EXPLORING CLASSICAL MECHANICS

Exploring Classical Mechanics

A Collection of 350+ Solved Problems for Students, Lecturers, and Researchers

Gleb L. Kotkin
Valeriy G. Serbo

Novosibirsk State University, Russia

Second revised and enlarged English edition

OXFORD

UNIVERSITY PRESS

OXFORD

UNIVERSITY PRESS

Great Clarendon Street, Oxford, OX2 6DP,
United Kingdom

Oxford University Press is a department of the University of Oxford.
It furthers the University's objective of excellence in research, scholarship,
and education by publishing worldwide. Oxford is a registered trade mark of
Oxford University Press in the UK and in certain other countries

First Edition published in 2020

Impression: 1

Published in the United States of America by Oxford University Press
198 Madison Avenue, New York, NY 10016, United States of America

British Library Cataloguing in Publication Data

Data available

Library of Congress Control Number: 2020937520

ISBN 978-0-19-885378-7 (hbk.)
ISBN 978-0-19-885379-4 (pbk.)

DOI: 10.1093/oso/9780198853787.001.0001

Printed and bound by
CPI Group (UK) Ltd, Croydon, CR0 4YY

Contents

Preface to the second English edition

This book was written by the working physicists for students of physics faculties of universities.

The first English edition of this book under the title *Collection of Problems in Classical Mechanics* was published by Pergamon Press in 1971 with the invaluable help by the translation editor D. ter Haar. This second English publication is based on the fourth Russian edition of 2010 and includes new problems from among those used in teaching at the physics faculty of Novosibirsk State University as well as the problems added in the publications in Spanish and French. As a result, this book contains 357 problems instead of the 289 problems that appeared in the first English edition.

We are grateful to A. V. Mikhailov for useful discussions of some new problems, to Z. K. Silagadze for numerous indications of misprints and inaccuracies in previous editions, and to O. V. Karpushina for an invaluable help in preparation of this manuscript.

In this edition, the main notations are:

m, e, $\mathbf{r}$, $\mathbf{p}$, and $\mathbf{M} = [\mathbf{r}, \mathbf{p}]$ – mass, charge, radius vector, momentum, and angular momentum of a particle, respectively;

L, H, E, and U – Lagrangian function, Hamiltonian function, energy, and potential energy of a system, respectively;

$\mathbf{E}$ and $\mathbf{B}$ – electric and magnetic field intensities, respectively;

φ and $\mathbf{A}$ – scalar and vector potentials, respectively, of the electromagnetic field;

c – velocity of light; and

$d\Omega$ – solid angle element.

For problems about the motion of particles in electromagnetic fields, we use Gaussian units, and in problems on electrical circuits, SI units.

From the Preface to the first English edition

This collection is meant for physics students. Its contents correspond roughly to the mechanics course in the textbooks by Landau and Lifshitz [1], Goldstein [4], or ter Haar [6]. We hope that the reading of this collection will give pleasure not only to students studying mechanics, but also to people who already know it. We follow the order in which the material is presented by Landau and Lifshitz, except that we start using the Lagrangian equations in § 4. The problems in §§ 1–3 can be solved using the Newtonian equations of motion together with the energy, linear momentum and angular momentum conservation laws.

As a rule, the solution of a problem is not finished with obtaining the required formulae. It is necessary to analyse the result, and this is by no means the "mechanical" part of the solution. It is also very useful to investigate what happens if the conditions of the problem are varied. We have, therefore, suggested further problems at the end of several solutions.

A large portion of the problems were chosen for the practical classes with students from the physics faculty of the Novosibirsk State University for a course on theoretical mechanics given by Yu. I. Kulakov. We want especially to emphasize his role in the choice and critical discussion of a large number of problems. We owe a great debt to I. F. Ginzburg for useful advice and hints which we took into account. We are very grateful to V. D. Krivchenkov whose active interest helped us to persevere until the end.

We are extremely grateful to D. ter Haar for his help in organizing an English edition of our book.

Problems

§1

Integration of one-dimensional equations of motion

1.1. Describe the motion of a particle in the following potential fields $U(x)$:

a) $U(x) = A(e^{-2\alpha x} - 2e^{-\alpha x})$ (Morse potential, Fig. 1a);

b) $U(x) = -\dfrac{U_0}{\cosh^2 \alpha x}$ (Fig. 1b);

c) $U(x) = U_0 \tan^2 \alpha x$ (Fig. 1c).

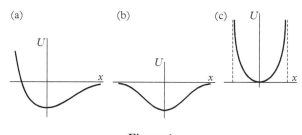

Figure 1

1.2. Describe the motion of a particle in the field $U(x) = -Ax^4$ for the case when its energy is equal to zero.

1.3. Give an approximate description of the motion of a particle in the field $U(x)$ near the tuning point $x = a$ (Fig. 2).

Hint: Use a Taylor expansion of $U(x)$ near the point $x = a$. Consider the cases $U'(a) \neq 0$ and $U'(a) = 0$, $U''(a) \neq 0$.

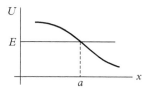

Figure 2

1.4. Determine how the period of a particle moving in the field in Fig. 3 tends to infinity as its energy E approaches U_m.

Exploring Classical Mechanics: A Collection of 350+ Solved Problems for Students, Lecturers, and Researchers. First Edition.
Gleb L. Kotkin and Valeriy G. Serbo, Oxford University Press (2020). © Gleb L. Kotkin and Valeriy G. Serbo 2020.
DOI: 10.1093/oso/9780198853787.001.0001

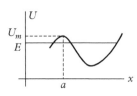 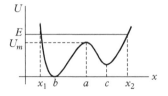

Figure 3   Figure 4

1.5. a) Estimate the period of the particle motion in the field $U(x)$ (Fig. 4), when its energy is close to U_m (i.e., $E - U_m \ll U_m - U_{\min}$).

b) Determine during which part of the period the particle is in the interval from x to $x + dx$.

c) Determine during which part of the period the particle has a momentum $m\dot{x}$ in the interval from p to $p + dp$.

d) In the plane x, $p = m\dot{x}$ represent qualitatively lines $E(x, p) = \text{const}$ for the cases $E < U_m$, $E = U_m$, $E > U_m$.

1.6. A particle of mass m moves along a circle of radius l in a vertical plane under the influence of the field of gravity (mathematical pendulum). Describe its motion for the case when its kinetic energy E in the lowest point is equal to $2mgl$.

Estimate the period of revolution of the pendulum for the case when $E - 2mgl \ll 2mgl$.

1.7. Describe the motion of a mathematical pendulum for an arbitrary value of the energy.

Hint: The time dependence of the angle the pendulum makes with the vertical can be expressed in terms of elliptic functions (e.g. see [1], § 37).

1.8. Determine the change in the motion of a particle moving along a section which does not contain turning points when the field $U(x)$ is changed by a small amount $\delta U(x)$.

Consider the applicability of the results obtained for the case of a section near the turning point.

1.9. Find the change in the motion of a particle caused by a small change $\delta U(x)$ in the field $U(x)$ in the following cases:

a) $U(x) = \frac{1}{2} m\omega^2 x^2$, $\delta U(x) = \frac{1}{3} m\alpha x^3$;

b) $U(x) = \frac{1}{2} m\omega^2 x^2$, $\delta U(x) = \frac{1}{4} m\beta x^4$.

1.10. Determine the change in the period of a finite orbit of a particle caused by the change in the field $U(x)$ by a small amount $\delta U(x)$.

1.11. Find the change in the period of a particle moving in a field $U(x)$ caused by adding to the field $U(x)$ a small term $\delta U(x)$ in the following cases:

a) $U(x) = \frac{1}{2}m\omega^2 x^2$ (a harmonic oscillator), $\delta U(x) = \frac{1}{4}m\beta x^4$;

b) $U(x) = \frac{1}{2}m\omega^2 x^2$, $\delta U(x) = \frac{1}{3}m\alpha x^3$;

c) $U(x) = A(e^{-2\alpha x} - 2e^{-\alpha x})$, $\delta U(x) = -Ve^{\alpha x}$ $(V \ll A)$.

1.12. The particle moves in the field $U(x) = \dfrac{U_0}{\cosh^2 \alpha x}$ with the energy $E > U_0$. Find the particle delay time at the motion from $x = -\infty$ to $x = +\infty$ in comparison with the free motion time with the same energy.

§2

Motion of a particle in three-dimensional fields

2.1. Describe qualitatively the motion of a particle in the field $U(r) = -\frac{\alpha}{r} - \frac{\gamma}{r^3}$ for different values of the angular momentum and of the energy.

2.2. Find the trajectories and the laws of motion of a particle in the field

$$U(r) = \begin{cases} -V, & \text{when } r < R, \\ 0, & \text{when } r > R \end{cases}$$

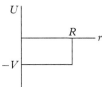

(Fig. 5, "spherical rectangular potential well") for different values of the angular momentum and of the energy.

Figure 5

2.3. Determine the trajectory of a particle in the field $U(r) = \frac{\alpha}{r} + \frac{\beta}{r^2}$. Give an expression for the change in the direction of velocity when the particle is scattered as a function of angular momentum and energy.

2.4. Determine the trajectory of a particle in the field $U(r) = \frac{\alpha}{r} - \frac{\beta}{r^2}$. Find the time it takes the particle to fall to the centre of the field from a distance r. How many revolutions around the centre will the particle then make?

2.5. Determine the trajectory of a particle in the field $U(r) = -\frac{\alpha}{r} + \frac{\beta}{r^2}$. Find the angle $\Delta\varphi$ between the direction of radius vector at two successive passages through the pericentre (i.e., when $r = r_{\min}$); also find the period of the radial oscillations, T_r. Under what conditions will the orbit be a closed one?

2.6. Determine the trajectory of a particle in the field $U(r) = -\frac{\alpha}{r} - \frac{\beta}{r^2}$. A field of this kind arises in the motion of a relativistic particle in the Coulomb field in the special theory of relativity; see [7], §42.1 for details.

Exploring Classical Mechanics: A Collection of 350+ Solved Problems for Students, Lecturers, and Researchers. First Edition.
Gleb L. Kotkin and Valeriy G. Serbo, Oxford University Press (2020). © Gleb L. Kotkin and Valeriy G. Serbo 2020.
DOI: 10.1093/oso/9780198853787.001.0001

2.7. For what values of the angular momentum M is it possible to have finite orbits in the field $U(r)$ for the following cases:

a) $U(r) = -\dfrac{\alpha e^{-\varkappa r}}{r}$; b) $U(r) = -Ve^{-\varkappa^2 r^2}$.

2.8. A particle falls from a finite distances towards the centre of the field $U(r) = -\alpha r^{-n}$. Will it make a finite number of revolutions around the centre? Will it take a finite time to fall towards the centre? Find the equation of the orbit for small r.

2.9. A particle in the field $U(r)$ flies off to infinity from a distance $r \neq 0$. Is the number of revolutions around the centre made by the particle finite for the following cases?

a) $U(r) = \alpha r^{-n}$ b) $U(r) = -\alpha r^{-n}$

2.10. How long will it take a particle to fall from a distance R to the centre of the field $U(r) = -\alpha/r$. The initial velocity of the particle is zero. Treat the orbit as a degenerate ellipse.

2.11. One particle of mass m moves along the x-axis from a long distance with velocity v towards the origin O of the coordinate system. Another particle of the same mass moves towards the origin O along the y-axis from a long distance with the same velocity magnitude. If the particles didn't interact, the second would pass through point O in time τ after the first one. However, they repulse from each other, and potential energy of interaction is $U(r) = \alpha/r$, where r is the distance between particles. Find the minimum distance between the particles.

2.12. Two particles with masses m_1 and m_2 move with velocities $\mathbf{v_1}$ and $\mathbf{v_2}$ from long distances along the crossing lines, the distance between which is equal to $\overrightarrow{AB} = \rho$. If particles didn't interact, particle 1 would pass the point of minimum distance A at time τ earlier than particle 2 would pass the point B. However, there is the force of attraction between the particles, which is given by the potential energy $U(r) = -\beta/r^2$.

a) At what relation between ρ and τ will particles collide?
b) At what distance from the point A will such a collision occur?

2.13. Determine the minimal distance between the particles, the one approaching from infinity with an impact parameter ρ and an initial velocity v and the other one initially at rest. The masses of the particles are m_1 and m_2, and the interaction law is $U(r) = \alpha/r^n$.

2.14. Determine in the centre of a mass system the finite orbits of two particles of masses m_1 and m_2, and an interaction law $U(r) = -\alpha/r$.

2.15. Determine the position of the focus of a beam of particles close to the beam axis, when the particles are scattered in a central field $U(r)$ under the assumption that a particle flying along the axis is turned back.

2.16. Find the inaccessible region of space for a beam of particles flying along the z-axis with a velocity v and being scattered by a field $U(r) = \alpha/r$.

2.17. Find the inaccessible region of space for particles flying with a velocity v from a point A in all directions and moving in a potential field $U(r) = -\alpha/r$.

2.18. Use the integral of motion $\mathbf{A} = [\mathbf{v}, \mathbf{M}] - \alpha \frac{\mathbf{r}}{r}$ (the Laplace vector – see [1], §15 and [7], §3.3) to find the orbit of a particle moving in the field $U(r) = -\alpha/r$.

2.19. The spacecraft is moving in a circular orbit of the radius R around the Earth. A body, whose mass is negligible in comparison with the mass of the spacecraft, is thrown from the spacecraft with relative velosity v, directed to the centre of the Earth. Find the orbit of the body.

   Hint: To find the orbit of the body, try using the Laplace vector.

   (This problem is formulated based on the real incident: during a spacewalk, the cosmonaut A. Leonov threw the plug from the camera in the direction of the Earth – see [8], §8.)

2.20. Determine in quadratures the change of the period T of radial oscillations of a particle moving in the central field $U(r)$ when this field is changed by a small amount $\delta U(r)$.

2.21. Show that the orbit of a particle in the field

$$U(r) = -\frac{\alpha}{r} e^{-r/D}$$

is a lowly precession ellipse when $r_{\max} \ll D$. Find the angular velocity of precession.

2.22. Find the precessional velocity of the orbit in the field $U(r) = -\alpha/r^{1+\varepsilon}$, when $|\varepsilon| \ll 1$.

2.23. Find the angular velocity of the orbit precession of a particle in the field

$$U(r) = \tfrac{1}{2} m\omega^2 r^2 + \frac{\beta}{r^4}$$

for $\beta \ll m\omega^2 a^6, m\omega^2 b^6$, where a and b are parameters of the unperturbed trajectory:

$$\left(\frac{r\cos\varphi}{a}\right)^2 + \left(\frac{r\sin\varphi}{b}\right)^2 = 1.$$

2.24. The particle slides on the surface of a smooth paraboloid of revolution whose axis (the z-axis) is directed straight up:

$$z = \frac{x^2 + y^2}{2l}.$$

Find the angular velocity of the orbit precession. The maximum and the minimum distances of a particle from the z-axis are a and b, where $a \ll l$.

2.25. Study the motion of the Earth–Moon system in the field of the Sun. Assume that the mass of the Moon is 81 times less than the mass of the Earth, and the distance from the Earth to the Moon ($r = 380$ thousand km) is a lot less than the average distances to the Sun ($R = 150$ million km).

a) For simplicity, taking that the plane of the Moon orbit coincides with the plane of the Earth orbit, show that the potential energy of the Earth–Moon system in the field of the Sun, averaged over a month, has the form

$$U(R) = -\frac{\alpha}{R} - \frac{\beta}{R^3},$$

where R is the distance from the Sun to the centre of mass of the Earth–Moon system. Determine the precession of perihelion for a 100-year period.

b) The plane of the Moon's orbit makes an angle of $\theta = 5°$ with the plane of the Earth's orbit. Determine the related average velocity of precession for the Moon's orbital plane.

2.26. Determine the angular velocity of the orbit precession of a particle in the field $U(r) = -\frac{\alpha}{r} + \delta U(r)$ if the orbit eccentricity e is much less than 1, assuming

$$\delta U(r) = \delta U(a) + (r - a)\delta U'(a) + \tfrac{1}{2}(r - a)^2 \delta U''(a),$$

where $a = \tfrac{1}{2}(r_{max} + r_{min})$ is the average orbit radius.

2.27. Determine the angular velocity of the orbit precession of a particle in the field $U(r) = -\frac{\alpha}{r} + \delta U(r)$ ($\delta U(r)$ is a small correction) up to second order in $\delta U(r)$ inclusively.

2.28. Find the equation of motion of the orbit of a particle moving in the field $U(r) = -\frac{\alpha}{r} + \frac{\gamma}{r^3}$, assuming $\frac{\gamma}{r^3}$ to be a small correction to the Coulomb field.

2.29. Show that the problem of the motion of two charged particles in a uniform electric field **E** can be reduced to the problem of the motion of the centre of mass and that of the motion of a particle in a given field.

2.30. Under what conditions can the problem of the motion of two charged particles in a constant uniform magnetic field **B** be separated into the problem of the centre of mass motion and the relative motion problem?

Take the vector potential of the magnetic field in the form

$$\mathbf{A}(\mathbf{r}_i) = \tfrac{1}{2}[\mathbf{B}, \mathbf{r}_i], \quad i = 1, 2.$$

2.31. Express the kinetic energy, the linear momentum, and the angular momentum of a system of N particles in terms of the Jacobi coordinates:

$$\boldsymbol{\xi}_n = \frac{m_1\mathbf{r}_1 + \ldots + m_n\mathbf{r}_n}{m_1 + \ldots + m_n} - \mathbf{r}_{n+1} \quad (n = 1\ldots N-1),$$

$$\boldsymbol{\xi}_N = \frac{m_1\mathbf{r}_1 + \ldots + m_N\mathbf{r}_N}{m_1 + \ldots + m_N}.$$

2.32. A particle with a velocity v at infinity collides with another particle of the same mass m which is at rest. Their interaction potential energy is $U(r) = \alpha/r^n$ and the collision is a central one. Find the point where the first particle comes to rest.

2.33. Prove that

$$\mathbf{MB} + \frac{e}{2c}[\mathbf{r},\mathbf{B}]^2,$$

is the integral of motion for a charged particle in a uniform constant magnetic field $\mathbf{B}$. Here $\mathbf{M} = m[\mathbf{r},\mathbf{v}]$, and c is the velocity of light.

2.34. Find the trajectory and the law of motion of a charged particle in the magnetic field $\mathbf{B}(\mathbf{r}) = g\mathbf{r}/r^3$ (the field of the magnetic monopole).

Such form has a field of a thin long solenoid outside its end at distances which are large compared to the solenoid's diameter, but small compared to its length.

2.35. Give a qualitative description of the motion and the shape of the orbit of a charged particle moving in the field of a magnetic dipole $\boldsymbol{\mu}$, in the plane perpendicular to the vector $\boldsymbol{\mu}$. Take the vector potential of magnetic dipole in the form $\mathbf{A}(\mathbf{r}) = [\boldsymbol{\mu},\mathbf{r}]/r^3$.

2.36. a) Give a qualitative description of the motion of a charged particle in the field $U(r) = \frac{1}{2}m\lambda r^2$, where r is a distance from the z-axis, for the case where there is a constant uniform magnetic field $\mathbf{B}$ parallel to the z-axis present.

b) Find the law of motion and the orbit of a charged particle moving in the field $U(r) = \alpha/r^2$ in a plane perpendicular to a constant uniform magnetic field $\mathbf{B}$.

2.37. A charged particle moves in the Coulomb field $U(r) = -\alpha/r$ in a plane perpendicular to a constant uniform magnetic field $\mathbf{B}$. Find the orbit of the particle. Study the case when $\mathbf{B}$ is small and the case when the field $U(r)$ is a small perturbation.

2.38. Describe the motion of two identical charged particles in a constant uniform magnetic field $\mathbf{B}$ for the case when their orbits lie in the same plane which is perpendicular to $\mathbf{B}$ and where we may consider their interaction energy $U(r) = e^2/r$ to be a small perturbation.

2.39. Show that the quantity

$$\mathbf{F}[\mathbf{v},\mathbf{M}] - \frac{\alpha}{r}\mathbf{Fr} + \tfrac{1}{2}[\mathbf{F},\mathbf{r}]^2$$

is a constant of motion in the field $U(\mathbf{r}) = -\frac{\alpha}{r} - \mathbf{Fr}$ where $\mathbf{F} = \mathbf{const}$.

Give the meaning of this integral of motion when F is small.

2.40. Study the effect of a small extra term $\delta U(\mathbf{r}) = -\mathbf{Fr}$, where $\mathbf{F} = \text{const}$, added to the Coulomb field on the finite orbit of a particle.

a) Find the average rate of change of the angular momentum, averaged over one period.

b) Find the time-dependence of the angular momentum, the size, and the orientation of the orbit for the case when the force $\mathbf{F}$ lies in the orbital plane.

c) Do the same as under b) for the case when the orientation of the force is arbitrary.

Hint: Write down the equations of motion for the vectors $\mathbf{M} = m[\mathbf{r}, \mathbf{v}]$ and $\mathbf{A} = [\mathbf{v}, \mathbf{M}] - \alpha \mathbf{r}/r$ averaged over one period and solve them.

2.41. Find the systematic displacement of a finite orbit of a charged particle moving in the field $U(r) = -\alpha/r$ under influence of weak constant uniform electric $\mathbf{E}$ and magnetic $\mathbf{B}$ fields.

a) Consider the limiting case when the magnetic field is perpendicular to the orbit plane and the electric field is in this plane.

b) Consider the general case.

2.42. Find the systematic change of the elliptic orbit of a particle in the field $U(r) = -\alpha/r$ under the influence of a small perturbation

$$\delta U(r, \theta) = -\beta r^2 (3 \cos^2 \theta - 1).$$

Only consider the case when the orbit plane passes through the z-axis. This problem is a simplified model of the satellite motion in the Earth field taking into account the gravitational field of the Moon near the Earth space.

2.43. Taking that the orbit of the Moon in the Earth field is an ellipse lying in the plane of the Earth orbit, find the systematic change of the Moon orbit under the influence of the perturbation

$$\delta U(r, \chi) = -\tfrac{1}{2} m \Omega^2 r^2 (3 \cos^2 \chi - 1),$$

where m is the Moon mass, Ω is the angular velocity of the Earth around the Sun, χ is the angle between Earth to Sun and Earth to Moon directions.

2.44. Find the systematic displacement of the finite orbit of a charged particle moving in the field $U(r) = -\alpha/r$ and in the field of the magnetic dipole $\boldsymbol{\mu}$, if the effect of the latter may be considered to be a small perturbation. Take the vector potential in the form $\mathbf{A}(\mathbf{r}) = [\boldsymbol{\mu}, \mathbf{r}]/r^3$.

2.45. Find the average precession rate of the orbit of a particle moving in the field $U(r) = -\alpha/r$ under the influence of a small additional "friction force" $\mathbf{F} = \beta \dot{\mathbf{v}}$ (such form has the force of radiation damping; in this case, $\beta = \frac{2}{3} \frac{q^2}{c^3}$, where q is the charge of the particle and c is the velocity of light; see [2], § 75).

§3

Scattering in a given field. Collision between particles

3.1. Find the differential cross-section for the scattering of particles with initial velocity parallel to the z-axes by smooth elastic surfaces of revolution $\rho(z)$ for the following cases:

a) $\rho = b\sin\frac{z}{a}, \quad 0 \leqslant z \leqslant \pi a$;

b) $\rho = Az^n, \quad 0 < n < 1$;

c) $\rho = b - \frac{a^2}{z}, \quad \frac{a^2}{b} \leqslant z < \infty$.

3.2. Find the surface of revolution which is such that the differential cross-section for elastic scattering by this surface coincides with the Rutherford scattering cross-section.

3.3. Find the differential cross-section for the scattering of particles by spherical "potential barrier":

$$U(r) = \begin{cases} V, & \text{when } r < a, \\ 0, & \text{when } r > a. \end{cases}$$

3.4. Find the cross-section for the process where a particle falls towards the centre of the field $U(r)$ when $U(r)$ is given by:

a) $U(r) = \frac{\alpha}{r} - \frac{\beta}{r^2}$,   b) $U(r) = \frac{\beta}{r^2} - \frac{\gamma}{r^4}$.

3.5. Calculate the cross-section for particles to hit a small sphere of radius R placed at the centre of the field $U(r)$ for the cases:

a) $U(r) = -\frac{\alpha}{r^n}, n \geqslant 2$;   b) $U(r) = \frac{\beta}{r^2} - \frac{\gamma}{r^4}$.

3.6. A uniform beam of meteors with velocity $\mathbf{v}_\infty$ flies towards the planet. What fraction of meteorites that fell on the planet falls on the part invisible from the beam side? Take a planet as a uniform ball of radius R and mass m_0.

Exploring Classical Mechanics: A Collection of 350+ Solved Problems for Students, Lecturers, and Researchers. First Edition.
Gleb L. Kotkin and Valeriy G. Serbo, Oxford University Press (2020). © Gleb L. Kotkin and Valeriy G. Serbo 2020.
DOI: 10.1093/oso/9780198853787.001.0001

3.7. Find the differential cross-section for the scattering of particles by the field $U(r)$:

a) $U(r) = \begin{cases} \frac{\alpha}{r} - \frac{\alpha}{R}, & \text{when } r < R, \\ 0, & \text{when } r > R; \end{cases}$

b) $U(r) = \begin{cases} \frac{1}{2} m\omega^2 (r^2 - R^2), & \text{when } r < R, \\ 0, & \text{when } r > R. \end{cases}$

3.8. Find the differential cross-section for the scattering of fast particles $(E \gg V)$ by the field $U(r)$ for the following cases:

a) $$U(r) = \begin{cases} V\sqrt{1 - (r^2/a^2)}, & \text{when } r < a, \\ 0, & \text{when } r > a; \end{cases}$$

b) $$U(r) = \begin{cases} V \ln (r/a), & \text{when } r < a, \\ 0, & \text{when } r > a; \end{cases}$$

c) $$U(r) = V \ln\left(1 + \frac{a^2}{r^2}\right);$$

d) $$U(r) = \begin{cases} V\left(1 - \frac{r^2}{R^2}\right), & \text{when } r < R, \\ 0, & \text{when } r > R. \end{cases}$$

3.9. Find the differential cross-section for small angle scattering in the field $U(r) = \frac{\beta}{r^4} - \frac{\alpha}{r^2}$.

3.10. Find the differential cross-section for the scattering of particles by the field $U(r) = -\alpha/r^2$.

3.11. Find the differential cross-section for the scattering of fast particles $(E \gg V)$ by the following fields $U(r)$:

a) $U(r) = V e^{-\varkappa^2 r^2}$; b) $U(r) = \dfrac{V}{1 + \varkappa^2 r^2}$.

Study in detail the limiting cases when the deflection angle is close to its minimum or maximum value.

3.12. A beam of particles with their velocities initially parallel to the z-axis is scattered by the fixed ellipsoid

$$\frac{x^2}{a^2} + \frac{y^2}{b^2} + \frac{z^2}{c^2} = 1.$$

Find the differential scattering cross-section for the following cases:

a) the ellipsoid is smooth and the scattering elastic;

b) the ellipsoid is smooth and the scattering inelastic;

c) the ellipsoid is rough and the scattering elastic.

3.13. Find the differential cross-section for small-angle scattering by the following fields $U(\mathbf{r})$ (**a** is a constant vector):

a) $U(\mathbf{r}) = \frac{\mathbf{ar}}{r^2}$; b) $U(\mathbf{r}) = \frac{\mathbf{ar}}{r^3}$.

3.14. Find the change in the differential cross-section for scattering of particles by the field $U(r)$ when $U(r)$ is varied by a small amount $\delta U(r)$ for the following cases:

a) $U(r) = \frac{\alpha}{r}$, $\delta U(r) = \frac{\beta}{r^2}$;

b) $U(r) = \frac{\alpha}{r}$, $\delta U(r) = \frac{\gamma}{r^3}$;

c) $U(r) = \frac{\beta}{r^2}$, $\delta U(r) = \frac{\gamma}{r^3}$.

3.15. Find the differential cross-section as function of the energy acquired by fast particles $(E \gg V_{1,2})$ due to their scattering in the field

$$U(r, t) = (V_1 + V_2 \sin \omega t) e^{-\varkappa^2 r^2}.$$

3.16. A particle with velocity V decays into two identical particles. Find the distribution of the secondary particles over the angle of divergence, that is, the angle between the directions at which two secondary particles fly off. The decay is isotropic in the centre of mass system and velocity of the secondary particles is v_0 in that system.

3.17. Find the energy distribution of secondary particles in the laboratory system, if the angular distribution in the centre of mass system is

$$\frac{dN}{N} = \frac{3}{8\pi} \sin^2 \theta_0 \, d\Omega_0,$$

where θ_0 is the angle between the velocity $\mathbf{V}$ of the original particle and the direction in which one of secondary particle flies off in the centre of mass system. The velocity of the secondary particles in the centre of mass system is v_0.

3.18. An electron moving at infinity with velocity v collides with another electron at rest; the impact parameter is ρ. Find the velocities of the two electrons after the collision.

3.19. Find the range of possible values for the angle between the velocity directions after a moving particle of mass m_1 has collided with a particle of mass m_2 at rest.

3.20. Find the differential cross-section for the scattering of inelastic smooth spheres by similar ones at rest.

3.21. Find the change in intensity of a beam of particles travelling through a volume filled with absorbing centres; their density n cm^{-3}, and the absorption cross-section σ.

3.22. Find the number of reactions occurring during a time dt in a volume element dV when two beams of particles with velocities $\mathbf{v}_1$, $\mathbf{v}_2$ and densities n_1, n_2, respectively, collide. The reaction cross-section is σ.

3.23. A particle of mass M moves in a volume filled with particles of mass m ($m \ll M$) which is at rest initially. The differential cross-section for scattering of M by m is $d\sigma = f(\theta) \, d\Omega$, and the collisions are assumed to be elastic.

a) Find the "friction force" acting on the particle M;

b) Find the average of the square of the angle Θ over which the particle M is deflected.

§4

Lagrangian equations of motion. Conservation laws

4.1. A particle, moving in the field $U(x) = -Fx$, travels from the point $x = 0$ to the point $x = a$ in a time τ. Find the law of motion of particle, assuming it to be of the form $x(t) = At^2 + Bt + C$, and determine the parameters A, B, and C such that the action is a minimum.

4.2. A particle moves in the xy-plane in the field

$$U(x, y) = \begin{cases} 0, & \text{when } x < 0, \\ V, & \text{when } x > 0, \end{cases}$$

and travels in a time τ from the point $(-a, 0)$ to the point (a, a). Find its position as a function of time, assuming that it satisfies the equations

$$x_{1,2}(t) = A_{1,2}t + B_{1,2},$$
$$y_{1,2}(t) = C_{1,2}t + D_{1,2}.$$

The indices 1 and 2 refer, respectively, to the left-hand $(x < 0)$ and right-hand $(x > 0)$ half-planes.

4.3. Prove by direct calculations the invariance of the Lagrangian equations of motion under the coordinate transformation

$$q_i = q_i(Q_1, Q_2, \ldots, Q_s, t), \quad i = 1, 2, \ldots, s.$$

4.4. What is the change in the Lagrangian function in order that the Lagrangian equations of motion retain their form under the transformation to new coordinates and "time":

$$q_i = q_i(Q_1, Q_2, \ldots, Q_s, \tau), \quad i = 1, 2, \ldots, s, \quad t = t(Q_1, Q_2, \ldots, Q_s, \tau)?$$

Exploring Classical Mechanics: A Collection of 350+ Solved Problems for Students, Lecturers, and Researchers. First Edition.
Gleb L. Kotkin and Valeriy G. Serbo, Oxford University Press (2020). © Gleb L. Kotkin and Valeriy G. Serbo 2020.
DOI: 10.1093/oso/9780198853787.001.0001

4.5. Write down the Lagrangian function and the equations of motion for a particle moving in a field $U(x)$, introducing a "local time" $\tau = t - \lambda x$.

4.6. How does the Lagrangian function

$$L = -\sqrt{1 - \left(\frac{dx}{dt}\right)^2}$$

transform when we change to the coordinates q and "time" τ through the equations

$$x = q\operatorname{ch}\lambda + \tau\operatorname{sh}\lambda$$
$$t = q\operatorname{sh}\lambda + \tau\operatorname{ch}\lambda$$

4.7. How do the energy and generalized momenta change under the following coordinate transformation

$$q_i = f_i(Q_1, \ldots, Q_s, t) \quad i = 1, \ldots, s?$$

4.8. How do the energy and generalized momenta which are conjugate to (a) the spherical polar and (b) the Cartesian coordinates transform under a change to a coordinate system which is rotating around the z-axis

a) $\varphi = \varphi' + \Omega t, \quad r = r'$

b) $x = x'\cos\Omega t - y'\sin\Omega t, \quad y = x'\sin\Omega t + y'\cos\Omega t$

4.9. How do the energy and momenta change when we change to a frame of reference which is moving with a constant velocity $\mathbf{V}$? Take the Lagrangian function L' in the moving frame of reference in either of two forms:

a) $L'_1 = L(\mathbf{r}' + \mathbf{V}t, \dot{\mathbf{r}}' + \mathbf{V}, t)$, where $L(\mathbf{r}, \dot{\mathbf{r}}, t)$ is the Lagrangian function in the original frame of reference;

b) $L'_2 = \frac{1}{2}m\mathbf{v}'^2 - U(\mathbf{r}' + \mathbf{V}t, t)$. Here L'_2 differs from L'_1 by the total derivative with respect to the time of the function

$$\mathbf{V}m\mathbf{r}' + \tfrac{1}{2}m\mathbf{V}^2 t.$$

4.10. Smooth rod OA of length l rotates around a point O in a horizontal plane with a constant angular velocity Ω (Fig. 6). The bead is fixed on the rod at a distance a from the point O. The bead is released, and after a while, it is slipping off the rod. Find its velocity at the moment when the bead is slipping?

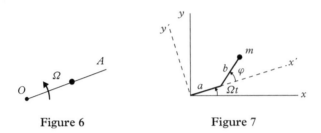

Figure 6   Figure 7

4.11. At the end of the rod of length a, rotating with constant angular velocity Ω in the plane of the figure, is pivotally attached another freely rotating rod of length b, at the end of which there is a particle of mass m (Fig. 7). Find the Lagrangian function $L(\varphi, \dot{\varphi}, t)$ of a system and the frequency of small oscillations of the second rod.

4.12. Consider an infinitesimal transformation of the coordinates and the time of the form

$$q_i' = q_i + \varepsilon f_i(q, t),$$
$$t' = t + \varepsilon h(q, t), \quad \varepsilon \to 0.$$

Let that the action is invariant under this transformation (with an accuracy to terms of the order of ε, inclusive),

$$\int_{t_1}^{t_2} L\left(q, \frac{dq}{dt}, t\right) dt = \int_{t_1'}^{t_2'} L\left(q', \frac{dq'}{dt'}, t'\right) dt'.$$

(We emphasize that in the left and right sides of this equation is the same function L, but from different arguments.) Prove that the quantity

$$\sum_i \frac{\partial L}{\partial \dot{q}_i}(\dot{q}_i h - f_i) - Lh$$

is an integral of motion.

4.13. Generalize the theorem of the preceding problem to the case when under the transformation of the coordinates and the time the action changes in the following way:

$$\int_{t_1}^{t_2} L\left(q, \frac{dq}{dt}, t\right) dt = \int_{t_1'}^{t_2'} \left\{ L\left(q', \frac{dq'}{dt'}, t'\right) + \varepsilon \frac{dF(q', t')}{dt'} \right\} dt',$$

where $F(q, t)$ is an arbitrary function of coordinates and time.

4.14. Find the integrals of motion if the action remains invariant under

    a) a translation;
    b) a rotation;
    c) a shift in the origin of the time;
    d) a screw shift;
    e) the transformation of problem 4.6.

4.15. Find the integrals of motion for a particle moving in

    a) a uniform field $U(\mathbf{r}) = -\mathbf{Fr}$;
    b) a field $U(\mathbf{r})$, where $U(\mathbf{r})$ is a homogeneous function,

$$U(\alpha\mathbf{r}) = \alpha^n U(\mathbf{r});$$

    specify, for what value of n the similarity transformation leaves the action invariant;
    c) the field of a travelling wave $U(\mathbf{r}, t) = U(\mathbf{r} - \mathbf{V}t)$, where $\mathbf{V}$ is a constant vector;
    d) a magnetic field specified by the vector potential $\mathbf{A}(\mathbf{r})$, where $\mathbf{A}(\mathbf{r})$ is a homogeneous function;
    e) an electromagnetic field rotating with a constant angular velocity Ω around the z-axis.

4.16. Find the time of the fall of the particle in the centre of the field $U(\mathbf{r}) = -(\mathbf{ar})^2/r^4$. At the initial moment, the coordinates and velocity of the particle are equal to $\mathbf{r}_0$ and $\mathbf{v}_0$.

4.17. Find the integral of motion corresponding to the Galilean transformations.
    Hint: Use the result of problem 4.13.

4.18. Find the integrals of motion of a particle moving in a uniform constant magnetic field $\mathbf{B}$ if the vector potential is given in the form:

    a) $\mathbf{A}(\mathbf{r}) = \frac{1}{2}[\mathbf{B}, \mathbf{r}]$;   b) $A_x = A_z = 0, A_y = xB$.

4.19. Find the integrals of motion for a particle moving in

    a) the field of magnetic dipole specified by the vector potential $\mathbf{A}(\mathbf{r}) = [\boldsymbol{\mu}, \mathbf{r}]/r^3$, where $\boldsymbol{\mu} = \mathbf{const}$;
    b) the field given by the vector potential $A_\varphi = \mu/r, A_r = A_z = 0$.

4.20. Find the equations of motion of a system with the following Lagrangian function:

a) $L(x, \dot{x}) = e^{-x^2 - \dot{x}^2} + 2\dot{x}e^{-x^2}\int_0^{\dot{x}} e^{-y^2}\, dy;$

b) $L(x, \dot{x}, t) = \frac{1}{2}e^{\alpha t}(\dot{x}^2 - \omega^2 x^2).$

4.21.

a) Write down the components of the acceleration vector for a particle in the spherical coordinate system.

b) Find the components of the acceleration in the orthogonal coordinate q_i, if the line element is given by the equation

$$ds^2 = h_1^2\, dq_1^2 + h_2^2\, dq_2^2 + h_3^2\, dq_3^2,$$

where $h_i(q_1, q_2, q_3)$ are the Lamé coefficients.

4.22. Write down the equations of motion of a point particle using arbitrary coordinates q_i which are connected with the Cartesian coordinates x_i by the relations:

a) $x_i = x_i(q_1, q_2, q_3),\quad i = 1, 2, 3;$
b) $x_i = x_i(q_1, q_2, q_3, t),\quad i = 1, 2, 3.$

4.23. Show that the Lagrangian function [25]

$$L = \frac{m}{2}(\dot{r}^2 + r^2\dot{\theta}^2 + r^2\dot{\varphi}^2 \sin^2\theta) - \frac{eg}{c}\dot{\varphi}\cos\theta,$$

where r, θ, and φ are the spherical coordinates, describes the motion of a charged particle in the magnetic field $\mathbf{B}(\mathbf{r}) = g\mathbf{r}/r^3$ (see problem 2.34). Find the integrals of motion.

4.24. Verify that one can use the Lagrangian functions

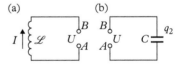

Figure 8

$$L_1 = \frac{\mathscr{L}\dot{q}_1^2}{2} - Uq_1, \quad L_2 = -\frac{q_2^2}{2C} + Uq_2,$$

to get the correct "equations of motion" for q_1 and q_2 and the correct energies. Here $\dot{q}_1 = I$ is the current flowing through the inductance $\mathscr{L}$ in the direction from A to B (Fig. 8a), q_2 the charge on the upper plate of the capacitor (Fig. 8b), and U the voltage between A and B ($U = \varphi_B - \varphi_A$).

4.25. Use the additivity property of the Lagrangian functions and the results of the preceding problem to find the Lagrangian functions and the Lagrangian equations of motion for the circuits of Fig. 9a–9c.

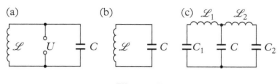

Figure 9

4.26. Find the Lagrangian functions for the following systems:

a) a circuit with a variable capacitor, the movable plate of which is connected to a pendulum of mass m (Fig. 10), and the capacitor of which is a known function $C(\varphi)$ of the angle φ the pendulum makes with the vertical. The mass of the capacitor plate may be ignored;

b) a core suspended from a spring with elastic constant k inside a solenoid with inductance $\mathscr{L}(x)$ which is a given function of the displacement x of a core (Fig. 10b).

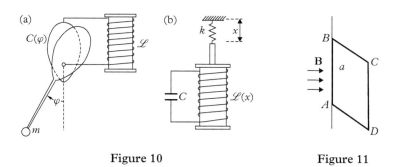

Figure 10   Figure 11

4.27. A perfectly conducting square frame can rotate around a fixed side $AB = a$ (Fig. 11). The frame is placed in a constant uniform magnetic field **B** at right angles to the AB axis. The inductance of the frame is $\mathscr{L}$ and the mass of the side CD is m; the masses of the other sides may be neglect.
Describe qualitatively the motion of the frame.

4.28. Use the method of the undetermined Lagrangian multipliers to obtain the equations of motion of a particle in the field of gravity when it is constrained to move

a) along a parabola $z = ax^2$ in a vertical plane;
b) along a circle of radius $r = l$ in a vertical plane.

Determine the forces of constraint.

4.29. A particle moves in the field of gravity along a straight line which is rotating uniformly in a vertical plane. Write down the equations of motion and determine the moment of the forces of constraint.

4.30. One can describe the influence of constraints and friction on the motion of a system by introducing generalized constraint and friction forces R_i into the equations of motion:

$$\frac{d}{dt}\frac{\partial L}{\partial \dot{q}_i} - \frac{\partial L}{\partial q_i} = R_i.$$

a) How does the energy of the system vary with time?
b) What is the transformation of the forces R_i which lives the equations of motion invariant under a transformation to new generalized coordinates

$$q_i = q_i(Q_1, \ldots, Q_s, t)?$$

4.31. Let the constraint equations be

$$\dot{q}_\beta = \sum_{n=r+1}^{s} b_{\beta n}\dot{q}_n, \quad \beta = 1, \ldots, r,$$

while the Lagrangian function $L(q_{r+1}, \ldots, q_s, \dot{q}_1, \ldots, \dot{q}_s, t)$ and the coefficients $b_{\beta n}$ do not depend on coordinates q_β.

Show that the equations of motion can be written in the form

$$\frac{d}{dt}\frac{\partial \tilde{L}}{\partial \dot{q}_n} - \frac{\partial \tilde{L}}{\partial q_n} + \sum_{\beta=1}^{r}\frac{\partial L}{\partial \dot{q}_\beta}\sum_{m=r+1}^{s}\left(\frac{\partial b_{\beta m}}{\partial q_n} - \frac{\partial b_{\beta n}}{\partial q_m}\right)\dot{q}_m = 0,$$

where $\tilde{L}(q_{r+1}, \ldots, q_s, \dot{q}_{r+1}, \ldots, \dot{q}_s, t)$ is the function obtained from L by using the constraint equations to eliminate the velocity $\dot{q}_1, \ldots, \dot{q}_r$.

4.32. A continuous string can be thought of as the limiting case of a system of N particles (Fig. 12) which are connected by an elastic thread, in the limit as $N \to \infty$, $a \to 0$, $Na = $ const. The Lagrangian function for a discrete system is

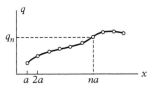

Figure 12

$$L(q_1, q_2, \ldots, q_N, \dot{q}_1, \dot{q}_2, \ldots, \dot{q}_N, t)$$

$$= \sum_{n=1}^{N+1} L_n(q_n, q_n - q_{n-1}, \dot{q}_n, t),$$

where q_n is the displacement of the nth particle from its equilibrium position.

a) Obtain the equation of motion for a continuous system as the limit case of the Lagrangian equations of motion of a discrete system.

b) Obtain an expression for the energy of a continuous system as the limiting case of the expression for the energy of a discrete system.

Hint: Introduce the coordinate x of a point on the string together with the expression obtained as a result of taking the limit as $a \to 0, n = x/a \to \infty$:

$$q(x, t) = \lim q_n(t), \quad \frac{\partial q}{\partial x} = \lim \frac{q_n(t) - q_{n-1}(t)}{a},$$

$$\mathscr{L}\left(x, q, \frac{\partial q}{\partial x}, \frac{\partial q}{\partial t}, t\right) = \lim \frac{L_n(q_n, q_n - q_{n-1}, \dot{q}_n, t)}{a}.$$

4.33. A charged particle moves in the potential field $U(\mathbf{r})$ and a constant magnetic field $\mathbf{B}(\mathbf{r})$, where $U(\mathbf{r})$ and $\mathbf{B}(\mathbf{r})$ are homogeneous functions of the coordinates of degrees k and n, respectively, that is,

$$U(\alpha\mathbf{r}) = \alpha^k U(\mathbf{r}), \quad \mathbf{B}(\alpha\mathbf{r}) = \alpha^n \mathbf{B}(\mathbf{r}).$$

Develop for this system the similarity principle, determining for what value of n it holds.

4.34. Generalize the virial theorem for a system of charged particles in a uniform constant magnetic field $\mathbf{B}$. The potential energy U of the system is a homogeneous function of the coordinates,

$$U(\alpha\mathbf{r}_1, \ldots, \alpha\mathbf{r}_s) = \alpha^k U(\mathbf{r}_1, \ldots, \mathbf{r}_s),$$

and the system moves in a bounded region of space with velocities which remain finite.

4.35. Three identical particles move along the same line and interact pairwise according to the law $U_{ik} = U(x_i - x_k)$, where x_i is the coordinate of ith particle. Prove that besides the obvious integrals of motion

$$P = m(\dot{x}_1 + \dot{x}_2 + \dot{x}_3),$$

$$E = \tfrac{1}{2} m(\dot{x}_1^2 + \dot{x}_2^2 + \dot{x}_3^2) + U_{12} + U_{23} + U_{31}$$

there is an additional integral of motion [23]

$$A = m\dot{x}_1\dot{x}_2\dot{x}_3 - \dot{x}_1 U_{23} - \dot{x}_2 U_{31} - \dot{x}_3 U_{21},$$

when the function $U(x)$ has the following form:

a) $U(x) = \dfrac{g^2}{x^2}$,

b) $U(x) = \dfrac{g^2 a^2}{\sinh^2 ax}$.

4.36. Consider the collision of the three particles described in the preceding problem. Assume $x_1 > x_2 > x_3$, let the distances between the particles be infinitely large at $t \to -\infty$, and let their velocities $v_i = \dot{x}_i$ $(t = -\infty)$ be such that $v_3 > v_2 > v_1$. Find $v_i' = \dot{x}_i$ $(t = +\infty)$.

§5

Small oscillations of systems with one degree of freedom

5.1. Find frequency ω of the small oscillations for particles moving in the following fields $U(x)$:

  a) $U(x) = V\cos\alpha x - Fx$;   b) $U(x) = V(\alpha^2 x^2 - \sin^2\alpha x)$.

5.2. Find the frequency of the small oscillations for the system depicted in Fig. 13. The system rotates with the angular velocity Ω in the field of gravity around a vertical axis.

5.3. A point charge q of mass m moves along a circle of radius R in a vertical plane and in the field of gravity. Another charge q is fixed at the lowest point of the circle (Fig. 14). Find the equilibrium position and the frequency of the small oscillations for the first point charge.

5.4. Describe the motion along a curve close to a circle for a point particle in the central field $U(r) = -\alpha/r^n$ $(0 < n < 2)$.

5.5. Find the frequency of the small oscillations of a spherical pendulum (a particle of mass m suspended a string of length l) if the angle of deflection from the vertical, θ oscillates about the value θ_0.

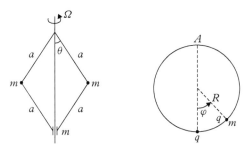

Figure 13   Figure 14

Exploring Classical Mechanics: A Collection of 350+ Solved Problems for Students, Lecturers, and Researchers. First Edition.
Gleb L. Kotkin and Valeriy G. Serbo, Oxford University Press (2020). © Gleb L. Kotkin and Valeriy G. Serbo 2020.
DOI: 10.1093/oso/9780198853787.001.0001

5.6. Find the correction to the frequency of the small oscillations of a diatomic molecule due to its angular momentum M.

5.7. Determine the free oscillations of the system shown in Fig. 15 for the case when the particle moves:

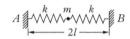

Figure 15

a) along the straight line AB;

b) at a right angle to AB.

How does the frequency depend on the tension of the springs in the equilibrium position?

5.8. Find the free oscillations of the system (Fig. 16) in a uniform field of gravity for the case when the particle can only move vertically.

5.9. Find the stable small oscillations of a pendulum when its point of suspension moves uniformly along a circle of radius a with an angular velocity Ω (Fig. 17). The pendulum length is l ($l \gg a$).

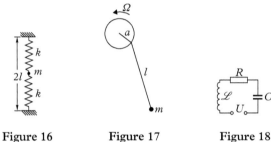

Figure 16   Figure 17   Figure 18

5.10. Find the stable oscillations in the voltage across a capacitor and the current in a circuit with an e.m.f. $U(t) = U_0 \sin \omega t$ (Fig. 18).

5.11. Determine the law of motion of an oscillator with friction which initially is at rest and which is acted upon by a force $F(t) = F \cos \gamma t$.

5.12. Determine the energy E acquired by an oscillator under the action of a force $F(t) = F e^{-(t/\tau)^2}$ during the total time its acts

a) if the oscillator was at rest at $t = -\infty$;

b) if the amplitude of the oscillation at $t = -\infty$ was a.

5.13. Describe the motion under the action of a force $F(t)$

a) of an unstable system described by the equation

$$\ddot{x} - \mu^2 x = \frac{1}{m} F(t);$$

b) of an oscillator with friction

$$\ddot{x} + 2\lambda\dot{x} + \omega_0^2 x = \frac{1}{m}F(t).$$

5.14. Find the differential cross-section for an isotropic oscillator to be exited to the energy ε by a fast particle $(E \gg V)$ if the interaction between the two particles is through the field

$$U(r) = Ve^{-\varkappa^2 r^2}.$$

The energy of the oscillator is zero initially.

5.15. An oscillator can oscillate only along the z-axis. Find the differential cross-section for the oscillator to be excited to an energy ε by a fast particle $E \gg V$, if the interaction between the particles is through the potential energy $U(r) = Ve^{-\varkappa^2 r^2}$. The particle moves along the z-axis with velocity $\mathbf{v}_\infty$, and the initial energy of the oscillator is ε_0.

5.16. A force $F(t)$, for which $F(-\infty) = 0$, $F(+\infty) = F_0$, acts upon an harmonic oscillator. Find the energy $E(+\infty)$ gained by the oscillator during the total time the force acts, and the amplitude of the oscillator at $t \to +\infty$, if it were at rest at $t \to -\infty$.

5.17. Find the energy acquired by an oscillator under the action of the force

$$F(t) = \begin{cases} \frac{1}{2}F_0 e^{\lambda t}, & \text{when } t < 0, \\ \frac{1}{2}F_0 (2 - e^{-\lambda t}), & \text{when } t > 0. \end{cases}$$

At $t \to -\infty$ the energy of the oscillator was E_0.

5.18. Estimate the change in the amplitude of the vibrations of an oscillator when a force $F(t)$ is switched on slowly over a period of τ so that $\omega\tau \gg 1$. Assume $F(t) = 0$ for $t < 0$, $F(t) = F_0$ for $t > \tau$, while $F^{(k)} \sim F_0/\tau^k$ $(k = 0, 1, \ldots, n+1)$ for $0 < t < \tau$ and $F^{(s)}(0) = F^{(s)}(\tau) = 0$ $(s = 1, 2, \ldots, n-1)$, while the nth derivative of the force has a discontinuity at $t = 0$ and at $t = \tau$.

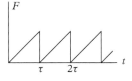

Figure 19

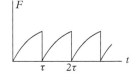

Figure 20

5.19. Find the stable oscillations of an oscillator which is acted upon by a periodic force $F(t)$ in following cases:

a) $F(t) = F \cdot (t/\tau - n)$, when $n\tau \leqslant t < (n+1)\tau$ (Fig. 19);

b) $F(t) = F \cdot (1 - e^{-\lambda t'})$, $t' = t - n\tau$, when $n\tau \leqslant t < (n+1)\tau$ (Fig. 20).

c) Find the stable current through the circuit of Fig. 16 in which there is a.m.f. $U(t) = (t/\tau - n)V$ for $n\tau \leqslant t < (n+1)\tau$. The internal resistance of the battery is zero.

5.20. An oscillator with eigen-frequency ω_0 and with a friction force acting upon it given by $f_{\text{fr}} = -2\lambda m\dot{x}$ has an additional force $F(t)$ acting upon it.

a) Find the average work A done by $F(t)$ when the oscillator is vibrating in a stable mode for the case when $F(t) = f_1 \cos \omega t + f_2 \cos 2\omega t$.

b) Repeat the calculation for the case when

$$F(t) = \sum_{n=-\infty}^{\infty} a_n e^{in\omega t}, \quad a_{-n} = a_n^*.$$

c) Find the average over a long time interval of the work done by the force $F(t) = f_1 \cos \omega_1 t + f_2 \cos \omega_2 t$ when the oscillator performs stable vibrations.

d) Find the total work done by the force

$$F(t) = \int_{-\infty}^{\infty} \psi(\omega) e^{i\omega t} \, d\omega, \quad \psi(-\omega) = \psi^*(\omega)$$

for the case where the oscillator was at rest at $t \to -\infty$.

5.21. A harmonic oscillator is in the field of the travelling wave that acts upon it with the force $F(x, t) = f(t - x/V)$, where x is the displacement of the oscillator from the equilibrium position and V is the wave velocity. Assuming x is small enough, find the connection between the energy ΔE and the momentum

$$\Delta p = \int_{-\infty}^{\infty} F(x(t), t) \, dt,$$

transferred to the oscillator. Consider the approximation quadratic in F only and assume $f(\pm\infty) = 0$.

§6

Small oscillations of systems with several degrees of freedom

In problems 6.1–6.22, we study free and forced oscillations of simple systems (with two or three degrees of freedom) using common methods. In problems 6.19– 6.22, systems with the degenerate frequencies are presented.

For more complex systems, it is helpful to use orthogonality of eigen-oscillations and the symmetry properties of the system. The corresponding theorems are given in problems 6.23 and 6.40 (see also [7], § 24 and § 25), whereas their application are illustrated, for instance, in problems 6.27–6.43 and 6.41–6.47, 6.49 (see also problems related to vibrations of molecules in 6.48, 6.50–6.54).

How small changes in the system can influence its motion will be studied using the perturbation theory. The general form of the perturbation theory for small oscillations is given in problem 6.35, whereas some specific examples can be found in problems 6.36, 6.38, 6.42, 6.43, 6.52 b. It is useful to note that the perturbation theory in quantum mechanics is constructed in a similar way.

Oscillations of systems in which gyroscopic forces act are studied in problems 6.37– 6.39 (see also problems 9.27–9.30 and [7], § 23).

6.1. Find the free oscillations of the system of Fig. 21 for the case when the particles move only vertically. Find the normal coordinates and express the Lagrangian function in terms of these coordinates.

6.2. Find the stable oscillations of the system described in the preceding problem, if the point of suspension moves vertically according to equation $a(t)$, where

a) $a(t) = a \cos \gamma t,$

b) $a(t) = a\left(\frac{t}{\tau} - n\right)$ for $n\tau \leqslant t < (n+1)\tau.$

6.3. Find the free small oscillations of the coplanar double pendulum (Fig. 22).

6.4. Find the normal oscillations for the double pendulum (Fig. 23) with the angle 60°. The angle is between the planes of vibrations of the upper particle with a mass of $3m$

Exploring Classical Mechanics: A Collection of 350+ Solved Problems for Students, Lecturers, and Researchers. First Edition.
Gleb L. Kotkin and Valeriy G. Serbo, Oxford University Press (2020). © Gleb L. Kotkin and Valeriy G. Serbo 2020.
DOI: 10.1093/oso/9780198853787.001.0001

and the lower particle with mass m. The length of each rod is l, the masses of the rods are negligible.

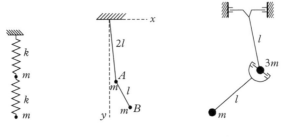

Figure 21   Figure 22   Figure 23

6.5. Find the free oscillations of the system described by the Lagrangian function

$$L = \tfrac{1}{2}\left(\dot{x}^2 + \dot{y}^2 - \omega_1^2 x^2 - \omega_2^2 y^2\right).$$

What is the trajectory of a point with Cartesian coordinate (x,y)?

6.6. Find the normal coordinates of the systems with the following Lagrangian functions:

a) $L = \tfrac{1}{2}\left(\dot{x}^2 + \dot{y}^2 - \omega_1^2 x^2 - \omega_2^2 y^2\right) + \alpha xy$;

b) $L = \tfrac{1}{2}\left(m_1\dot{x}^2 + m_2\dot{y}^2 - x^2 - y^2\right) + \beta \dot{x}\dot{y}$.

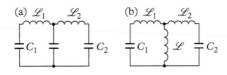

Figure 24

6.7. Find the eigen-oscillations of a system of coupled circuits: a) Fig. 24a;   b) Fig. 24b.

Figure 25   Figure 26

6.8. Find the eigen-oscillations of a system of particles which are connected by springs (Fig. 25). The particles can move only along the straight line AB. Find the free oscillations of the system.

6.9. Find the free oscillations of the system of Fig. 26 where the particles can move only along the straight line AB

   a) if at $t = 0$ one of the particles moves with velocity v while the second particle is at rest and the displacements from the equilibrium position of both particles are zero;

   b) if at $t = 0$ one of the particles is displaced from its equilibrium position over a distance a, while the other is at rest at its equilibrium position and both particles are at rest.

6.10. Determine the flux of energy flow from one particle to the other in the preceding problem.

6.11. Find the free oscillations of the system of Fig. 26 if each of the particles is acted upon by a frictional force which is proportional to the particle's velocity.

6.12. Find the free small oscillations of the coplanar double pendulum of Fig. 27 if in the initial moment the upper pendulum is vertical, while the bottom pendulum is displaced by an angle $\beta \ll 1$, and their velocities are zero. The pendulum masses are M and m, respectively ($M \gg m$).

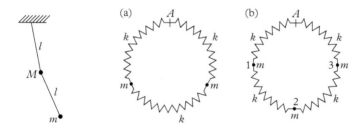

Figure 27   Figure 28

6.13. Find the stable oscillations of the system of Fig. 26 if the point A moves according to the relation $a \cos \gamma t$ in the direction of a straight line AB.

6.14. Find the stable oscillations of a system of two particles which move along the ring[1] (Fig. 28a) for the case when the point A moves along the ring according to the relation $a \cos \gamma t$. Study how the amplitude of the oscillations depends on the frequency of the applied force.

6.15. Three particles with mass m each, connected by springs, can move along the ring (Fig. 28b). Find the stable oscillations of the system if point A moves along the ring according to the law $a \cos \gamma t$.

---

[1] In this and similar problems, the ring is assumed to be smooth and fixed.

6.16. Find the stable oscillations of the system of Fig. 26 when point A moves according to the relation $a\cos\gamma t$. The particles are acted upon by the frictional forces proportional to their velocities.

6.17. Find the motion of the system of Fig. 25 when in the initial moment the particle are at rest in the equilibrium positions, and point A moves according to the relation $a\cos\gamma t$. The masses of all particles are equal ($m_1 = m_2 = m$).

6.18. Determine the stable oscillations of the particle of Fig. 29 moving in a variable field $U(\mathbf{r}, t) = -\mathbf{F}(t)\mathbf{r}$, where the vector $\mathbf{F}(t)$ lies in the plane of the figure, for the following cases:

  a) $\mathbf{F}(t) = \mathbf{F}_0 \cos\gamma t$,

  b) the vector $\mathbf{F}(t)$ rotates with constant absolute amplitude magnitude with a frequency γ.

6.19. Find the eigen-oscillations of three identical particles connected by identical springs and moving along the ring (Fig. 30).

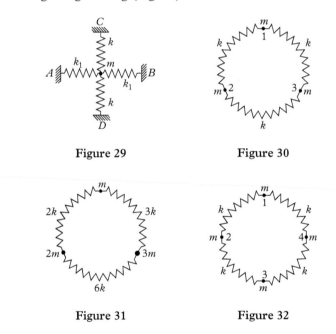

Figure 29   Figure 30

Figure 31   Figure 32

Determine the normal coordinates which reduce the Lagrangian function to a diagonal form.

6.20. Find the free oscillations of the system considered in the preceding problem, if at $t = 0$ one of the particles is displaced from its equilibrium position. The initial velocities of all particles are zero.

6.21. Find the eigen-oscillations of the system of three particles which can move along the ring (Fig. 31).

6.22. Find the normal coordinates of the system of four identical particles on the ring (Fig. 32).

6.23. Consider a system described by the Lagrangian function[1]

$$L = \frac{1}{2}\sum_{i,j}(m_{ij}\dot{x}_i\dot{x}_j - k_{ij}x_ix_j).$$

Let their eigen-oscillations be described by the following equation:

$$x_i^{(l)}(t) = A_i^{(l)}\cos(\omega_l t + \varphi_l).$$

Prove that the amplitudes corresponding to the oscillations with different frequencies ω_l and ω_s satisfy the relation

$$\sum_{i,j}A_i^{(s)}m_{ij}A_j^{(l)} = \sum_{i,j}A_i^{(s)}k_{ij}A_j^{(l)} = 0.$$

6.24. Consider a system described by the Lagrangian function

$$L = \frac{1}{2}\sum_{i,j}(m_{ij}\dot{x}_i\dot{x}_j - k_{ij}x_ix_j)$$

which also has the different eigen-frequencies $\Omega_1 < \Omega_2 < \; < \ldots < \Omega_N$. The linear relationship

$$\sum_i a_ix_i = 0, \quad a_i = \text{const}$$

is imposed upon the system. Prove that all eigen-frequencies ω_l of the system with such relation lie between Ω_l:

$$\Omega_1 \leqslant \omega_1 \leqslant \Omega_2 \leqslant \omega_2 \leqslant \ldots \leqslant \omega_{N-1} \leqslant \Omega_N.$$

---

[1] Since the product $\dot{x}_i\dot{x}_j$ is symmetric with respect to the replacement of $i \leftrightarrow j$, then the matrix m_{ij} can always be chosen as the symmetric one, $m_{ij} = m_{ji}$. The same holds for the matrix $k_{ij} = k_{ji}$.

6.25. We can write the stable oscillations of the system described by the Lagrangian function

$$L = \frac{1}{2}\sum_{i,j}(m_{ij}\dot{x}_i\dot{x}_j - k_{ij}x_ix_j) + \sum_i x_if_i\cos\gamma t,$$

in the form $x_i(t) = \sum_l \lambda^{(l)}A_i^{(l)}\cos\gamma t$ (see problem 6.23.) (Why?)
   Express the coefficients $\lambda^{(l)}$ in terms of the f_i and the $A_i^{(l)}$.
   Study the γ-dependence of the $\lambda^{(l)}$.
   Show that $\lambda^{(s)} = 0$, if $\sum_i f_i A_i^{(s)} = 0$ for the sth normal oscillation.

6.26. A system of particles connected by springs can perform small oscillations. One of them (the particle A) be acted upon by the force $F(t) = F_0\cos\gamma t$ along the x-direction, while the other (the particle B) performs the stable oscillations, during which the projection of its displacement on the x'-direction has the form $x'_B = C\cos\gamma t$.
   Show that when force F acts upon the particle B along the x'-axis, there will be the oscillations $x_A = C\cos\gamma t$ of the particle A (the reciprocity theorem).

6.27. Find the eigen-oscillations of the system of four particles moving along the ring (Fig. 33). All springs are identical, and the masses of particles 1 and 3 are m, while those of particles 2 and 4 are M.

6.28. Find the eigen-oscillations of the system of four particles of Fig. 34; the particles move along the ring.

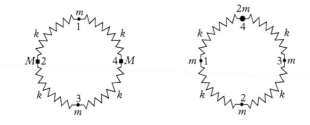

Figure 33   Figure 34

6.29. a) Find the normal oscillations of the system of Fig. 35. All particles and springs are identical. The tension in the springs at equilibrium is $f = kl$, where l is the equilibrium distance between the particles.
   Hint: Several of the eigen-vibrations are obvious. The determination of the others can be simplified by using the relations of problem 6.23.
   b) Find the eigen-oscillations of the system of four identical particles of Fig. 32 for the case where the mass of particle 5 is equal to zero and the springs in this place are

connected. The elasticity coefficients and tensions at equilibrium of all the springs are the same as before.

6.30. Three particles with different masses m_i ($i = 1, 2, 3$) move along the ring (Fig. 36). At what values of the elasticity coefficients k_i will the degeneracy of frequencies occur in this system?

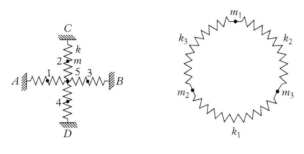

Figure 35   Figure 36

6.31. Which of the eigen-vibrations of the system of Fig. 30 remain practically unchanged when the following small changes are made in the system:

a) the elasticity coefficient of spring 1–2 is changed by a small amount δk;

b) a small mass δm is added to particle 3;

c) a small mass δm_1 is added to particle 1 and δm_2 to particle 2?

6.32. Describe the free oscillations of the system of the preceding problem for the case a) and b) if initially the particles 1 and 3 are displaced over equal distances in opposite directions so as to decrease their mutual distance. All velocity are initially zero.

6.33. The system of Fig. 32 has degenerate frequencies; therefore, its eigen-vibrations are not uniquely determined. Even a small change in the mass of particles or a small change in the elasticity of the springs can lead to *the removal of degeneration*.
  Find the eigen-vibrations of the system of Fig. 32 which are practically the same as the eigen-vibrations of the system which is obtained

a) by adding identical masses to the first and the second particles;

b) by changing the elasticity coefficients of the springs 1–2 and 3–4;

c) by adding an extra mass to the first particle.

6.34. The particles 1 and 3 of the system described in the problem 6.35b are at time $t = 0$ displaced by the same amount from their equilibrium positions in opposite direction so as to decrease their distance apart. Initially all velocities are equal zero. Describe the free oscillations of the system.

6.35. Consider a system described by the Lagrangian function

$$L_0 = \frac{1}{2} \sum_{i,j} (m_{ij}\dot{x}_i\dot{x}_j - k_{ij}x_ix_j).$$

Now consider the new system which has been obtained by the small change in the original Lagrangian function:

$$\delta L = \frac{1}{2} \sum_{i,j} (\delta m_{ij}\dot{x}_i\dot{x}_j - \delta k_{ij}x_ix_j).$$

Find the small change of the eigen-frequencies of the new system if all eigen-frequencies of the original system are non-degenerate.

6.36. Find the small changes in the eigen-frequencies of the system of Fig. 34 when a small mass δm is added to the first particle so that $\varepsilon = \delta m/m \ll 1$.

6.37. Determine the free oscillations of an anisotropic charged oscillator moving in the potential field

$$U(\mathbf{r}) = \frac{1}{2} m(\omega_1^2 x^2 + \omega_2^2 y^2 + \omega_3^2 z^2)$$

and in a uniform constant magnetic field $\mathbf{B}$ which is parallel to the z-axis. Consider in particular the following limit cases:

 a) $|\omega_B| \ll |\omega_1 - \omega_2|$,
 b) $|\omega_B| \gg \omega_{1,2}$,
 c) $\omega_1 = \omega_2 \gg |\omega_B|$

(here $\omega_B = eB/(mc)$).

6.38. Determine the free oscillations of an anisotropic harmonic oscillator moving in the potential field

$$U(\mathbf{r}) = \frac{1}{2} m(\omega_1^2 x^2 + \omega_2^2 y^2 + \omega_3^2 z^2)$$

and in a weak constant uniform magnetic field $\mathbf{B} = (B_x, 0, B_z)$, considering the effect of the magnetic field to be a small perturbation.

6.39. A mathematical pendulum is part of an electric chains (Fig. 37). A constant uniform magnetic field **B** is applied at right angles to the plane of the figure. Find the eigen-vibrations of the system.

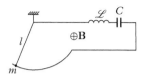

6.40. Suppose that a system, making small oscillations (and hence its Lagrangian function $L(x, \dot{x})$ as well), does not change its form with respect to the replacement

Figure 37

$$x_i \to \sum_j S_{ij} x_j; \quad \dot{x}_i \to \sum_j S_{ij} \dot{x}_j, \quad i, j = 1, 2, \ldots, N.$$

Here the constant coefficients $S_{ij} = S_{ji}$ satisfy the condition[1]

$$\sum_j S_{ij} S_{jk} = \delta_{ik}.$$

Prove that

a) if the eigen-oscillation $x_i = A_i \cos(\omega t + \varphi)$ is not degenerate, then the amplitude A_i is either symmetric or anti-symmetric with respect to the given transformation, that is, either $\sum_j S_{ij} A_j = +A_i$ or $\sum_j s_{ij} A_j = -A_j$;

b) if the frequency is degenerate, one can choose the eigen-oscillation to be either symmetric or anti-symmetric;

c) if the system is acted upon by an external force which is symmetric (or anti-symmetric) with respect to this transformation, then the anti-symmetric (or the symmetric) eigen-oscillations is not excited. (This is one example of the so-called selection rules.)

6.41. Using symmetry considerations, find the eigen-oscillations of the system of Fig. 38.

6.42. Determine the free oscillations of the system of Fig. 35 when at $t = 0$ particles 1 and 4 are displaced over equal distances in the horizontal direction towards each other in such a way that the distance between them decreases. At $t = 0$ the velocities of all particles equal zero. The tension of the springs is $f = kl_1$; $l - l_1 \ll l$, where l is the equilibrium distance between the particles (cf. problem 6.32).

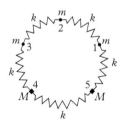

Figure 38

---

[1] This condition means that double application of such a replacement returns the system to its original state. For example, this is valid for reflections with respect to a plane of the system symmetry or rotations at $180°$ relative to the symmetry axis.

6.43. Find corrections to the frequencies of the eigen-oscillations of the systems of four particles on the ring (Fig. 32). Such corrections arise from the small changes in the particles' masses: by δm_1 for the particle 1 and by δm_2 for the particle 2.

6.44. Using symmetry considerations, determine the vectors of the eigen-oscillations for the system of Fig. 39. All the masses of the particles and the springs are identical.

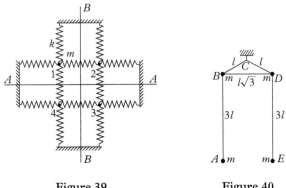

Figure 39   Figure 40

6.45. Find the eigen-oscillations of the "scales" of Fig. 40. The suspension of a rigid frame BCD is implemented with a short flexible thread allowing any turns of the frame around the point C. The length of the rods $BC = CL = l$, $BD = l\sqrt{3}$, the length of the threads $AB = DE = 3l$. The identical particles are fixed at points A, B, D, and E. The masses of the rods and threads can be neglected.

6.46. Find the eigen-oscillations of the system of eight particles which are attached to the fixed frame by springs (Fig. 41). The elasticity coefficients k, the tensions f and the lengths l of springs are identical.

6.47. The frame shown in Fig. 41 oscillates along the direction AA according to the equation $a\cos\gamma t$. At what values of the frequency γ will the resonant amplification of the oscillations be possible?

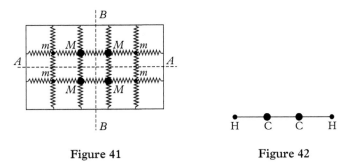

Figure 41   Figure 42

6.48. Find the eigen-oscillations of a linear symmetric molecule of acetylene C_2H_2 (Fig. 42), assuming that the potential energy of the molecule depends on the distance between neighbouring atoms as well as on angles HCC.

6.49. Two identical particles are attached to a fixed frame by springs (Fig. 43). The system is symmetric with respect to the *CF*-axis. What information about the eigen-oscillations of the system can be obtained if the elasticities and the tensions of the springs are unknown?

6.50. Classify the eigen-oscillations of a molecule of ethylene C_2H_4 by their symmetry properties with respect to axes *AB* and *CD* (Fig. 44). In the equilibrium position all atoms of the molecule are located in the same plane.

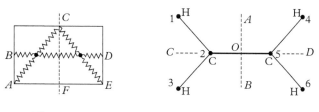

Figure 43   Figure 44

6.51. Determine the eigen-oscillations in a plane of a molecule which has the shape of an equilateral triangle. Assume that the potential energy depends only on the distances between atoms and that all atoms are the same. The angular momentum is equal to zero up to terms of first order in the amplitudes of the oscillations.

6.52. A molecule AB_3 has the shape of an equilateral triangle with atom A in its centre and atoms B in its vertices (the molecule of boron chloride BCl_3 is an example of such a model).

   a) Using symmetry considerations, determine the degree of degeneracy of the eigen-frequencies of the molecule.

   b) Consider the oscillations which leave the molecule as an equilateral triangle, and the oscillations which output atoms of the plane. Determine how the frequencies of such oscillations will change when one of the atoms B (its mass m) is replaced by an isotope whose mass $m + \delta m$ is close to m. The mass of atom A is m_A.

6.53. Use symmetry arguments to determine the degree of degeneracy of various frequencies for the case of a "molecule" consisting of four identical "atoms" which has the form of a regular tetrahedron at equilibrium.

6.54. A molecule of methane CH_4 has the form of a regular tetrahedron with fourth hydrogen atoms in vertices and one carbon atom in its centre.

a) Determine the degree of degeneracy of the eigen-frequencies of the molecule.

b) How many eigen-frequencies will experience resonance amplification when this molecule is under the influence of a uniform alternating electric field? (Such are, for instance, the electromagnetic waves of the infrared range whose wavelength is several orders of magnitude greater than the size of the molecule.) Note that the hydrogen and carbon atoms have charges of opposite signs.

How does the amplitude of the oscillations of the carbon atom depend on the orientation of the molecule with respect to the electric field?

§7

Oscillations of linear chains

Chain of particles connected by springs are the simplest models used in theory of solids (e.g. see, [16]). The electrical analogues of such lines are r.f. lines employed in in radio engineering.

7.1. Determine the eigen-vibrations of a system of N identical particles with masses m connected by identical springs with elastic constant k and moving along a straight line (Fig. 45).

Hint: Express the eigen-vibrations in the form of a superposition of travelling waves.

7.2. Repeat this for the system of Fig. 46 with one free end.

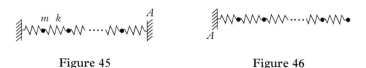

Figure 45   Figure 46

7.3. Determine the eigen-vibrations of the system of N plane pendulums suspended from each other (Fig. 47). The masses of all pendulums are the same; the lengths are equal to Nl, $(N-1)l$, ..., $2l$, l, counting from the top.

7.4. Find the free oscillations of N particles which are connected by springs and which can move along the ring (Fig. 48). All particles and the elastic constants of the springs are the same. Let the motion be that of a wave travelling along the ring. Check whether the energy flux equals the product of the linear energy density and the group velocity.

7.5. Determine the free oscillations of a system of particles moving along a straight line for the following cases (consider the hint for problem 7.1):

Exploring Classical Mechanics: A Collection of 350+ Solved Problems for Students, Lecturers, and Researchers. First Edition.
Gleb L. Kotkin and Valeriy G. Serbo, Oxford University Press (2020). © Gleb L. Kotkin and Valeriy G. Serbo 2020.
DOI: 10.1093/oso/9780198853787.001.0001

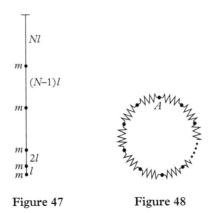

Figure 47   Figure 48

a) $2N$ particles, alternating with masses m and M connected by springs of elastic constant k (Fig. 49);

b) $2N$ particles with masses m connected by springs with alternating elastic constants k and K (Fig. 50);

Figure 49   Figure 50

c) $2N + 1$ particles of mass m connected by springs with alternating elastic constants k and K (Fig. 51).

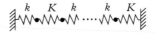

Figure 51

7.6. a) Find the stable oscillations of the system described in problem 7.1 if the point A moves according to $a \cos \gamma t$ (Fig. 45).

b) Let the left point, at which the spring of the same system is fixed, moves according to $x_0 = a \cos \gamma t$. Under what law should move the point A to make the oscillation represent a wave travelling along the chain towards the point A?

What will be in this case the flux of energy along the chain?

c) Consider the same question as in a) for the system of Fig. 46.

7.7. The same questions as in a) and b) of the preceding problem, but for the system of Fig. 49.

7.8. Determine the eigen-vibrations of a system of N particles which move along a straight line for the following cases:

   a) $m_i = m \neq m_N$, $i = 1, 2, \ldots, N - 1$; the elastic constants of all the springs are the same (Fig. 52). Study the cases when $m_N \gg m$ and $m_N \ll m$;

   b) $k_i = k \neq k_{N+1}$, $i = 1, 2, \ldots, N$; all the masses are the same (Fig. 53). Study the cases when $k_{N+1} \gg k$ and $k_{N+1} \ll k$.

        Figure 52   Figure 53

7.9. a) N pendulums are connected by springs and move only in the vertical plane passing through the horizontal suspension line (Fig. 54). Find the eigen-vibrations of the system, if all the pendulums and the springs are the same and, in the position of equilibrium, the length of the spring is equal to the distance between the suspension points of neighbouring pendulums.

   b) Find the forced oscillations of the system of Fig. 54 for the case when the last particle is acted upon by the driving force $F(t) = F \sin \gamma t$ which is directed parallel to the suspension line.

   c) $2N$ identical pendulums are connected by identical springs and move only in vertical planes which are perpendicular to the circular line of suspension (Fig. 55). The distance between neighbouring suspension points is a. The length of each of the springs in the unstretched state is b. Study how the stability of small oscillations near the vertical line depends on the parameter $b - a$. Consider the radius of the circular suspension line R to be large enough so that small quantities l/R, a/R, b/R can be neglected.

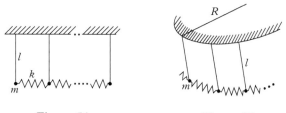

        Figure 54   Figure 55

7.10. a) An AC voltage source $U \cos \gamma t$ is connected to one end of the artificial line (Fig. 56). Which chain Z with resistance R and inductance $\mathcal{L}_0$ (or do we need capacity?) must be connected to the other end of the line to produce the oscillations in the form of a *travelling wave* (i.e. in this case the voltage on each of the capacitors will differ from that of the neighbour's only by a definite phase shift)?

   b) Do the same for the artificial line of Fig. 57.

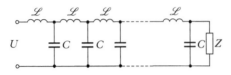

Figure 56

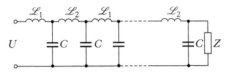

Figure 57

7.11. Consider the elastic rod to be the limiting case of the system of N particles of Fig. 45 in the limit $N \to \infty$, $a \to 0$, where m and a are, respectively, the mass of particles and the distance between neighbouring particles at equilibrium, while Nm and Na kept constant. Write down the equations of motion for the oscillations of the rod as the limiting case of the equations of motion of the discrete system.

Hint: Introduce the coordinate of a point of the rod at equilibrium $\xi = na$ and consider the following quantities

$$x(\xi, t) = \lim x_n(t), \qquad \frac{\partial x}{\partial \xi} = \lim \frac{x_n(t) - x_{n-1}(t)}{a}$$

with the limit $a \to 0$.

7.12. Write down the equations of motion for the oscillations of the rod in the preceding problem taking into account the first non-vanishing correction due to a finite distance a between the neighbouring particles.

§8

Non-linear oscillations

8.1. Determine the distortion in the oscillations of a harmonic oscillator which is caused by the presence of anharmonic terms in the potential energy for the following cases:
a) $\delta U(x) = \frac{1}{4} m\beta x^4$; b) $\delta U(x) = \frac{1}{3} m\alpha x^3$.

8.2. Determine the distortion in the oscillations of a harmonic oscillator which is caused by the presence of anharmonic term $\delta T = \frac{1}{2} m\gamma x \dot{x}^2$ in the kinetic energy.

8.3. Determine the anharmonic corrections to the oscillations of a pendulum whose point of suspension moves along a circle (Fig. 17; $l \gg a$).

8.4. Determine the oscillations of a harmonic oscillator when there is a force

$$f_1 \cos\omega_1 t + f_2 \cos\omega_2 t$$

acting upon it, taking anharmonic corrections into account for the case when $\delta U(x) = \frac{1}{3} m\alpha x^3$.

8.5. A pendulum consists of a particle of mass m in the field of gravity g suspended on a spring with the elasticity constant k (Fig. 58). The length of the spring in the free state is l_0. Find the anharmonic corrections to the oscillations of the pendulum.
  Use the Cartesian coordinates of the displacement of the particle from the equilibrium position.

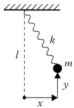

Figure 58

8.6. Find the amplitude of the stable oscillations of an anharmonic oscillator which satisfies the equation of motion

$$\ddot{x} + 2\lambda\dot{x} + \omega_0^2 x + \beta x^3 = f\cos\omega t$$

a) in the resonance region, $|\omega - \omega_0| \ll \omega_0$;
b) in the region where there is resonance with the tripled frequency of the force, $|3\omega - \omega_0| \ll \omega_0$ (frequency tripling).

Exploring Classical Mechanics: A Collection of 350+ Solved Problems for Students, Lecturers, and Researchers. First Edition.
Gleb L. Kotkin and Valeriy G. Serbo, Oxford University Press (2020). © Gleb L. Kotkin and Valeriy G. Serbo 2020.
DOI: 10.1093/oso/9780198853787.001.0001

8.7. a) Determine the amplitude and   phase of the stable vibration of an oscillation under conditions of parametric resonance:

$$\ddot{x} + 2\lambda\dot{x} + \omega_0^2(1 + h\cos 2\omega t)x + \beta x^3 = 0$$

provided

$$h \ll 1, \quad |\omega - \omega_0| \ll \omega_0, \quad \beta x^2 \ll \omega_0^2.$$

b) Determine the amplitude of the third harmonic in the stable vibration.

8.8. Determine the vibrations of the oscillator

$$\ddot{x} + \omega_0^2(1 + h\cos 2\omega t)x = 0$$

provided

$$h \ll 1, \quad |\omega - \omega_0| \ll \omega_0$$

a) in the region where instability for parametric resonance occurs;
b) close to the region of instability.

8.9. Let the frequency of a harmonic oscillator $\omega(t)$ change as is indicated in Fig. 59. Find the region where instability against parametric resonance occurs.

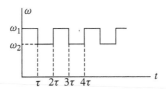

Figure 59

8.10. Determine how amplitudes will change over time for the weakly coupled oscillators whose Lagrangian function is

$$L = \tfrac{1}{2}m(\dot{x}^2 + \dot{y}^2 - \omega^2 x^2 - 4\omega^2 y^2 + 2\alpha x^2 y).$$

8.11. Find the frequency of the small free oscillations of a pendulum whose point of suspension performs vertical oscillations with a high frequency γ ($\gamma \gg \sqrt{g/l}$).

8.12. Find the effective potential energy for the following cases:

a) a particles of mass m moves in the field

$$U(\mathbf{r}) = \frac{\alpha}{|\mathbf{r} - \mathbf{a}\cos\omega t|} - \frac{\alpha}{|\mathbf{r} + \mathbf{a}\cos\omega t|}$$

at a distance $r \gg a$;

b) a harmonic oscillator moving in the field

$$U(\mathbf{r}) = 2\alpha\,\frac{\mathbf{ar}}{r^3}\,\cos\omega t.$$

8.13. Determine the motion of a fast particle entering the field $U(\mathbf{r}) = a(x^2 - y^2)\sin kz$ at a small angle to the z-axis $(k^2 E \gg a)$.

8.14. Determine velocity of the orbit centre's displacement for a charged particle in the weakly inhomogeneous magnetic field

$$B_x = B_y = 0, \quad B_z = B(x), \quad \varepsilon = \frac{B'(x)}{B(x)}r \ll 1, \quad r = \frac{mvc}{eB(x)},$$

where r is the orbit radius.

8.15. The following problem represents a mechanical model of phase transitions of the second kind.

The iron ball of mass m can oscillate along the y-axis. It is connected to the spring whose potential energy has the form[1] $U(y) = -Cy^2 + By^4$. Using an electromagnet one can excite vibrations of the ball according to the equation $y(t) = y_0\cos\gamma t$, where γ is considerably greater than the frequency of the eigen-vibration of the ball.[2]

Find how the frequency of small eigen-oscillations of the ball depends on the parameter $T = y_0^2$.

---

[1] Such is, for example, the potential energy of the system shown in Fig. 15 where a ball can move only in the direction of the y-axis perpendicular to the AB line, and the length of the unstretched springs l_0 is larger then l. Therein, if $|y| \ll l$, we have
$$C = k(l_0 - l)/l, \quad B = kl_0/(4l^3).$$

[2] Under the action of high-frequency force $f(t) = -(y_0\gamma^2/m)\cos\gamma t$, the amplitude of the forced vibration tends to y_0 in limit $\gamma \to \infty$, and the corrections to the amplitude and the higher harmonics are small $\sim 1/\gamma^2$.

§9

Rigid-body motion. Non-inertial coordinate systems

9.1. Find the normal oscillations of a uniform ring of radius R, suspended on a string of length R in the gravity field g (Fig. 60). Consider only small oscillations in the plane of the ring.

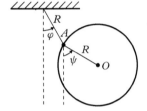

9.2. Masses m and M are located at the vertices of a square with side lengths $2a$ (Fig. 61a). Find the components of the inertia tensor

    a) relative to the x-, y-, and z-axes;

Figure 60

    b) relative to the the x' and y'-axes which are the diagonals of the square and the z-axis.

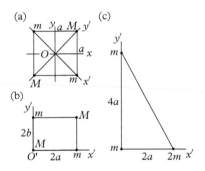

Figure 61

9.3. Find the principal axes and the principal moments of inertia for the following systems:

    a) masses m and M at the vertices of a rectangle with side lengths $2a$ and $2b$ (Fig. 61b).

    b) masses m and $2m$ at the vertices of a right-angled triangle with side lengths $2a$ and $4a$ (Fig. 61c).

Exploring Classical Mechanics: A Collection of 350+ Solved Problems for Students, Lecturers, and Researchers. First Edition.
Gleb L. Kotkin and Valeriy G. Serbo, Oxford University Press (2020). © Gleb L. Kotkin and Valeriy G. Serbo 2020.
DOI: 10.1093/oso/9780198853787.001.0001

9.4. Give an expression for the moment of inertia $I_{\mathbf{n}}$ with respect to an axis parallel to a unit vector $\mathbf{n}$ and passing through the centre of mass of the body in terms of the components of the inertia tensor.

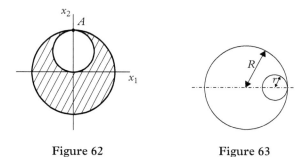

Figure 62   Figure 63

9.5. The cloakroom number tag is made of uniform thin plate. The tag is shaped as a disc of radius R with a hole in the shape of circle of radius $R/2$. The centre of this hole is at a distance $R/2$ from the tag's edge (Fig. 62). Find the frequency of small oscillations of the tag, suspended in the gravity field g and able to rotate around point A.

9.6. Determine the principal moments of inertia of a uniform ball of radius R inside of which there is a cavity in the form of a ball of radius r (Fig. 63).

9.7. Express the components of the mass quadrupole moment tensor,

$$D_{ik} = \int (3x_i x_k - r^2 \delta_{ik}) \rho \, dV$$

where ρ is the density, in terms of the components I_{ik} of the inertia tensor.

9.8. Determine the frequency of small vibrations of a uniform hemisphere which lies on on a smooth horizontal surface in the field of gravity.

9.9. The period of the Earth's rotation around its axis increases due to the action of tidal forces and may become equal to the period of circulation of the Moon around the Earth, that is, to the month. What will be the period of the Earth's rotation at the moment when it becomes equal to the month? Consider for simplicity that the Earth's axis of rotation is perpendicular to the plane of the Earth and the Moon orbits. For numerical estimates, consider the Earth as a uniform ball of radius $a = 6.4$ thousand km and mass M which is 81 times greater than the mass of the Moon m; the distance from the Earth to the Moon equals $R = 380$ thousand km.

9.10. Two identical uniform balls, rotating with equal angular velocities ω, slowly approach closer and then rigidly connect with each other. Determine the motion of the newly formed body. Find which part of the initial kinetic energy has been converted into heat. Prior to connecting, the angular velocities of the balls were directed:

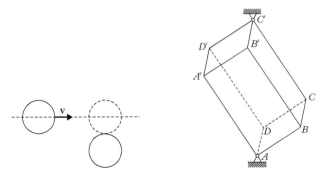

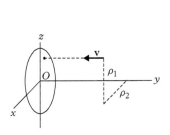

Figure 64

Figure 65

a) perpendicular to the line of the balls' centres and parallel to each other;

b) one along the line of the balls' centres, and the other, perpendicular to this line.

9.11. A uniform ball of radius r and mass m rolls, without slipping, on a horizontal plane at velocity v. At the moment when it touches another motionless ball, both balls become rigidly connected (Fig. 64). The plane is smooth; therefore, upon connecting, the balls freely slide on it. What are the forces with which the balls act upon the plane? The acceleration of gravity is large enough so that the balls are in constant contact with the plane.

9.12. A uniform rectangular parallelepiped has two spherical hinges attached to its opposite vertices A and C'. The parallelepiped rotates freely around the diagonal AC' with an angular velocity Ω (Fig. 65). Find the forces acting upon the hinges.

9.13. A particle moving parallel to the Oy-axis with velocity $\mathbf{v}$ and with impact parameters ρ_1 and ρ_2 is incident upon a uniform ellipsoid of rotation (with semi-axes $a = b$ and c) (Fig. 66) and sticks to it. Describe the motion of the ellipsoid assuming its mass to be much larger than that of the incident particle.

9.14. A gyrocompass is a rapidly rotating disc whose spinning axis is confined to a horizontal plane (Fig. 67). Study the motion of the gyrocompass at latitude α. The angular velocity of the Earth's rotation equals Ω.

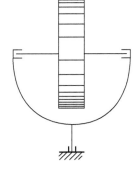

Figure 66

Figure 67

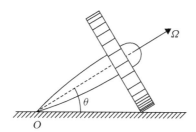

Figure 68

9.15. A top with a fixed fulcrum O touches the horizontal plane with the edge of its disc (Fig. 68). Before the touchdown, the top was spinning around its axis with an angular velocity Ω (assume velocity of precession to be small).

Find the angular velocity of the top when slipping of the disc vanishes. There were no nutations at the time of the touchdown.

9.16. An isotropic ellipsoid of revolution of mass m moves in the gravitational field produced by a fixed point of mass M. Use the spherical coordinates of the centre of mass and the Euler angles as generalized coordinates and determine the Lagrangian function of the system. Assume the size of the ellipsoid to be small compared to the distance from the centre of the field.

Hint: The potential energy of the system is approximately equal to

$$U(\mathbf{R}) = m\varphi(R) + \frac{1}{6}\sum_{i,k=1}^{3} D_{ik}\frac{\partial^2\varphi(R)}{\partial X_i\partial X_k},$$

where $\mathbf{R} = (X_1, X_2, X_3)$ is the radius vector of the centre of the ellipsoid, D_{ik} is the mass quadrupole tensor (see problem 9.7), and $\varphi(R) = -GM/R$ is the potential of the gravitational field (cf. [2], §42).

9.17. Determine the angular velocity of precession for the Earth's axis caused by the attraction forces of the Sun and the Moon. For simplicity, consider the Earth as a uniform ellipsoid of rotation whose equatorial axis a is greater than its polar axis c, so that $\frac{a-c}{a} \approx \frac{1}{300}$. Assume the Earth's and the Moon's orbits to be the circumferences lying in the same plane. The tilt of the Earth's axis to the plane of the Earth and Moon's orbits equals $67°$.

9.18. Write down the equations of motion for the components of the angular momentum along moving coordinate axes. These axes coincide with the principal axes of the inertia tensor. Integrate these equations for the case of the free motion of a symmetric top.

9.19. Use the Euler equations to study the stability of rotations around the principal axes of the moment of inertia tensor of an asymmetric top.

9.20. A uniform ball of radius a moves in the field of gravity along the inner surface of a vertical cylinder of radius b without slipping. Find the motion of the ball.

9.21. a) A plane disc, symmetric around its axis, rolls in the field of gravity over a smooth horizontal plane without friction. Find its motion in the form of quadratures.

Answer in detail the following questions:

Under what conditions does the angle of inclination of the disc to the plane remain constant?

If the disc rolls in such a way that its axis has a fixed (horizontal) direction in space, at what angular velocity will the rotation around this axis be stable?

b) A disc rolls without slipping over a horizontal surface. Find the equations of motion for this case and answer the same questions as in a).

c) Do the same for a disc which rolls without slipping over the horizontal plane and without rotation around the vertical axis.[1]

d) A disc rotates without slipping around its diameter which is at right angle to an inclined plane the disc is placed upon. The plane makes a small angle α with the horizontal plane. Find the displacement of the disc over a long time period.

9.22. a) Find in terms of quadratures the law of motion of an inhomogeneous ball which is slipping without friction on a horizontal plane in the field of gravity. The mass distribution of the ball is symmetrical with respect to the axis passing through the geometric centre and the centre of mass of the ball.

b) Find the equations of motion of the ball described in 9.22a if it rolls without slipping over a horizontal plane.

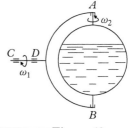

Figure 69

9.23. A particle is dropped from a height of h with zero initial velocity. Find its displacement from the vertical in the direction of the East and the South. For simplicity, consider the Earth as a uniform ball of radius R.

9.24. A vessel partially filled with gradually solidifying epoxy resin, is rotating in the field of gravity with an angular velocity ω_2 around the *AB*-axis, which in turn rotates around the fixed *CD*-axis with an angular velocity ω_1 (Fig. 69). What form will it take when the surface of the resin completely solidifies?[2]

9.25. A particle moves in a central field $U(r)$. Find the equation for its trajectory and describe its motion in a coordinate system which rotates uniformly with an angular velocity $\boldsymbol{\Omega}$ parallel to its angular momentum **M**.

9.26. Find the small oscillations of the particle with mass m which is fastened to a frame by springs with elastic constants k_1 and k_2. The frame rotates in its own plane with an angular velocity $\boldsymbol{\Omega}$ (Fig. 70). The particle moves in the plane of the frame.

---

[1] This means that the cohesion of the disc to the plane at the "point" of contact is so firm that the area of contact neither slips nor rotates. The energy loss due to rolling friction can be neglected.

[2] Problem by V. S. Kuzmin and M. P. Perelroizen.

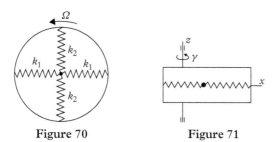

Figure 70   Figure 71

9.27. A smooth paraboloid

$$z = \frac{x^2}{2a} + \frac{y^2}{2b}$$

rotates in the field of gravity around the vertical z-axis with an angular velocity ω. At what value of ω will the lower position of a particle inside the paraboloid become unstable? The acceleration of gravity is $\mathbf{g} = (0, 0, -g)$.

9.28. A particle of mass m is fixed inside a frame by the strings of the lengths l, the elasticity constants k and the tension f (when the frame is at rest). The frame rotates with an angular velocity γ around the z-axis, which is shifted by a distance a from the centre of the frame (Fig. 71).

   Determine the equilibrium distance of the particle from the axis of rotation and investigate the stability of this equilibrium positions.

   Consider the following cases:

   a) the particle can only move along the springs;
   b) any displacements of the particle are possible.

9.29. Two stars move along the circular orbits around their centre of mass. Consider the frame of reference where the stars are motionless. Find the points at which a light particle in this system is also motionless. Investigate the stability of these "equilibrium positions". (Ignore the points lying on a straight line connecting the stars.)

§10

The Hamiltonian equations of motion. Poisson brackets

10.1. Let the Hamiltonian function H of a system of particles be invariant under an infinitesimal translation (rotation). Prove the linear (angular) momentum conservation law.

10.2. Use the Euler angles θ, φ, and ψ to find the Hamiltonian function of a symmetric top when there is no external field acting on the top.

10.3. Determine the Hamiltonian function of an anharmonic oscillator, if the Lagrangian function is given by the equation

$$L = \tfrac{1}{2}\left(\dot{x}^2 - \omega^2 x^2\right) - \alpha x^3 + \beta x \dot{x}^2.$$

10.4. Describe the motion of a particle with the Hamiltonian function

$$H(x, p) = \tfrac{1}{2}\left(p^2 + \omega_0^2 x^2\right) + \tfrac{1}{4}\lambda(p^2 + \omega_0^2 x^2)^2.$$

10.5. Do the same for $H(x, p) = A\sqrt{p} - xF$.

10.6. Find the equations of motion for the case when the Hamiltonian function is of the form $H(\mathbf{r}, \mathbf{p}) = \frac{c|\mathbf{p}|}{n(\mathbf{r})}$ (a beam of light).
   Find the trajectory if $n(\mathbf{r}) = ax$.

10.7. Find the Lagrangian function for the case when the Hamiltonian function is
   a) $H(\mathbf{r}, \mathbf{p}) = \frac{\mathbf{p}^2}{2m} - \mathbf{pa}$ ($\mathbf{a} = \text{const}$);   b) $H(\mathbf{r}, \mathbf{p}) = \frac{c|\mathbf{p}|}{n(\mathbf{r})}$.

10.8. Describe the motion of a charged particle in a uniform constant magnetic field $\mathbf{B}$ by solving the Hamiltonian equations of motion; take the vector potential in the form

$$A_x = A_z = 0, \quad A_y = xB.$$

Exploring Classical Mechanics: A Collection of 350+ Solved Problems for Students, Lecturers, and Researchers. First Edition.
Gleb L. Kotkin and Valeriy G. Serbo, Oxford University Press (2020). © Gleb L. Kotkin and Valeriy G. Serbo 2020.
DOI: 10.1093/oso/9780198853787.001.0001

10.9. Study qualitatively the motion of a charged particle in a non-uniform magnetic field described by the vector potential $\mathbf{A}(\mathbf{r}) = (0, hx^2, 0)$. Compare the result with the drift approximation.

10.10. Show that the problem of motion of two particles with opposite charges (e and $-e$) in a uniform constant magnetic field is reducible to the problem of the motion of one particle in a given potential field [24].

In problems 10.11 to 10.14 we are dealing with the motion of electrons in a metal or semiconductor. Electrons in a solid form a system of particles which interact both with themselves and with the ions which form the crystal lattice. Their motion is described by quantum mechanics. In solid state theory one is often able to reduce the problem of many interacting particles which form the solid to the problem of the motion of separate free particles (the so-called quasi-particles: electrons or holes, depending on the sign of their charge) for which, however, the momentum dependence $\varepsilon(\mathbf{p})$ ("dispersion law") of the energy is complicated. For instance, for "holes" in silicon crystal

$$\varepsilon(\mathbf{p}) = \frac{1}{2m}\left[-4p^2 \pm \sqrt{1.2p^4 + 17(p_x^2 p_y^2 + p_x^2 p_z^2 + p_y^2 p_z^2)}\right],$$

where the coordinate axes are chosen to coincide with the crystal symmetry and m is the electron mass.

In many cases it turns out that it is possible to consider the motion of the quasi-particles using classical mechanics (e.g. [16]). The function $\varepsilon(\mathbf{p})$ is periodic with a period equal to the period of the so-called reciprocal lattice.[1] Otherwise one can assume that $\varepsilon(\mathbf{p})$ has an arbitrary form.

10.11. It is well known that $\varepsilon(\mathbf{p})$ is a periodic function of $\mathbf{p}$ with a period which is equal to that of the reciprocal lattice, multiplied by $2\pi\hbar$; for instance, for a simple cubic lattice with lattice constant a, the period of $\varepsilon(\mathbf{p})$ is equal to $2\pi\hbar/a$).

Describe the motion of an electron in a uniform electric field $\mathbf{E}$.

Hint to problems 10.12–10.14: In these problems it is convenient to introduce the kinematic momentum $\mathbf{p} = \mathbf{P} - \frac{e}{c}\mathbf{A}$ besides the generalized momentum $\mathbf{P}$ (here $\mathbf{A}$ is the vector potential of the magnetic field).

10.12. Obtain the equation of motion using the Hamiltonian function

$$H(\mathbf{P}, \mathbf{r}) = \varepsilon\left(\mathbf{P} - \frac{e}{c}\mathbf{A}\right) + e\varphi$$

where the electron charge e is taken to be negative, $e < 0$.

---

[1] For instance, for a crystal the lattice of which has a smallest period a in the x-direction, we have $\varepsilon(p_x, p_y, p_z) = \varepsilon\left(p_x + \frac{2\pi\hbar}{a}, p_y, p_z\right)$, where $\hbar$ is the Planck constant.

10.13. a) Determine the integrals of motion of an electron moving in a solid in uniform constant magnetic field **B**. What does the "orbit" in momentum space look like?

b) Prove that the projection of the electron orbit in a uniform constant magnetic field onto a plane at right angles to **B** in coordinate space can be obtained by rotation and change of scale of the orbit in momentum space.

10.14. Express the period of revolution of an electron in a uniform constant magnetic field **B** in terms of the area of $S(E, p_B)$ of the section cut off by the plane $p_b = \mathbf{p}\mathbf{B}/B = $ const of the surface $\varepsilon(\mathbf{p}) = E$ in momentum space.

10.15. Evaluate the Poisson brackets:

a) $\{M_i, x_j\}, \{M_i, p_j\}, \{M_i, M_j\}$;

b) $\{\mathbf{ap}, \mathbf{br}\}, \{\mathbf{aM}, \mathbf{br}\}, \{\mathbf{aM}, \mathbf{bM}\}$;

c) $\{\mathbf{M}, \mathbf{rp}\}, \{\mathbf{p}, r^n\}, \{\mathbf{p}, (\mathbf{ar})^2\}$,

   where x_i, p_i, and M_i are the Cartesian components of the vectors, while **a** and **b** are the constant vectors.

10.16. Evaluate the Poisson brackets $\{A_i, A_j\}$, where

$$A_1 = \tfrac{1}{4}(x^2 + p_x^2 - y^2 - p_y^2), \quad A_2 = \tfrac{1}{2}(xy + p_x p_y),$$
$$A_3 = \tfrac{1}{2}(xp_y - yp_x), \quad A_4 = x^2 + y^2 + p_x^2 + p_y^2.$$

10.17. Evaluate the Poisson brackets $\{M_i, \Lambda_{jk}\}, \{\Lambda_{jk}, \Lambda_{il}\}$, where $\Lambda_{ik} = x_i x_k + p_i p_k$.

10.18. Show that the Poisson bracket $\{M_z, \varphi\} = 0$, where φ is an arbitrary scalar function of the coordinates and momenta of a particle, $\varphi = \varphi(\mathbf{r}^2, \mathbf{p}^2, \mathbf{rp})$.

Show also that the Poisson bracket $\{M_z, \mathbf{f}\} = [\mathbf{n}, \mathbf{f}]$, where **n** is the unit vector along the z-axis and **f** is the vector function of the coordinates and momenta of a particle, that is, $\mathbf{f} = \mathbf{r}\varphi_1 + \mathbf{p}\varphi_2 + [\mathbf{r}, \mathbf{p}]\varphi_3$ and $\varphi_i = \varphi_i(\mathbf{r}^2, \mathbf{p}^2, \mathbf{rp})$.

10.19. Evaluate the Poisson brackets $\{\mathbf{f}, \mathbf{aM}\}, \{\mathbf{fM}, \mathbf{lM}\}$, where **f** and **l** are the vector functions of **r** and **p** while **a** is the constant vector.

10.20. Evaluate the Poisson bracket $\{M_\zeta, M_\xi\}$, where M_ζ and M_ξ are the components of the angular moment along the Cartesian ζ- and ξ-axes which are fixed in a rotating rigid body.

10.21. Write down the equations of motion for the components M_α of the angular momentum along the axes fixed in a freely rotating rigid body. The Hamiltonian function is of the form

$$H = \tfrac{1}{2}\sum_{\alpha, \beta}(I^{-1})_{\alpha\beta}M_\alpha M_\beta.$$

10.22. In this problem we consider the model of the electron and nuclear paramagnetic resonance (see [16], Ch. IX). The Hamiltonian function of the magnetized ball in a uniform magnetic field **B** has the form

$$H = \frac{M^2}{2I} - \gamma \mathbf{MB},$$

where I is the moment of inertia of the ball and γ is a gyromagnetic ratio. Write down the equations of motion for the vector **M** and find the law of its motion in the following cases:

a) $\mathbf{B} = (0, 0, B_0)$;

b) $\mathbf{B} = (B_1 \cos \omega t, B_1 \sin \omega t, B_0)$ and $\mathbf{M} = (0, 0, M_0)$ in the initial time.

10.23. Evaluate the Poisson brackets $\{v_i, v_j\}$ for a particle in a magnetic field. Here v_i are the components of the particle velocity in the Cartesian coordinates.

10.24. Prove that the value of any function $f(q(t), p(t))$ of the coordinates and momenta of a system at time t can be expressed in terms of the values of the q and p at $t = 0$ as

$$f(q(t), p(t)) = f + \frac{t}{1!}\{H, f\} + \frac{t^2}{2!}\{H, \{H, f\}\} + \ldots,$$

where $f = f(q(0), p(0))$ and $H = H(q(0), p(0))$ is the Hamiltonian function. (Assume that the series converges.)

Apply this formula to evaluate $q(t)$, $p(t)$, $q^2(t)$, and $p^2(t)$ for the following cases:

a) a particle moving in a uniform field of force;

b) a harmonic oscillator.

10.25. Prove the equalities

a) $\{f(q_1, p_1), \Phi(\varphi(q_1, p_1), q_2, p_2, \ldots)\} = \dfrac{\partial \Phi}{\partial \varphi}\{f, \varphi\}$;

b) $\{f(q_1, p_1, q_2, p_2), \Phi(\varphi(q_1, p_1, q_2, p_2), q_3, p_3, \ldots)\} = \dfrac{\partial \Phi}{\partial \varphi}\{f, \varphi\}$;

c) $\{f(q, p), \Phi(\varphi_1(q, p), \varphi_2(q, p)\ldots)\} = \sum_i \dfrac{\partial \Phi}{\partial \varphi_i}\{f, \varphi_i\}$.

10.26. a) The Hamiltonian function depends on the variables q_1 and p_1 only through the function $f(q_1, p_1)$

$$H = H(f(q_1, p_1), q_2, p_2, \ldots, q_N, p_N).$$

Prove that the function $f(q_1, p_1)$ is an integral of motion.

b) Find the integrals of motion for a particle in the field $U(\mathbf{r}) = \frac{\mathbf{ar}}{r^3}$ (use the spherical coordinates).

10.27. A particle in the field $U(r) = -\alpha/r$ is known to have the integral of motion (see [1], §15 and [7], §3.3)

$$\mathbf{A} = [\mathbf{v}, \mathbf{M}] - \alpha \frac{\mathbf{r}}{r}.$$

a) Evaluate the Poisson brackets $\{A_i, A_j\}$, $\{A_i, M_j\}$,

b) Evaluate the Poisson brackets

$$\{H, \mathbf{J}_{1,2}\}, \quad \{\mathcal{J}_{1i}, \mathcal{J}_{2j}\}, \quad \{\mathcal{J}_{1i}, \mathcal{J}_{1j}\}, \quad \{\mathcal{J}_{2i}, \mathcal{J}_{2j}\}$$

for the finite motion ($E < 0$), if the vectors $\mathbf{J}_{1,2}$ are

$$\mathbf{J}_{1,2} = \frac{1}{2} \left(\mathbf{M} \pm \sqrt{\frac{m}{-2E}} \, \mathbf{A} \right).$$

Compare these Poisson brackets with the Poisson brackets for the components of the angular momentum $\mathbf{M}$.

Express the Hamiltonian function H in terms of $\mathbf{J}_1$ and $\mathbf{J}_2$.

§11

Canonical transformations

11.1. Determine the canonical transformation defined by the following generating functions:

a) $F(q, Q, t) = \frac{1}{2}m\omega(t)q^2 \cot Q.$

Write down the equations of motion for a harmonic oscillator with frequency $\omega(t)$ in terms of the variables Q and P.

b) $F(q, Q, t) = \frac{1}{2}m\omega\left[q - \frac{F(t)}{m\omega^2}\right]^2 \cot Q.$

Write down the equations of motion for a harmonic oscillator which is acted upon by an external force $F(t)$ in terms of the variables Q and P.

11.2. Determine the generating function $\Psi(p, Q)$ which produces the same canonical transformation as the generating function $F(q, P) = q^2 e^P$.

11.3. What is the condition that a function $\Phi(q, P)$ can be used as a generating function for a canonical transformation?
Consider in particular the example

$$\Phi(q, P) = q^2 + P^2.$$

11.4. Prove that a rotation in (q, p) space is a canonical transformation for a system with one degree of freedom.

11.5. Consider the small oscillations of an anharmonic oscillator with the Hamiltonian function

$$H = \frac{1}{2}\left(p^2 + \omega^2 x^2\right) + \alpha x^3 + \beta x p^2$$

under the assumption that $|\alpha x| \ll \omega^2$, $|\beta x| \ll 1$.

Exploring Classical Mechanics: A Collection of 350+ Solved Problems for Students, Lecturers, and Researchers. First Edition.
Gleb L. Kotkin and Valeriy G. Serbo, Oxford University Press (2020). © Gleb L. Kotkin and Valeriy G. Serbo 2020.
DOI: 10.1093/oso/9780198853787.001.0001

Find the parameters a and b for the canonical transformation produced by generating function $\Phi = xP + ax^2 P + bP^3$ such that the new Hamiltonian function does not contain any anharmonic terms up to first-order terms in $\alpha\omega^{-2}Q$ and βQ. Determine $x(t)$.

11.6. Determine the parameters a and b for the canonical transformation produced by the generating function

$$\Phi = xP + ax^3 P + bxP^3,$$

in such a way that that small oscillation of an anharmonic oscillator described by the Hamiltonian function

$$H = \tfrac{1}{2}\left(p^2 + \omega_0^2 x^2\right) + \beta x^4$$

can be reduced to harmonic oscillations in terms of the new variables Q and P. Neglect terms of second order in $\beta\omega^{-2}Q^2$ in the new Hamiltonian function.

11.7. Small oscillations of the coupled oscillators are described by the Hamiltonian function

$$H = \frac{1}{2m}\left[p_1^2 + (m\omega_1 x)^2 + p_2^2 + (m\omega_2 y)^2\right] - m\alpha x^2 y.$$

Neglecting the terms of the second degree in α, find the parameters a, b, and c for the canonical transformation produced by the generating function

$$\Phi(x, y, P_X, P_Y) = xP_X + yP_Y + axyP_X + bP_X^2 P_Y + cx^2 P_Y$$

such that in the new variables the system is reduced to two independent oscillators. Find $x(t)$ and $y(t)$ with the first-order corrections in α.

11.8. Prove that the following transformation is canonical:

$$x = X\cos\lambda + \frac{P_Y}{m\omega}\sin\lambda, \qquad y = Y\cos\lambda + \frac{P_X}{m\omega}\sin\lambda,$$
$$p_y = -m\omega X\sin\lambda + P_Y\cos\lambda, \qquad p_x = -m\omega Y\sin\lambda + P_X\cos\lambda.$$

Determine the new Hamiltonian function, $H'(P, Q)$, if the old Hamiltonian function is

$$H(p, q) = \frac{p_x^2 + p_y^2}{2m} + \tfrac{1}{2}m\omega^2(x^2 + y^2)$$

(cf. with problem 11.18). Describe the motion of a two-dimensional oscillator at $Y = P_Y = 0$.

11.9. Use the transformation of the preceding problem to reduce the Hamiltonian function of an isotropic harmonic oscillator in a magnetic field described by the vector potential $\mathbf{A}(\mathbf{r}) = (0, xB, 0)$, to a sum of squares and determine its motion.

11.10. Use the canonical transformation to diagonalize the Hamiltonian function of an anisotropic charged harmonic oscillator with potential energy

$$U(\mathbf{r}) = \tfrac{1}{2} m(\omega_1^2 x^2 + \omega_2^2 y^2 + \omega_3^2 z^2),$$

which is situated in a uniform constant magnetic field determined by the vector potential $\mathbf{A}(\mathbf{r}) = (0, xB, 0)$.

11.11. Apply the canonical transformation of problem 11.8 to pairs of normal coordinates corresponding to standing waves in the system of particles on the ring considered in problem 7.4 to obtain the coordinates corresponding to the travelling waves.

11.12. Prove that the following transformation is a canonical one:

$$x = \frac{1}{\sqrt{m\omega}}(\sqrt{2P_1}\sin Q_1 + P_2),$$

$$p_x = \tfrac{1}{2}\sqrt{m\omega}(\sqrt{2P_1}\cos Q_1 - Q_2),$$

$$y = \frac{1}{\sqrt{m\omega}}(\sqrt{2P_1}\cos Q_1 + Q_2),$$

$$p_y = \tfrac{1}{2}\sqrt{m\omega}(-\sqrt{2P_1}\sin Q_1 + P_2)$$

Find the Hamiltonian equations of motions for a particle in a magnetic field described by the vector potential $\mathbf{A}(\mathbf{r}) = \left(-\tfrac{1}{2}yB, \tfrac{1}{2}xB, 0\right)$ in terms of the new variables introduced through the previously described transformation with $\omega = \frac{eB}{mc}$.

11.13. What is the meaning of the canonical transformation produced by the generating function $\Phi(q, P) = \alpha qP$?

11.14. Prove that a gauge transformation of the potentials of the electromagnetic field is a canonical transformation for coordinates and momenta of charged particles, and find the corresponding generating function.

11.15. It is well known that replacing the Lagrangian function $L(q, \dot{q}, t)$ by

$$L'(q, \dot{q}, t) = L(q, \dot{q}, t) + \frac{dF(q, t)}{dt},$$

where $F(q, t)$ is an arbitrary function of the coordinates and time lives the Lagrangian equations of motion invariant. Prove that this transformation is a canonical one and find its generating function.

11.16. Find the generating function for the canonical transformation which consist in changing $q(t)$ and $p(t)$ to

$$Q(t) = q(t+\tau) \text{ and } P(t) = p(t+\tau),$$

where τ is constant for the following cases:

a) a free particle,
b) motions in a uniform field of force, $U(q) = -Fq$;
c) a harmonic oscillator.

11.17. Discuss the physical meaning of the canonical transformations produced by the following generating functions:

a) $\Phi(\mathbf{r}, \mathbf{P}) = \mathbf{r}\mathbf{P} + \delta\mathbf{a}\mathbf{P}$;
b) $\Phi(\mathbf{r}, \mathbf{P}) = \mathbf{r}\mathbf{P} + \delta\varphi[\mathbf{r}, \mathbf{P}]$;
c) $\Phi(q, P, t) = qP + \delta\tau H(q, P, t)$;
d) $\Phi(\mathbf{r}, \mathbf{P}) = \mathbf{r}\mathbf{P} + \delta\alpha(\mathbf{r}^2 + \mathbf{P}^2)$,

where $\mathbf{r}$ is the Cartesian radius vector while $\delta\mathbf{a}$, $\delta\varphi$, $\delta\tau$, and $\delta\alpha$ are infinitesimal parameters.

11.18. Prove that the canonical transformation produced by the generating function

$$\Phi(x, y, P_X, P_Y) = xP_X + yP_Y + \varepsilon(xy + P_XP_Y),$$

where $\varepsilon \to 0$ is a rotation in phase space.

11.19. Write down the generating function for infinitesimal canonical transformations corresponding to

(a) a screw motion;
b) a Galilean transformation;
c) a change to a rotating system of reference.

11.20. A canonical transformation is produced by the generating function $\Phi(q, P) = qP + \lambda W(q, P)$, where $\lambda \to 0$.
Determine up to first-order terms the change in value of an arbitrary function $f(q, p)$ when we change arguments:

$$\delta f(q, p) = f(Q, P) - f(q, p).$$

11.21. For the case when the Hamiltonian function is

$$H(\mathbf{r}, \mathbf{p}) = \frac{p^2}{2m} + \frac{\mathbf{a}\mathbf{r}}{r^3} \quad (\mathbf{a} = \text{const}),$$

determine the Poisson bracket $\{H, \mathbf{r}\mathbf{p}\}$ and use the result to obtain an integral of the equations of motion. Here, it is convenient to use the results of the preceding problem and problem 11.13.

11.22. Determine how the $\mathbf{r}$ and $\mathbf{p}$ dependence of $\mathbf{M}$, $\mathbf{p}^2$, $\mathbf{p}\mathbf{r}$, and $H(\mathbf{r}, \mathbf{p}, t)$ change under the canonical transformations of problem 11.17.

11.23. Show that if

$$\{W_1(q, p), W_2(q, p)\} = 0,$$

the result of applying two successive infinitesimal canonical transformations which are produced by the generating functions

$$\Phi_i(q, P) = qP + \lambda_i W_i(q, P), \quad \lambda_i \to 0, \quad i = 1, 2,$$

will be independent of the order in which they are taken, up to and including second-order terms.

11.24. Determine the canonical transformation which is the result of N successive infinitesimal transformations produced by the generating function

$$\Phi(q, P) = qP + \frac{\lambda}{N} W(q, P), \quad \lambda = \text{const}, \quad N \to \infty;$$

a) $W(\mathbf{r}, \mathbf{P}) = [\mathbf{r}, \mathbf{P}]\mathbf{a}, \ \mathbf{a} = \text{const};$ \quad b) $W(x, y, P_X, P_Y) = A_i,$

where A_i is defined in problem 10.16.

Hint: Construct—and solve for the different concrete forms of W—differential equations which $Q(\lambda)$ and $P(\lambda)$ must satisfy.

11.25.

a) What is the change with time in the volume, the volume in momentum space, and the volume in phase-space which are occupied by a group of particles which move freely along the x-axis? At $t = 0$ the particle coordinates are lying in the interval $x_0 < x < x_0 + \Delta x_0$, and their momenta in the range $p_0 < p < p_0 + \Delta p_0$.

b) Do the same for particles which move along the x-axis between two walls. Collisions with the walls are absolutely elastic. The particles do not interact with one another.

c) Do the same for a group of harmonic oscillators.

d) Do the same for a group of harmonic oscillators with friction.

e) Do the same for a group of anharmonic oscillators.

f) We shall describe the particle distribution in phase space at time t by the distribution function $w(x, p, t)$ which is such that $w(x, p, t)\,dx\,dp$ is the number of particles with coordinates in the interval from x to $x + dx$ and momenta in the range from p up to $p + dp$. Determine the distribution functions of a group of free particles and of a group of harmonic oscillators, if at $t = 0$

$$w(x, p, 0) = \frac{1}{2\pi\,\Delta p_0\,\Delta x_0}\exp\left\{-\frac{(x - X_0)^2}{2(\Delta x_0)^2} - \frac{(p - P_0)^2}{2(\Delta p_0)^2}\right\}.$$

11.26. For the variable

$$a = \frac{m\omega x + ip}{\sqrt{2m\omega}}e^{i\omega t},$$

a) find the Poisson bracket $\{a^*, a\}$. Express the Hamiltonian function of a harmonic oscillator

$$H_0 = \frac{p^2}{2m} + \tfrac{1}{2}m\omega^2 x^2$$

in terms of variables a and a^*.

b) Prove that $Q = a$ and $P = ia^*$ are the canonical variables. Find a new Hamiltonian function $H_0'(Q, P)$.

c) For an oscillator with the anharmonic addition to potential energy $\delta U = \tfrac{1}{4}m\beta x^4$, average the Hamiltonian function $H'(Q, P)$ over the period of fast oscillations $2\pi/\omega$.

Use the averaged Hamiltonian function to determine the slow changes in variables Q and P.

d) Study the variation of the amplitude of oscillations for a harmonic oscillator under the action of the nonlinear resonant force

$$H = \frac{p^2}{2m} + \tfrac{1}{2}m\omega^2 x^2 + m^2\omega^2\alpha x^4\cos 4\omega t.$$

11.27. Study the variation of the amplitude of oscillations for the system of three harmonic oscillators with the weak nonlinear coupling

$$H = \frac{1}{2m}(p_x^2 + p_y^2 + p_z^2) + \tfrac{1}{2}m(\omega_1^2 x^2 + \omega_2^2 y^2 + \omega_3^2 z^2 + \alpha xyz),$$

when $|\omega_1 - \omega_2 - \omega_3| \ll \omega_1$, $|\alpha x| \ll \omega_1^2$. Consider in more detail the cases when in the initial moment $|y| \ll |x|$, $z = 0$, $\dot{y} = \dot{z} = 0$.

Use the same method as in the preceding problem.

11.28. The Hamiltonian function of an anharmonic oscillator with the parametric excitation has the form

$$H = \frac{p^2}{2m} + \tfrac{1}{2} m\omega^2 (1 + h\cos 2\gamma t) x^2 + \tfrac{1}{4} m\beta x^4.$$

Introduce the canonical variables

$$a = \frac{m\omega x + i p}{\sqrt{2m\omega}} e^{i\gamma t}, \quad P = i a^*.$$

a) Determine the new Hamiltonian function $H'(a, P, t)$ and average it over the period of fast oscillations $2\pi/\gamma$.

b) Study the change in an amplitude of oscillations in the resonance region $|\gamma - \omega| \ll h\omega$, $h \ll 1$ when in the initial moment quantity a is close to zero.

11.29. Check that the transformation

$$x = Q\cos\gamma t + \frac{1}{m\omega} P\sin\gamma t,$$
$$p = -m\omega Q\sin\gamma t + P\cos\gamma t$$

is canonical. Determine the new Hamiltonian function $H'(Q, P, t)$ for an oscillator with the parametric excitation:

$$H = \frac{p^2}{2m} + \tfrac{1}{2} m\omega^2 x^2 (1 - h\cos 2\gamma t).$$

b) Average $H'(Q, P, t)$ by the period $2\pi/\gamma$ and describe qualitatively the motion of a point on the phase plane QP. Take $h \ll 1$, $\varepsilon = 1 - \gamma/\omega \ll 1$.

11.30. Consider the motion of two weakly coupled oscillators

$$H = H_0 + V, \quad H_0 = \frac{1}{2m}(p_x^2 + p_y^2) + \tfrac{1}{2} m(\omega_1^2 x^2 + \omega_2^2 y^2),$$
$$V = m\beta xy \sin(\omega_1 - \omega_2)t, \quad \beta \ll \omega_1^2 \sim \omega_2^2.$$

In the plane of the variable x, $p_x/(m\omega_1)$ change the variable to the coordinate system X, $P_X/(m\omega_1)$ which rotates clockwise with an angular velocity ω_1. Make a similar change for the variables $y, p_y/(m\omega_2)$. Prove that the variables X, Y, and P_X, P_Y are the canonical variables.

Determine the new Hamiltonian function $H'(X, Y, P_X, P_Y, t)$ and average it over the time t_{av} such that

$$\frac{1}{\omega_{1,2}} \ll t_{av} \ll \frac{\omega_{1,2}}{\beta},$$

For the period $t \lesssim \omega_{1,2}/\beta$, study the change over time of the amplitudes of oscillations for the variables x and y.

b) Do the same for $V = m\beta xy\sin(\omega_1 + \omega_2)t$.

§12

The Hamilton–Jacobi equation

12.1. Describe the motion of a particle moving in the field $U(\mathbf{r})$ by using the Hamilton–Jacobi equation for the following cases:

a) $U(\mathbf{r}) = -Fx$;

b) $U(\mathbf{r}) = \frac{1}{2}\left(m\omega_1^2 x^2 + m\omega_2^2 y^2\right)$.

12.2. Describe the motion of a particle which is scattered in the field $U(\mathbf{r}) = \mathbf{ar}/r^3$. Express the equation of the trajectory in terms of quadratures; express it analytically for the case when $E\rho^2 \gg a$, where ρ is the impact parameter. Before the scattering the velocity of the particle is parallel to the vector $(-\mathbf{a})$.

12.3. Find the cross-section for the small-angle scattering of particles with velocities before the scattering anti-parallel to the z-axis for the cases where the scattering is caused by the following fields $U(\mathbf{r})$:

a) $U(\mathbf{r}) = \dfrac{a\cos\theta}{r^2}$;

b) $U(\mathbf{r}) = \dfrac{b\cos^2\theta}{r^2}$;

c) $U(\mathbf{r}) = \dfrac{b(\theta)}{r^2}$.

12.4. Find the cross-section for a particle to fall into the centre of one of the following fields $U(\mathbf{r})$:

$$\text{a) } U(\mathbf{r}) = \frac{\mathbf{ar}}{r^3}; \quad \text{b) } U(\mathbf{r}) = \frac{\mathbf{ar}}{r^3} + \frac{\lambda}{r};$$

$$\text{c) } U(\mathbf{r}) = \frac{\mathbf{ar}}{r^3} - \frac{\gamma}{r^4}; \quad \text{d) } U(\mathbf{r}) = \frac{b(\theta)}{r^2}.$$

Assuming that all directions of $\mathbf{a}$ are equally probable, evaluate the averages of the cross-section obtained.

Exploring Classical Mechanics: A Collection of 350+ Solved Problems for Students, Lecturers, and Researchers. First Edition.
Gleb L. Kotkin and Valeriy G. Serbo, Oxford University Press (2020). © Gleb L. Kotkin and Valeriy G. Serbo 2020.
DOI: 10.1093/oso/9780198853787.001.0001

12.5. Find the cross-section for particles to hit a sphere of radius R placed at the centre of the field $U(\mathbf{r}) = \mathbf{ar}/r^3$.

12.6. Describe the motion of particles which are scattered by and fall towards the centre of the field $U(\mathbf{r})$ for the following cases:

a) $U(\mathbf{r}) = \dfrac{a \cos \theta}{r^2}$;

b) $U(\mathbf{r}) = -\dfrac{a(1 + \sin \theta)}{r^2}$.

The particle velocity before scattering is parallel to the z-axis.

Give an expression for the trajectories in terms of quadratures and analytically for the case when $E\rho^2 \gg a$.

For the first field, find also an analytic expression for the trajectory of a particle falling into the centre at $E\rho^2 \ll a$.

12.7. Describe the motion of a particle falling towards the centre of the field $U(\mathbf{r}) = \mathbf{ar}/r^3$ for the case when at infinity the particle moved along the straight line $y = \rho$, $x = -z \tan \alpha$, where ρ is the impact parameter. The vector $\mathbf{a}$ is along the z-axis and the initial spherical coordinates of the particle are $r = \infty, \theta = \pi - \alpha, \varphi = 0$. Express the trajectory in terms of quadratures and analytically for the case when $\alpha^2 < \dfrac{2E\rho^2}{a} \ll 1$.

12.8. a) Determine in terms of quadratures the finite orbit of a particle moving in the field $U(\mathbf{r}) = \dfrac{a \cos \theta}{r^2} - \dfrac{\alpha}{r}$ for the case when $M_z = 0$, where the z-axis is taken along the direction of $\mathbf{a}$.

b) Do the same for the field $U(\mathbf{r}) = \dfrac{a \cos \theta}{r^2} + \dfrac{\gamma}{r^4}$.

12.9. Under what condition is the orbit in the preceding problem a closed one?

12.10. Describe qualitatively the motion of a particle in the field

$$U(\mathbf{r}) = \frac{\mathbf{ar}}{r^3} - \frac{\alpha}{r}.$$

12.11. Find the values of the angular momentum M_z for which the orbits in the following fields $U(\mathbf{r})$ are finite:

a) $U(\mathbf{r}) = \dfrac{\gamma}{r^4} - \dfrac{b \cos^2 \theta}{r^2}$;

b) $U(\mathbf{r}) = \dfrac{b \cos^2 \theta}{r^2} - \dfrac{\alpha}{r}$.

Describe the orbits in both cases.

12.12. Describe the motion in terms of parabolic coordinates for a particle moving in the field $U(\mathbf{r})$ for the following cases:

a) $U(\mathbf{r}) = -\frac{\alpha}{r}$;

b) $U(\mathbf{r}) = -\frac{\alpha}{r} - \mathbf{F}\mathbf{r}$.

In b), consider only finite orbits it terms of quadratures.

12.13. A particle starts from the origin at an angle α to the z-axis inside a smooth elastic ellipsoid of revolution:

$$\frac{x^2}{a^2} + \frac{y^2}{a^2} + \frac{z^2}{c^2} = 1.$$

Find the region inside the ellipsoid which can not be reached by the particle.

12.14. A particle moves in a smooth horizontal plane in the area bounded by an elastic smooth wall having the form of ellipse

$$\frac{x^2}{a^2} + \frac{y^2}{b^2} = 1.$$

The particle starts to move from a point on the x-axis at the angle α to it:

$$(x,y) = (x_0, 0), \quad (\dot{x}, \dot{y}) = v(\cos\alpha, \sin\alpha).$$

Find the region inside the ellipse which can not be reached by the particle.
Hint: Use the elliptic coordinates ζ, φ, determined by the equations

$$x = c \cosh\zeta \cos\varphi, \quad y = c \sinh\zeta \sin\varphi,$$
$$0 \le \zeta < \infty, \quad -\pi \le \varphi < \pi,$$

where c is the parameter of transformation, $c^2 = a^2 - b^2$.

12.15. Describe in terms of quadratures the trajectory of a particle moving in the field two Coulomb centres,

$$U(\mathbf{r}) = \frac{\alpha}{r_1} - \frac{\alpha}{r_2}$$

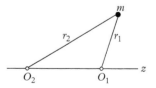

Figure 72

(Fig. 72). At infinity the velocity of the particle was parallel to the axis O_2O_1z. Describe the motion of a particle falling onto a "dipole" formed by these centres.

12.16. A short magnetic lens is produced by a field governed by the vector potential

$$A_\varphi = \tfrac{1}{2} rB(z), \quad A_r = A_z = 0,$$

where $B(z)$ is non-vanishing in the interval $|z| < a$. A beam of electrons close to the z-axis incident onto the lens from point $(0, 0, z_0)$. Find the point $(0, 0, z_1)$ onto which the beam is focused. Assume that $z_{0,1} \gg a$.

Hint: Find the solution of the Hamilton–Jacobi equation in the form of expansion of $S(r, \varphi, z, t)$ in powers of r:

$$S(r, \varphi, z, t) = -Et + p_\varphi \varphi + f(z) + r\psi(z) + \tfrac{1}{2} r^2 \sigma(z) + \ldots$$

12.17. A magnetic lens is produced by a field governed by the vector potential

$$A_\varphi = \tfrac{1}{2} rB(z), \quad A_r = A_z = 0,$$

where

$$B(z) = \frac{B}{1 + \varkappa^2 z^2}.$$

A beam of electrons close to the z-axis is incident from the point $(0, 0, z_0)$. Find the points where it will be focused.

Hint: Find a solution of the Hamilton–Jacobi equation as an expansion in r.

12.18. How can one find the action as a function of the coordinates and time, if the solution of the Hamilton–Jacobi equation is known?

12.19. Formulate and prove the theorem about the integrating the equations of motion using the complete solution of the equation

$$\frac{\partial S}{\partial t} + H\left(-\frac{\partial S}{\partial p}, p, t\right) = 0,$$

where $H(q, p, t)$ is the Hamiltonian function (Hamilton–Jacobi equation in the momentum representation).

12.20. Use the Hamilton–Jacobi equation in the momentum representation to find the trajectory and law of motion of a particle moving in a uniform field.

§13

Adiabatic invariants

13.1. A particle of mass m is suspended from a thread which passes through a small ring A (Fig. 73). Determine the average force exerted upon the ring by the thread when the pendulum performs small oscillations. Find the change in the energy of the pendulum when the ring is slowly displaced vertically.

13.2. A particle moves in a rectangular potential well of width l. Consider the collisions of the particle with the "wall" of the well to find how the energy of the particle will change when l is changed slowly.

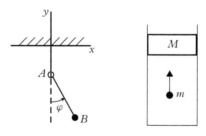

Figure 73   Figure 74

13.3. A ball of mass m moves between a heavy piston of mass $M \gg m$ and the bottom of the cylinder in the gravitation field g (Fig. 74). Equilibrium distance from the bottom of the cylinder to the piston is equal to X_0. Considering that the velocity of the ball is much more than the piston's velocity, determine the law of motion of the piston, averaged over the period of the ball's motion. Find the frequency of small oscillations of the piston. Assume the collisions to be elastic. (This is a model of a "gas" consisting of a single molecule.)

13.4. A small ball is jumping up and down on an elastic plate in a lift. What is the change in the maximum height the ball reaches if the acceleration of the lift is slowly changed? How does the height vary if the plate is raised slowly?

Exploring Classical Mechanics: A Collection of 350+ Solved Problems for Students, Lecturers, and Researchers. First Edition.
Gleb L. Kotkin and Valeriy G. Serbo, Oxford University Press (2020). © Gleb L. Kotkin and Valeriy G. Serbo 2020.
DOI: 10.1093/oso/9780198853787.001.0001

13.5. How does the energy of a particle moving in the field $U(x,\alpha)$ change when the parameters of the field change slowly for the following cases:

a) $U(x,\alpha) = A(e^{-2\alpha x} - 2e^{-\alpha x})$;   b) $U(x,\alpha) = -\dfrac{U_0}{\cosh^2 \alpha x}$;

c) $U(x,\alpha) = U_0 \tan^2 \alpha x$;   d) $U(x,\alpha) = \alpha |x|^n$.

Hint: Use the formula $T = 2\pi \dfrac{\partial I}{\partial E}$ ([1], § 49).

13.6. A particle moves down an inclined plane AB (Fig. 75) and is reflected elastically by a wall at the point A. How does the maximum height the particle reaches change when the angle α changes slowly?

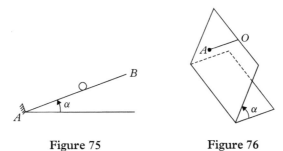

Figure 75   Figure 76

13.7. A pendulum is placed upon an inclined plane OA (Fig. 76). How does its amplitude change, when the angle α is changes slowly?

13.8. Determine the adiabatic invariant for a mathematical pendulum for the case when the amplitude of the oscillations is not small.

13.9. Two elastic small balls of small radius and with masses m and M, respectively, ($m \ll M$) move along the straight line OA (Fig. 77). At O the particle m is reflected elastically by a wall. Assume that the velocity of the lighter particle at $t = 0$ is much larger that of the heavier particle and determine the motion of the heavier particle averaged over period of motion of lighter particle.

13.10. In this problem we consider a simplified model of an ion H_2^+. Two particles of mass M each and one of mass m ($m \ll M$), located between them, move along a straight line AB (Fig. 78). The lighter particle is attracted with the constant force f to each of the heavier particles and is reflected elastically at collisions with the them. Assume the velocity of the lighter particle to be much higher than that of the heavier particles. Find the frequency of small oscillations of the distance between the heavier particles averaging on the fast motion of the lighter one.

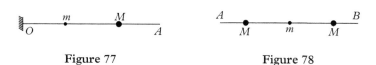

Figure 77   **Figure 78**

13.11. Use the method of successive approximations to solve the equations for P and Q of problem 11.1a for the case when the frequency changes slowly, $|\dot{\omega}| \ll \omega^2$, $|\ddot{\omega}| \ll |\dot{\omega}|\omega$, up to and including terms of first order in $\dot{\omega}/\omega^2$.

What is the advantage of the variables P, Q over p, q in this case?

13.12. Check that up to terms of first order in $\dot{\omega}/\omega^2$ the expression

$$q = \frac{1}{\sqrt{\omega}} e^{i \int \omega dt}$$

satisfies the equation

$$\ddot{q} + \omega^2(t)q = 0.$$

13.13. A force $F(t)$ acts upon a harmonic oscillator. Find the time-dependence of the adiabatic invariant $I = \frac{1}{2\pi} \oint p \, dq$.

13.14. Find a connection between the volume and the pressure of a "gas" consisting of particles which move parallel to the edges and inside an elastic cube, when the size of the cube changes slowly.

13.15. A particle moves inside an elastic parallelepiped. How does its energy change

a) if the size of the parallelepiped changes slowly;

b) if the parallelepiped rotates slowly?

13.16. A particle moves inside an elastic sphere, the radius of which changes slowly. How does its energy change, and how does the angle at which it hits the sphere change?

13.17. Determine the change in the energy and in the trajectory of a particle performing a finite motion in the field $U(\mathbf{r})$ in the following cases:

a) $U(r) = -\dfrac{\gamma}{r^n}$ $(0 < n < 2)$;

b) $U(r) = \dfrac{\mathbf{a}r}{r^3} + \dfrac{\gamma}{r^4}$,

when the coefficient γ changes slowly.

13.18. Determine the change in the energy of a particle moving in the central field $U(r)$ when a small extra field $\delta U(r)$ is slowly "switched on".

13.19. Find the time-dependence of the energy of a system two coupled harmonic oscillators with the Lagrangian function

$$L = \tfrac{1}{2}(\dot{x}^2 + \dot{y}^2 - \omega_1^2 x^2 - \omega_2^2 y^2) + \alpha xy$$

when ω_1 changes slowly.

What is the change in the orbit of the point (x, y)?

13.20. Let the coupling between the oscillators in the preceding problem be small: $\alpha \ll \omega_{1,2}^2$. Prove that if we are far from degeneracy, when $\omega_1 = \omega_2$, the adiabatic invariants calculated neglecting the coupling are conserved, but that they change rapidly when we pass slowly through the region of degeneracy.

13.21. In which range of frequency $\omega_1(t)$ will the adiabatic invariants of the harmonic oscillators in problem 13.19 change sharply, if the coupling is $\delta U(x, y) = \beta x^2 y$?

13.22. Determine the shortest distance a particle approaches the edge of a dihedral angle α after being reflected elastically from its faces. The angle at which particle is incident on one of the faces at a distance of l from the edge is φ_0.

Solve this problem by two methods: either by using a reflection method and solving it directly or by using adiabatic invariant assuming that α and φ_0 are small.

13.23. Define the boundaries of the region in which a particle moves between two elastic surfaces $y = 0$ and $y = \dfrac{a \cosh \alpha x}{\sqrt{\cosh 2\alpha x}}$. The particle starts at the origin at the angle φ to the y-axis in the plane xy ($\alpha, \varphi \ll 1$). Also determine the period of the oscillations along the x-axis.

13.24. What is the change in the radius and the centre of the orbits of a charged particle moving in a uniform magnetic field which is slowly changing its strength? Take the vector potential in one of the following two forms

   a) $\mathbf{A}(\mathbf{r}) = (0, xB, 0)$;
   b) $A_r = A_z = 0, A_\varphi = \tfrac{1}{2} rB$.

13.25. Calculate the adiabatic invariants for a charged isotropic harmonic oscillator moving in a uniform magnetic field. Take the vector potential in the form $A_r = A_z = 0$, $A_\varphi = \tfrac{1}{2} rB(t)$.

13.26. a) Determine the adiabatic invariants for a charged anisotropic harmonic oscillator with the potential energy

$$U(\mathbf{r}) = \tfrac{1}{2} m(\omega_1^2 x^2 + \omega_2^2 y^2 + \omega_3^2 z^2)$$

moving in a uniform magnetic field $\mathbf{B}$ parallel to the z-axis. Take the vector potential in the form $\mathbf{A}(\mathbf{r}) = (0, xB, 0)$.

b) Let initially $\mathbf{B} = 0$ and let the trajectory of the oscillator fill the rectangle $|x| \leqslant a$, $|y| \leqslant b$. Find the motion of the oscillator after the magnetic field slowly increases its strength up to a value which is so high that $\omega_B = eB/(mc) \gg \omega_{1,2}$.

c) Let the magnetic field be weak ($\omega_B \ll \omega_1 - \omega_2 > 0$) and let the oscillator originally oscillate near along the x-axis. What happens to its motion when the value of ω_1 decreases slowly to reach a value $\omega_1' < \omega_2$ such that $\omega_B \ll \omega_2 - \omega_1'$?

13.27. A particle performs a finite motion in a plane at right angles to a magnetic dipole $\boldsymbol{\mu}$. What is the change in the energy of the particle when the magnitude $\boldsymbol{\mu}$ changes slowly?

13.28. Determine the period of the oscillations along the z-axis of an electron in a magnetic trap. The magnetic field of the trap is symmetric with respect to the z-axis, and we have

$$B_\varphi = 0, \quad B_z = B_z(z), \quad B_r = -\tfrac{1}{2} r B_z'(z),$$

Consider the following cases:

a) $B_z(z) = B_0\left(1 + \lambda \tanh^2 \frac{z}{a}\right)$;

b) $B_z(z) = B_0\left(1 + \frac{z^2}{a^2}\right)$.

13.29. How does the energy and the period of oscillations of an electron moving in the magnetic trap of the preceding problem change when the field parameters B_0, λ, and a change slowly?

13.30. Determine the change in the energy of a particle moving in a central field $U(r)$ when a weak iniform magnetic field $\mathbf{B}$ is slowly "switched on".

13.31. The number of single-valued integrals of motion is known to increase when the motion becomes degenerate. Find the integrals of motion when a particle moves in the field

$$U(x, y) = \tfrac{1}{2} m\omega^2 (x^2 + 4y^2).$$

13.32. Find the action and angle variables for the following systems:

a) a harmonic oscillator;

b) a particle moving in the field $U(x) = \begin{cases} \infty, & \text{when} \quad x < 0, \\ xF, & \text{when} \quad x > 0. \end{cases}$

13.33. Use the generating function

$$S(x, P) = \int\limits_{0}^{x} \sqrt{2m|E - U(x)|}\, dx,$$

to perform a canonical transformation for the case of a particle moving in the periodic field

$$U(x) = \begin{cases} 0, & \text{when } na < x < \left(n + \dfrac{1}{2}\right)a, \\ V, & \text{when } \left(n + \dfrac{1}{2}\right)a < x < (n+1)a, \end{cases} \qquad n = 0, \pm 1, \pm 2, \ldots,$$

when $E > V$, and where E in the expression for $S(x, P)$ can be derive as a function of P from the equation

$$P = \int\limits_{0}^{a} \sqrt{2m|E - U(x)|}\, dx.$$

Answers and solutions

§1

Integration of one-dimensional equations of motions

1.1. a) The energy of the particle E is determined by the initial values $x(0)$ and $\dot{x}(0)$. The motion follows from the energy conservation law

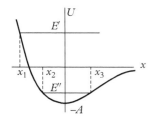

Figure 79

$$\tfrac{1}{2}m\dot{x}^2 + U(x) = E. \tag{1}$$

When $E \geqslant 0$, the particle can move in the region $x \geqslant x_1$: the motion is infinite ($E = E'$ in Fig. 79). When $E < 0$ ($E = E''$), the particle moves in the interval $x_2 \leqslant x \leqslant x_3$: the motion is finite. The turning points are determined from (1) through $U(x_i) = E$:

$$x_1 = \frac{1}{\alpha} \ln \frac{\sqrt{A(A+E)} - A}{E}, \qquad \text{when} \quad E > 0,$$

$$x_1 = -\frac{\ln 2}{\alpha}, \qquad\qquad\qquad \text{when} \quad E = 0, \tag{2}$$

$$x_{2,3} = \frac{1}{\alpha} \ln \frac{A \mp \sqrt{A(A - |E|)}}{|E|}, \quad \text{when} \quad E < 0.$$

From (1) we obtain

$$t = \sqrt{\frac{m}{2}} \int_{x(0)}^{x} \frac{dx}{\sqrt{E - U(x)}}. \tag{3}$$

whence

$$x(t) = \frac{1}{\alpha} \ln \frac{A - \sqrt{A(A - |E|)}\cos(\alpha t \sqrt{2|E|/m} + C)}{|E|}, \qquad \text{when } E < 0, \tag{4}$$

Exploring Classical Mechanics: A Collection of 350+ Solved Problems for Students, Lecturers, and Researchers. First Edition.
Gleb L. Kotkin and Valeriy G. Serbo, Oxford University Press (2020). © Gleb L. Kotkin and Valeriy G. Serbo 2020.
DOI: 10.1093/oso/9780198853787.001.0001

$$x(t) = \frac{1}{\alpha} \ln \left[\frac{1}{2} + \frac{A\alpha^2}{m} (t+C)^2 \right], \qquad \text{when} \quad E = 0, \qquad (5)$$

$$x(t) = \frac{1}{\alpha} \ln \frac{\sqrt{A(A+E)} \cosh(\alpha t \sqrt{2E/m} + C) - A}{E}, \qquad \text{when} \quad E > 0. \qquad (6)$$

The C constants are defined by the initial values of $x(0)$. For example, in (4), when $\dot{x}(0) > 0$

$$C = \arccos \frac{A - |E| e^{\alpha x(0)}}{\sqrt{A - (A - |E|)}}.$$

Turning points (2) are also easily found from (4) to (6).

When $E < 0$ the motion is periodic according to (4) with a period $T = \frac{\pi}{\alpha} \sqrt{\frac{2m}{|E|}}$.
When E is close to the minimum value of $U(x)$, $U_{\min} = U(0) = -A$ (that is, when $\varepsilon = \frac{A - |E|}{A} \ll 1$), the period

$$T \approx T_0 \left(1 - \frac{\varepsilon}{2} \right), \quad T_0 = \frac{\pi}{\alpha} \sqrt{\frac{2m}{A}}$$

is nearly independent of E. In this case we can write (4) in the form

$$x(t) = -\frac{1}{\alpha} \ln(1 - \varepsilon) + \frac{1}{\alpha} \ln \left[1 - \sqrt{\varepsilon} \cos \left(\frac{2\pi}{T} t + C \right) \right] \approx -\frac{\sqrt{\varepsilon}}{\alpha} \cos \left(\frac{2\pi}{T_0} t + C \right). \qquad (7)$$

The particle now performs harmonic oscillations near the point $x = 0$ with an amplitude $\sqrt{\varepsilon}/\alpha$ determined by the difference $E - U_{\min}$, and with a frequency which is independent of the energy. This kind of motion for an energy E close to $U_{\min}$ occurs in any field $U(x)$, in which potential energy near the point of minimum $x = a$ has a non-zero second derivative $U''(a) \neq 0$. (For more information, see § 5, and [7], § 19.)

When $E \geqslant 0$ the particle coming to the right reaches the turning point x_1 (see (2)), turns back, and goes to infinity. Its velocity approaches the value $\sqrt{2E/m}$ from above.

$$\text{b)} \quad x(t) = \frac{1}{\alpha} \operatorname{arsinh} \left[\sqrt{\frac{-|E| + U_0}{|E|}} \sin(\alpha t \sqrt{2|E|/m} + C) \right], \qquad \text{when} \quad E < 0,$$

$$x(t) = \pm \frac{1}{\alpha} \operatorname{arsinh} \left[\sqrt{\frac{E + U_0}{E}} \sinh(\alpha t \sqrt{2E/m} + C) \right], \qquad \text{when} \quad E > 0,$$

$$x(t) = \pm \frac{1}{\alpha} \text{arsinh}(\alpha t \sqrt{2U_0/m} + C), \quad \text{when} \quad E = 0;^{1}$$

c) $$x(t) = \frac{1}{\alpha} \arcsin \left[\sqrt{\frac{E}{E + U_0}} \sin \left(\alpha t \sqrt{2(U_0 + E)/m} + C \right) \right].$$

Explain why in some of the formulae two signs occur?

1.2.

$$x(t) = \frac{x_0}{1 \pm t x_0 \sqrt{2A/m}}, \quad x_0 = x(0).$$

The sign in the denominator is the opposite of that of $\dot{x}(0)$. Let for definiteness $x(0) > 0$. When $\dot{x}(0) > 0$, the particle reaches infinity after a time $\sqrt{m/(2Ax_0^2)}$. Of course, for real system the particle only reaches finite, though long, distance, corresponding to the distance over which the specified field $U(x)$ has the given form.

When $\dot{x}(0) < 0$, the particle asymptotically approaches the point $x = 0$.

1.3. Near the turning point $U(x) \approx E - (x - a)F$, where $F = -U'(a)$, that is, the motion of the particle can be considered to occur under the action of a uniform constant force F. Assuming that $x(0) = a$, we get

$$x(t) = a + \frac{Ft^2}{2m}.$$

The further away from the point $x = a$ we go, the less accurate this formula becomes.

It takes a time $\tau \propto s$ to traverse a short path length s if it is far from a turning point. However, if this path length is at the turning point, $\tau \approx \sqrt{2ms/|F|}$, that is, $\tau \propto \sqrt{s}$.

If $U'(a) = 0$ (see Fig. 80), one must extend the expansion of $U(x)$ to the next term:

$$U(x) = E + \tfrac{1}{2}U''(a)(x - a)^2.$$

In this case

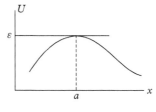

$$x(t) = a + se^{\pm \lambda t} \text{ with } s = x(0) - a, \lambda^2 = -\frac{U''(a)}{m},$$

Figure 80

where m is the particle mass, and where the sign in the index is determined by the direction of the velocity at $t = 0$. It takes an infinite time to reach the turning point.

---

[1] $\text{arsinh} x = \ln(x + \sqrt{x^2 + 1})$.

1.4. If $U''(a) \neq 0$, then $T \propto \ln \varepsilon$, where $\varepsilon = U_m - E$. If

$$U''(a) = \ldots = U^{(n-1)}(a) = 0, \quad U^{(n)}(a) \neq 0,$$

we have $t \propto \varepsilon^{-(n-2)/(2n)}$.

1.5. a) When $\varepsilon = E - U_m$ is small, the particle moves most slowly near the point $x = a$. Therefore, the entire period of movement T can be estimated by the time T_1 of the (back and forth) passage of the small interval δ in the neighbourhood of this point $a - \delta < x < a + \delta$:

$$T_1 = \sqrt{2m} \int_{a-\delta}^{a+\delta} \frac{dx}{\sqrt{E - U(x)}} \approx T.$$

In the neighbourhood of $x = a$, we present $U(x)$ in the form

$$U(x) = U_m - \tfrac{1}{2} k(x-a)^2,$$

where $k = -U''(a)$. If ε is small enough, you can choose δ such that velocity v on the boundary of the interval is a lot more then the minimum one (at $x = a$),

$$\tfrac{1}{2} mv^2 \sim \tfrac{1}{2} k\delta^2 \gg \varepsilon,$$

and at the same time to be $\delta \ll L = x_2 - x_1$, i.e.

$$\sqrt{\frac{\varepsilon}{k}} \ll \delta \ll L = x_2 - x_1.$$

As a result,

$$T_1 = 2\sqrt{\frac{m}{k}} \ln \frac{2k\delta^2}{\varepsilon}. \tag{1}$$

The time T_2 spent by a particle along the intervals $x_1 < x < a - \delta$ and $a + \delta < x_2$ satisfies the relation

$$T_2 \lesssim \frac{L}{v} \sim \sqrt{\frac{m}{k}} \frac{l}{\delta}.$$

When ε decreases, T_1 increases so that, for sufficiently small ε, $T_2 \ll T_1$, and we can use (1) to estimate the period of the motion. This formula is asymptotically exact. As $\varepsilon \to 0$, its relative error tends to zero as $1/\ln \varepsilon$. But with the same logarithmic accuracy we can replace δ by L in (1) and omit the multiplier 2 under the sign of logarithm:

$$T = 2\sqrt{\frac{m}{k}}\ln\frac{kL^2}{\varepsilon}. \tag{2}$$

If $U''(a) = 0$, but $U^{(4)} = -K \neq 0$, we have

$$T = 4\left(\frac{6m^2}{\varepsilon K}\right)^{1/4}\int\limits_0^\infty \frac{dx}{\sqrt{1+x^4}} = 11.6\left(\frac{m^2}{\varepsilon K}\right)^{1/4}$$

and the relative error tends to zero as $\varepsilon^{1/4}$ at $\varepsilon \to 0$.

b) If we observe the motion of the particle over a time which is much larger compared to the period T, the probability to find the particle between x and $x + dx$ is

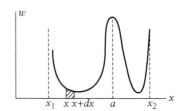

Figure 81

$$w(x)\,dx = 2\frac{dt}{T} = \frac{\sqrt{2m}\,dx}{T\sqrt{E-U(x)}},$$

where $2\,dt$ is the a entire time which the particle spends on an interval dx over the period. The dependence of the probability density $w(x)$ on x is shown in Fig. 81.

The probability $w(x)\,dx$ is represented by the shaded area (the total area under the curve is unity). For sufficiently small ε the area under the central maximum gives the main contribution T_1/T to the total area under the curve. Despite the fact that $w(x) \to \infty$ as $x \to x_{1,2}$, the contribution from the regions near the turning points is relatively small.

c)

$$\widetilde{w}(p)\,dp = \frac{1}{T}\sum_k\left|\frac{dt_k}{dp}\right|dp = \frac{1}{T}\sum_k\frac{dp}{\left|\dfrac{dU(x_k)}{dx}\right|},$$

where $x_k = x_k(p)$ is the various roots of the equation $\frac{p^2}{2m} + U(x) = E$.

The graph $\widetilde{w}(p)$ is shown in Fig. 82, where

$$p_1 = \sqrt{2m(E-U_m)}, \quad p_2 = \sqrt{2m[E-U(c)]}, \quad p_3 = \sqrt{2m[E-U(b)]}.$$

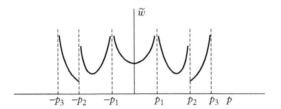

Figure 82

d) The lines $E(x, p) = \text{const}$ (the phase trajectories of the particle) are shown in Fig. 83, where curves are numbered in ascending order of energies. For $U(c) \leqslant E < U_m$, the phase trajectory 2 is double-connected. The arrows indicate the direction of the motion of the point representing the particle state.

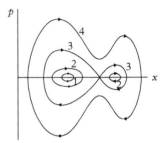

1.6. We put the value of the potential energy at the lowest point equal to zero. If $E = 2mgl$ we have

Figure 83

$$\varphi(t) = -\pi + 4\arctan\left(e^{\pm t\sqrt{g/l}}\tan\frac{\varphi(0) + \pi}{4}\right),$$

where φ is the angle between the pendulum and the vertical. The sign in the index is the same as that of $\dot\varphi(0)$. The pendulum asymptotically approaches its upper position.

In the case $0 < E - 2mgl \ll 2mgl$ the pendulum rotates, slowly crossing its upper position. One can estimate the period of rotation applying the result (2) of the preceding problem:

$$T = \sqrt{\frac{l}{g}}\ln\frac{\varepsilon_0}{E - 2mgl}; \quad \varepsilon_0 = 4\pi^2 mgl.$$

1.7. Once again, we put the value of the potential energy at the lowest point equal to zero, then the energy is equal to

$$E = \tfrac{1}{2}ml^2\dot\varphi^2 + mgl(1 - \cos\varphi).$$

Let at the moment t_0 the angle $\varphi(t_0) = 0$ and for definiteness $\dot\varphi(t_0) > 0$. By introducing $k = \sqrt{E/2mgl}$, we have

$$t = \frac{1}{2}\sqrt{\frac{l}{g}}\int_0^\varphi \frac{d\varphi}{\sqrt{k^2 - \sin^2\frac{\varphi}{2}}} + t_0. \tag{1}$$

At $k < 1$ the pendulum oscillates within the interval $-\varphi_m \leqslant \varphi \leqslant \varphi_m$ and $k = \sin \frac{\varphi_m}{2}$. The substitution $\sin \xi = \frac{1}{k} \sin \frac{\varphi}{2}$ is reduced the integral (1) to the form[1]
$t = \sqrt{\frac{l}{g}} F(\xi, k) + t_0$.
Hence

$$\varphi = 2 \arcsin[k \operatorname{sn}(u, k)], \quad u = (t - t_0) \sqrt{\frac{g}{l}}.$$

The period of oscillations is

$$T = 4 \sqrt{\frac{l}{g}} K \left(\sin \frac{\varphi_m}{2} \right).$$

In the extreme cases (cf. with the problem 1.4)

$$T = 2\pi \sqrt{\frac{l}{g}} \left(1 + \frac{\varphi_m^2}{16} \right), \qquad \text{when } \varphi_m \ll 1,$$

$$T = 4 \sqrt{\frac{l}{g}} \ln \frac{8}{\pi - \varphi_m}, \qquad \text{when } \pi - \varphi_m \ll 1.$$

If $k > 1$, the pendulum does not oscillate but rotates and from (1) we obtain

$$t = \frac{1}{k} \sqrt{\frac{l}{g}} F \left(\frac{\varphi}{2}, \frac{1}{k} \right) + t_0, \quad \varphi = 2 \operatorname{Arcsin} \operatorname{sn} \left(u, \frac{1}{k} \right), \quad u = k(t - t_0) \sqrt{\frac{g}{l}}.$$

[1] Functions

$$F(\xi, k) = \int_0^\xi \frac{d\alpha}{\sqrt{1 - k^2 \sin^2 \alpha}}, \quad E(\xi, k) = \int_0^\xi \sqrt{1 - k^2 \sin^2 \alpha} \, d\alpha$$

are the so-called *incomplete elliptic integrals* of the first and second kind, while functions

$$K(k) = F \left(\frac{\pi}{2}, k \right), \quad E(k) = E \left(\frac{\pi}{2}, k \right)$$

are the so-called *complete elliptic integrals* of the first and second kind, respectively. If $u = F(\xi, k)$, then ξ is expressed in one of the elliptic Jacobi functions, namely by the elliptic sine: $\sin \xi = \operatorname{sn}(u, k)$.
  Here is also the formulae for the two limiting cases:

$$K(k) = \frac{\pi}{2} \left(1 + \frac{k^2}{4} \right), \quad \text{when } k \ll 1; \quad K(k) = \frac{1}{2} \ln \frac{16}{1 - k^2}, \quad \text{when } 1 - k \ll 1.$$

Tables and formulae of these functions can be found, for instance, in [11].

The rotation period of the pendulum is

$$T = \frac{2}{k}\sqrt{\frac{l}{g}}\,\mathrm{K}\!\left(\frac{1}{k}\right).$$

In particular, if $E - 2mgl \ll 2mgl$ we get

$$T = \sqrt{\frac{l}{g}}\,\ln\frac{\varepsilon_0}{E - 2mgl},$$

where $\varepsilon_0 = 32mgl$. This result differs from the rather crude estimation made in the preceding problem in the value of ε_0, that is in a quantity which does not depend on $E - 2mgl$.

1.8. The motion in the field $U(x) + \delta U(x)$ is governed by the equality

$$t = \sqrt{\frac{m}{2}}\int_b^x \frac{dx}{\sqrt{E - U(x) - \delta U(x)}}, \tag{1}$$

where we have assumed that at $t = 0$, $x = b$. Expanding the integrand in (1) in powers of $\delta U(x)$, we obtain

$$t = t_0(x) + \delta t(x), \tag{2}$$

where

$$t_0(x) = \sqrt{\frac{m}{2}}\int_b^x \frac{dx}{\sqrt{E - U(x)}}, \tag{3}$$

$$\delta t(x) = \frac{1}{2}\sqrt{\frac{m}{2}}\int_b^x \frac{\delta U(x)\,dx}{[E - U(x)]^{3/2}}. \tag{4}$$

Let the orbit for the case when $\delta U(x) = 0$, which is determined by the equation $t = t_0(x)$, be $x = x_0(t)$. We then have from (2)

$$x = x_0(t - \delta t(x)), \tag{5}$$

where in small correction $\delta t(x)$ we substitute for x the function $x = x_0(t)$. Expanding (5) in terms of δt, we finally obtain

$$x = x_0(t) - x_0'(t)\delta t(x_0(t)). \tag{6}$$

Near a turning point $x = x_1$ one has to be careful, as the expansion (2) becomes inapplicable since the correction is $\delta t(x) \to \infty$ as $x \to x_1$.

It is remarkable, however, that formula (6) turns out to be correct up to the turning point if

$$|\delta U'(x)| \ll |F|, \quad F = -U'(x_1). \tag{7}$$

This is because although with the approach to a turning point δt increases, the dependence $x(t)$ near the extremum turns out to be weak.

It is obvious that close to x_1 the unperturbed motion has the form

$$x_0(t) = x_1 + \frac{F}{2m}(t - t_1)^2. \tag{8}$$

Adding δU shifts the turning point on δx_1, according to equation

$$U(x_1 + \delta x_1) + \delta U(x_1 + \delta x_1) = E,$$

hence $\delta x_1 = \delta U(x_1)/F$. Taking into account the perturbation δU similarly to (8) we have

$$x(t) = x_1 + \delta x_1 + \frac{F}{2m}(t - t_1 - \delta t_1)^2 \tag{9}$$

(because of (7), the correction to F is neglected). Make sure that the calculation of the formula (6) results in (9).

Divide the integration region in (4) into two parts: from b to c and from c to x, where c lies near x_1. In the second area we can put $\delta U = \delta U(x_1)$ and $U(x) = E - (x - x_1)F$. Then

$$\delta t = \frac{\sqrt{m}\,\delta U(x_1)}{\sqrt{2F^3(x - x_1)}} + \delta t_0, \tag{10}$$

$$\delta t_0 = \frac{1}{2}\sqrt{\frac{m}{2}} \int_b^c \frac{\delta U(x)\,dx}{(E - U)^{3/2}} - \frac{\sqrt{m}\,\delta U(x_1)}{\sqrt{F^3(c - x_1)}}.$$

Substituting (10) and (8) into (6) and neglecting $(\delta t_0)^2$, we obtain (9) with $\delta t_1 = \delta t_0$.

1.9. a) Here we can use the results of the preceding problem. For the unperturbed motion with $b = 0$ and $x_1 = a$ we have

$$x_0(t) = a\sin\omega t, \quad E = \tfrac{1}{2}m\omega^2 a^2.$$

In this case $|\delta U/U| \lesssim \varepsilon = \frac{\alpha a}{\omega^2} \ll 1$. For the correction we have

$$\delta t(x) = \frac{\alpha}{3\omega^3}\left(\sqrt{a^2 - x^2} + \frac{a^2}{\sqrt{a^2 - x^2}} - 2a\right),$$

$$\delta t(x_0(t)) = \frac{\varepsilon}{3\omega}\left(\cos\omega t + \frac{1}{\cos\omega t} - 2\right).$$

Substituting these expressions into (6) of the preceding problem, we get

$$x(t) = a\sin\omega t - \tfrac{1}{3}\varepsilon a\,(\cos^2\omega t + 1 - 2\cos\omega t).$$

Up to and including first-order terms in ε, we have

$$x(t) = a\sin\left(\omega t + \tfrac{2}{3}\varepsilon\right) - \tfrac{1}{2}\varepsilon a - \tfrac{1}{6}\varepsilon a\cos 2\omega t$$

(cf. problem 8.1 b).

b) Acting in the same way as previously described, we obtain

$$x(t) = a\sin\omega t + \varepsilon a\left(\tfrac{3}{2}\omega t\cos\omega t - \tfrac{9}{8}\sin\omega t - \tfrac{1}{8}\sin 3\omega t\right), \quad \varepsilon = \frac{\beta a^2}{4\omega^2} \ll 1. \qquad (1)$$

This result has a relative accuracy $\sim \varepsilon^2$ in one period, and after $1/\varepsilon$ periods the formula (1) becomes completely inapplicable. Taking into account the periodic character of the motion, we can extend the result (1) over a larger period of time. With accuracy up to and including first-order terms of ε, the formula (1) is transformed into a clearly periodic form

$$x(t) = a\left(1 - \tfrac{9}{8}\varepsilon\right)\sin\left[\omega\left(1 + \tfrac{3}{2}\varepsilon\right)t\right] - \tfrac{1}{8}\varepsilon a\sin\left[3\omega\left(1 + \tfrac{3}{2}\varepsilon\right)t\right]. \qquad (2)$$

The corrections, which were not taken into account in (1), have lead to the frequency change of the order of $\varepsilon^2\omega$, so that (2) has the relative accuracy $\sim \varepsilon$ during $1/\varepsilon$ periods (cf. problem 8.1 and, for more detail, see [7], § 29.1).

1.10. The change in the period is

$$\delta T = \sqrt{2m}\left[\int_{x_1+\delta x_1}^{x_2+\delta x_2} \frac{dx}{\sqrt{E - U(x) - \delta U(x)}} - \int_{x_1}^{x_2}\sqrt{E - U(x)}\,dx\right]. \qquad (1)$$

One can not expand the integrand (1) in terms of $\delta U(x)$ since the requirements of the theorem about differentiation improper integrals with respect to a parameter are violated, since the resulting integral diverges. However, we can expand the integrand in terms of $\delta U(x)$ up to terms of fist order in $\delta U(x)$ if we rewrite δT in the form

$$\delta T = 2\sqrt{2m}\,\frac{\partial}{\partial E}\left[\int_{x_1+\delta x_1}^{x_2+\delta x_2}\sqrt{E-U(x)-\delta U(x)}\,dx - \int_{x_1}^{x_2}\sqrt{E-U(x)}\,dx\right], \qquad (2)$$

whence

$$\delta T = -\sqrt{2m}\,\frac{\partial}{\partial E}\int_{x_1}^{x_2}\frac{\delta U(x)\,dx}{\sqrt{E-U(x)}} = -\frac{\partial}{\partial E}(T\langle\delta U\rangle), \qquad (3)$$

where

$$\langle\delta U\rangle = \frac{1}{T}\int_{0}^{T}\delta U[x(t)]\,dt \qquad (4)$$

is time-average value of δU.

The time spent near the turning points contributes little to the period, provided, of course, that $U'(x_{1,2}) \neq 0$ (cf. problem 1.3). Equation (3) can therefore give a good approximation.

We note that sometimes even small extra term $\delta U(x)$ may strongly affect the particle motion (e.g. see, problem 1.11 b, c).

Higher-order terms in the expansion of δT in terms of δU can be obtained by similar means:

$$T = \sqrt{2m}\sum_{n=0}^{\infty}\frac{(-1)^n}{n!}\,\frac{\partial^n}{\partial E^n}\int_{x_1}^{x_2}\frac{[\delta U(x)]^n\,dx}{\sqrt{E-U(x)}}. \qquad (5)$$

The formal expression (5) may be an asymptotic or even a convergent series.

1.11. a) According to (5) of the preceding problem the correction to the period $2\pi/\omega$ is

$$\delta T = -\frac{3\pi\beta E}{2m\omega^5}.$$

This correction is small for sufficiently small E.

b) In Fig. 84 we have given the potential energies $U(x)$ and $U(x)+\delta U(x)$. When $E > U_m = m\omega^6/(6\alpha^2)$ the extra term clearly means that the particle can move to infinity. When E is close to U_m,

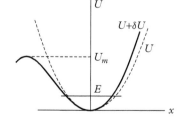

Figure 84

the period of oscillation increases without bound (as $|\ln(U_m - E)|$; see problem 1.4) so that one cannot determine the change in period just by a few terms in series (5) of the preceding problem.

When $E \ll U_m$ the correction to the period is

$$\delta T = \frac{5\pi}{18\omega} \frac{E}{U_m}.$$

c)

$$\delta T = \frac{3\pi A V \sqrt{m}}{\sqrt{2}\alpha |E|^{5/2}};$$

the formula is applicable when $|E| \gg |U_m| \approx \sqrt{8AV}$, $(E < 0)$.

1.12. The particle delay time is

$$\tau = \int\limits_{-\infty}^{+\infty} \left(\frac{1}{v} - \frac{1}{v_0}\right) dx = \frac{1}{\alpha v_0} \ln \frac{E}{E - U_0},$$

where $v = \sqrt{\frac{2}{m}|E - U(x)|}$, $v_0 = \sqrt{\frac{2E}{m}}$ (cf. problem 1.1b).

§2

Motion of a particle in three-dimensional fields

2.1. To study the orbits of the particle we use the energy and angular momentum conservation laws:

$$\frac{m\dot{\mathbf{r}}^2}{2} + U(r) = E, \tag{1}$$

$$m[\mathbf{r}, \dot{\mathbf{r}}] = \mathbf{M}. \tag{2}$$

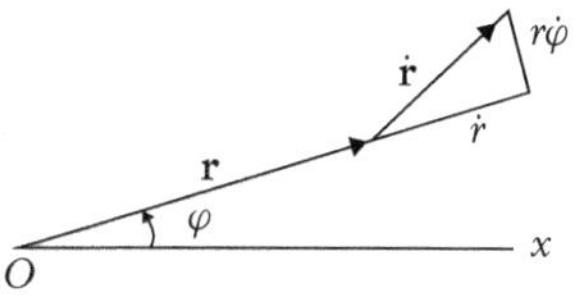

Figure 85

It follows from (2) that the orbit lies in a plane. Introducing the polar coordinates in that plane (see Fig. 85) we get

$$\tfrac{1}{2} m\dot{r}^2 + \tfrac{1}{2} mr^2\dot{\varphi}^2 + U(r) = E, \tag{3}$$

$$mr^2\dot{\varphi} = M. \tag{4}$$

Using (4) to eliminate $\dot{\varphi}$ from (3), we find

$$\frac{m\dot{r}^2}{2} + U_{\text{eff}}(r) = E, \tag{5}$$

where

$$U_{\text{eff}}(r) = U(r) + \frac{M^2}{2mr^2}.$$

The radial motion can thus be considered as one-dimensional motion in the field $U_{\text{eff}}(r)$.
For a qualitative discussion of the character of the orbits we use curves of

$$U_{\text{eff}}(r) = -\frac{\alpha}{r} - \frac{\gamma}{r^3} + \frac{M^2}{2mr^2} \tag{6}$$

for different values of M (Fig. 86).

Exploring Classical Mechanics: A Collection of 350+ Solved Problems for Students, Lecturers, and Researchers. First Edition.
Gleb L. Kotkin and Valeriy G. Serbo, Oxford University Press (2020). © Gleb L. Kotkin and Valeriy G. Serbo 2020.
DOI: 10.1093/oso/9780198853787.001.0001

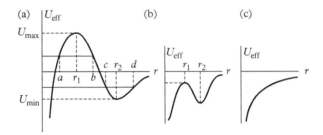

Figure 86

When $12\alpha\gamma m^2 < M^4$, then U_{eff} has two extrema at

$$r_{1,2} = \frac{M^2 \mp \sqrt{M^4 - 12\alpha\gamma m^2}}{2m\alpha}.$$

The maximum value of $U_{\text{eff}}(r_1) = U_{\text{max}}$ is positive when $M^4 > 16\alpha\gamma m^2$ (Fig. 86a) and negatively when $12\alpha\gamma m^2 < M^4 < 16\alpha\gamma m^2$ (Fig. 86b); in both these cases we have $U_{\text{eff}}(r_2) = U_{\text{min}} < 0$.

If $M^4 < 12\alpha\gamma m^2$, then the function $U_{\text{eff}}(r)$ is monotone (Fig. 86c).

Let us consider the case a) in somewhat more detail. If $E > U_{\text{max}}$ the particle coming from infinity falls in the centre of the field. The quantity of $\dot{\varphi}$ then increases according to (4). This is all we need to sketch the orbit (see Fig. 87a).

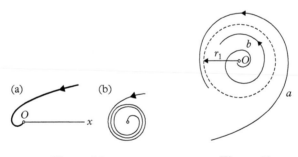

Figure 87   Figure 88

At large distances such that $\frac{\gamma}{r^3} \ll \frac{\alpha}{r}$, the main term in $U(r)$ is the term $(-\alpha/r)$ and the orbit differs little from the hyperbola. (See problem 2.8 on the form of a trajectory at $r \to 0$.)

If the energy E is close to U_{max} the particle pass very slowly through the range of values of r which are close to r_1. At the same time the radius vector turns round with an angular velocity $\dot{\varphi} \approx \frac{M}{mr_1^2}$ so that the particle may perform many revolutions around the centre before it passes through this range of values (Fig. 87b).

If $E = U_{max}$ the particle approaches the point of $r = r_1$ (cf. problem 1.3). The orbit is a spiral approaching the circle of radius r_1 and centre O (Fig. 88 curve a). If, on the other hand, the particle starts in the region $r < r_1$ and is moving away from the centre, the orbit also approaches this circle, but this time from the inside (Fig. 88 curve b). Finally, motion along the circle $r = r_1$ is possible when $E = U_{max}$. However, the motion along this circle is unstable one, as any change in the values of E or M will bring onto the orbit which moves away from the circle.

If $0 < E < U_{max}$ a particle coming from infinity will be reflected from the potential barrier $U_{eff}(r)$ and again move off to infinity. Examples of such orbits are given in Fig. 89 (curves a and b). If the energy is close to U_{max} the particle makes many revolutions around the centre before the radial velocity $\dot{r}$ changes sign. The closer the energy is to zero (for a fixed value of M this corresponds to larger values of the impact parameter) the less twisted the orbit of the particle.

When $E < U_{max}$ the case is also possible of the particle falling towards the centre of the field, when it is moving in the region $r < a$. The orbit for this case is given in Fig. 90.

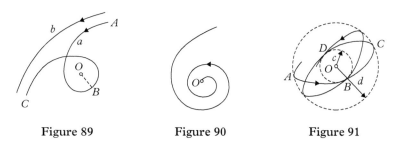

| Figure 89 | Figure 90 | Figure 91 |

When $U_{min} < E < 0$ the particle can also perform radial oscillations in region $c \leqslant r \leqslant d$ (Fig. 91). If the energy is close to zero, the amplitude of the radial oscillations will be large; their period can also become large and the radius vector may perform several revolutions during one radial oscillation. When the energy close to U_{min} the orbit close to a circle with radius r_2 while the angle over which the radius vector turns during one period of the radial oscillation depends on the quantities α, γ, M (compare problem 5.4). When $E = U_{min}$ the particle moves along of the previously mentioned circle.

One can analyse the motion of the particle in a similar way for the other cases.

What are the particular features of the orbit when $M^4 = 12\alpha\gamma m^2$?

Let us now consider the orbits in more detail. One can obtain the equation for the orbits from the (4) and (5). From (5) we get

$$\dot{r} = \frac{dr}{dt} = \pm\sqrt{\frac{2}{m}[E - U_{eff}(r)]}, \tag{7}$$

or

$$t = \pm\sqrt{\frac{m}{2}} \int \frac{dr}{\sqrt{E - U_{eff}(r)}} + C. \tag{8}$$

Using (4) to eliminate dt from (7), we find the equation of the orbit

$$\varphi = \pm \frac{M}{\sqrt{2m}} \int \frac{dr}{r^2 \sqrt{E - U_{\text{eff}}(r)}} + C_1. \tag{9}$$

Let us consider the case $M^4 > 12\alpha\gamma m^2$. If the particle moves towards the centre, we must take the lower sign in (7) and, of course, in (8). If at $t = 0, r = r_0$, we can rewrite (8) in the form

$$t = -\sqrt{\frac{m}{2}} \int_{r_0}^{r} \frac{dr}{\sqrt{E - U_{\text{eff}}(r)}}. \tag{10}$$

Equality (10) is an implicit equation for r as the function of t. If the orbit passes through the point $r = r_0, \varphi = \varphi_0$, the equation of the orbit becomes

$$\varphi = -M \int_{r_0}^{r} \frac{dr}{r^2 |p_r|} + \varphi_0, \tag{11}$$

where we have once again taken the lower sign and where

$$|p_r| = \sqrt{2m[E - U_{\text{eff}}(r)]}.$$

In particular, if the velocity of the particle at infinity is makes an angle ψ with the x-axis, we must put $r_0 = \infty, \varphi_0 = \pi - \psi$.

If $E > U_{\text{max}}$, (10) and (11) completely determine law of motion and the particle orbit.

However, if $0 < E < U_{\text{max}}$, these equations correspond only to the section AB of the orbit (see Fig. 89, curve a). At the point B the radial component of the velocity $\dot{r}$ vanishes and then changes sign. The section BC of the orbit is thus described by (9) with the upper sign, and we must redetermine the constant. It is helpful to write (9) in the form

$$\varphi = M \int_{r_{\text{min}}}^{r} \frac{dr}{r^2 |p_r|} + C_1. \tag{12}$$

As long as C_1 is not determined, we can choose the lower limit of the integral arbitrarily. According to (12), we now have

$$C_1 = \varphi(r_{\text{min}}). \tag{13}$$

Determining $\varphi(r_{\text{min}})$ from (11), we get the following equation for the section BC of the orbit:

$$\varphi = \left(\int_{r_{\text{min}}}^{r} - \int_{r_0}^{r_{\text{min}}} \right) \frac{M \, dr}{r^2 |p_r|} + \varphi_0. \tag{14}$$

Similarly we can determine the time dependence of r along the section BC

$$t = \left(\int\limits_{r_{\min}}^{r} - \int\limits_{r_0}^{r_{\min}} \right) \sqrt{\frac{m}{2}} \frac{dr}{\sqrt{E - U_{\text{eff}}(r)}}. \tag{15}$$

If $U_{\min} < E < 0$, $a < r_0 < b$, $\dot{r}(0) < 0$, $\varphi|_{t=0} = \varphi_0$, (11) describes the section AB of the orbit in Fig. 91. The section BC is described by the equation

$$\varphi = M \int\limits_{a}^{r} \frac{dr}{r^2 |p_r|} + \varphi_1, \tag{16}$$

where the angle φ_1 can be obtained by putting $r = a$ in (11). The equation for the section CD is

$$\varphi = -M \int\limits_{b}^{r} \frac{dr}{r^2 |p_r|} + \varphi_2, \tag{17}$$

where φ_2 is defined from (16) with $r = b$, and so on. Substituting the values φ_1 and φ_2 in (16) and (17), we get the equations for BC and CD in the form

$$\varphi = M \left(\int\limits_{a}^{r} - \int\limits_{r_0}^{a} \right) \frac{dr}{r^2 |p_r|} + \varphi_0, \tag{18}$$

$$\varphi = M \left(\int\limits_{b}^{r} + \int\limits_{a}^{b} - \int\limits_{r_0}^{a} \right) \frac{dr}{r^2 |p_r|} + \varphi_0 = M \left(-\int\limits_{a}^{r} + 2\int\limits_{a}^{b} - \int\limits_{r_0}^{a} \right) \frac{dr}{r^2 |p_r|} + \varphi_0. \tag{19}$$

One verify easily that the equation that section of the orbit which corresponds to the nth radial oscillation, taken the section AB to be the first, has the form[1]

---

[1] Equation (20) for the orbit can be written in the form

$$\cos \gamma (\varphi + \alpha) = \left(\gamma M \int\limits_{a}^{r} \frac{dr}{r^2 |p_r|} \right),$$

where

$$\frac{\pi}{\gamma} = M \int\limits_{a}^{b} \frac{dr}{r^2 |p_r|}, \quad \alpha = M \int\limits_{r_0}^{a} \frac{dr}{r^2 |p_r|} - \varphi_0.$$

$$\varphi = M \left(\pm \int\limits_a^r + 2(n-1) \int\limits_a^b - \int\limits_{r_0}^a \right) \frac{dr}{r^2 |p_r|} + \varphi_0. \tag{20}$$

In the equations given here we have assumed that the angle φ changes continuously and we have not introduced the limitation $0 \leqslant \varphi < 2\pi$. There is an infinity of values of φ corresponding to a given value of r, corresponding to different values of n and different signs in (20): φ is a many-valued function of r; on the other hand, r is a single-value function of φ.

All other cases can be treated in a similar fashion.

2.2. Outside the sphere of radius R the particle moves with a speed $\sqrt{2E/m}$ and inside with a speed $\sqrt{2(E+V)/m}$. Depending on the values of E and M, we get different kinds of orbits.

When $\dfrac{M^2}{2mR^2} - V < E < \dfrac{M^2}{2mR^2}$, the particle can either move inside the sphere and be reflected at its surface (Fig. 92a), or, provided $E > 0$, we can have an infinite orbit (which can be a straight line, see Fig. 93b). When $\dfrac{M^2}{2mR^2} < E$, we can have a reflected orbit (Fig. 92b).

What is the shape of the orbit when $E = \dfrac{M^2}{2mR^2} - V$?

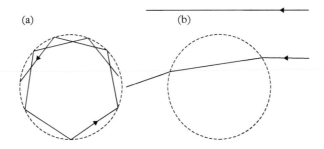

Figure 92

2.3. To determine the equation of the orbit we use the equations

$$\varphi = \int \frac{M\,dr}{r^2 \sqrt{2m(E - U_{\text{eff}})}}, \quad U_{\text{eff}} = U(r) + \frac{M^2}{2mr^2} \tag{1}$$

and we then get[1]

$$r = \frac{p}{e \cos \gamma (\varphi - \psi) - 1},$$ (2)

where

$$p = \frac{2}{\alpha}\left(\beta + \frac{M^2}{2m}\right), \quad e = \sqrt{1 + \frac{4E}{\alpha^2}\left(\beta + \frac{M^2}{2m}\right)}, \quad \gamma = \sqrt{1 + \frac{2m\beta}{M^2}},$$ (3)

$E > 0$, and ψ is an arbitrary constant.

The orbit is the curve which is obtained from a hyperbola by reducing the polar angles by a factor γ (Fig. 93). The constant ψ determines the orientation of the orbit.

The direction of the asymptotes is determined by the condition $r \to \infty$, or $e \cos(\varphi_{1,2} - \psi) = 1$. The direction of the velocity before and after the scattering are at the angle

$$\pi - (\varphi_1 - \varphi_2) = \pi - \frac{2}{\gamma}\arccos\frac{1}{e} = \pi - \frac{2}{\gamma}\arctan\sqrt{\frac{4E}{\alpha^2}\left(\beta + \frac{M^2}{2m}\right)}.$$

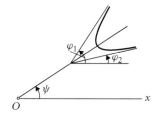

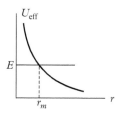

Figure 93   **Figure 94**

2.4. It is useful first to study the character of the orbit using a graph $U_{\text{eff}}(r)$. This graph is given in Fig. 94 for the case where $\beta < M^2/2m$. In that case all orbits are infinite and lie in the region $r \geqslant r_m$ when $E > 0$. The equation of the orbits are the same as in problem 2.3, (2), except that in equations (3) we must substitute β for $(-\beta)$.

---

[1] If we write the integral for φ in the form

$$\frac{\tilde{M}}{M}\varphi = \int \frac{\tilde{M}\,dr}{r^2\sqrt{2m\left(E - \frac{\tilde{M}^2}{2mr^2} - \frac{\alpha}{r}\right)}},$$

where $\tilde{M}^2 = M^2 + 2m\beta$, the integral is reduced to the corresponding integral in the Kepler problem (see [1], § 15 and [7], § 3.4).

The main difference with the orbits found in problem 2.3 occurs because now $\gamma < 1$. An example of an orbit is shown in Fig. 95. (The point of the inflection A is determined by the condition $dU(r)/dr = 0$, i.e. $r = 2\beta/\alpha$.)

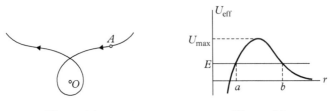

Figure 95   Figure 96

The form of $U_{\text{eff}}(r)$ for the case $\beta > M^2/2m$ is shown in Fig. 96.
If

$$E > U_{\max} = \frac{\alpha^2}{4(\beta - M^2/(2m))},$$

the particle flying from infinity will fall into the centre of the field. The equation of the orbit can be obtained from the equation of problem 2.3. We must then replace β by $(-\beta)$ and ψ by $\psi + \pi/2\gamma$, and also use the formulae

$$\sin ix = i \sinh x, \quad \sqrt{-x} = i\sqrt{x}.$$

As a result, we get

$$r = \frac{p'}{e' \sinh \gamma'(\varphi - \psi) + 1}, \tag{1}$$

$$p' = \frac{2}{\alpha}\left(\beta - \frac{M^2}{2m}\right), \quad e' = \sqrt{\frac{4E}{\alpha^2}\left(\beta - \frac{M^2}{2m}\right) - 1}, \quad \gamma' = \sqrt{\frac{2m\beta}{M^2} - 1}. \tag{2}$$

The orbit for this case is shown in Fig. 97a. We note that as $r \to 0$, $\varphi \to \infty$. This means that a particle coming from infinity and falling into the centre of the field makes an infinite number of revolutions around the centre.

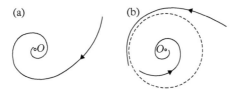

Figure 97

If $E < U_{max}$, we see from Fig. 96 that the particle can either move in the region $b \leqslant r < \infty$ (scattering), or in the region $0 < r \leqslant a$ (falling into the centre.) We obtain the equation for the orbit, using the equation $\cos ix = \cosh x$, and for the second case again the substitution of $\psi + \pi/\gamma$ for ψ:

$$r = \frac{p'}{1 \mp e'' \cosh \gamma'(\varphi - \psi)}, \quad e'' = \sqrt{1 - \frac{4E}{\alpha^2}\left(\beta - \frac{M^2}{2m}\right)}. \tag{3}$$

For the case $E = U_{max}$, it is not possible to use (2) of the problem 2.3, as we have assumed in its derivation that $e \neq 0$, and we must start again from the integral (1). We find

$$r = \frac{p'}{1 + c\exp(-\gamma'\varphi)},$$

that is

$$r = \frac{p'}{1 \pm \exp[-\gamma'(\varphi - \psi)]} \quad \text{or} \quad r = p'$$

depending on the initial value of r. The orbit is either a spiral starting at infinity and approaching asymptotically the circle with radius $r = p'$, or a spiral starting near the centre and approaching the same circle asymptotically, or it is the circle itself (Fig. 97b).

Finally, for the case $\beta = M^2/(2m)$, it is also simplest to start again from the original integral. In that case, scattering occurs and the equation for the orbit is

$$r = \frac{\alpha/E}{1 - m\alpha^2(\varphi - \psi)^2/(2M^2E)}.$$

The time it takes the particle to fall into the centre of the field is found from the equation

$$t = \sqrt{\frac{m}{2}} \int_0^r \frac{dr}{\sqrt{E - U_{eff}}}.$$

For instance, when the orbit has the form given by (2), the time to fall into the centre from a distance r is given by

$$t = \frac{1}{E}\sqrt{\frac{m}{2}} \left(\sqrt{Er^2 - \alpha r + \beta - M^2/(2m)} - \sqrt{\beta - M^2/(2m)} \right) +$$

$$+ \frac{\alpha}{2E}\sqrt{\frac{m}{2E}} \left(\text{arsinh}\frac{(2Er/\alpha) - 1}{e'} + \text{arsinh}\frac{1}{e'} \right).$$

2.5. The equation of the orbit is

$$r = \frac{p}{1 + e\cos\gamma(\varphi - \psi)},$$

where p, e, and γ are defined in problem 2.3. When $E < 0$, the orbit is the finite one, and[1]

Figure 98

$$T_r = \frac{\pi\alpha\sqrt{m}}{\sqrt{2}|E|^{3/2}}, \quad \Delta\varphi = \frac{2\pi}{\gamma}.$$

The orbit is closed if γ is the rational number. In Fig. 98, to show the orbit for $\gamma \approx 5$.

2.6. When $\beta < M^2/(2m)$ we have

$$r = \frac{\tilde{p}}{1 - \tilde{e}\cos\tilde{\gamma}(\varphi - \psi)}, \quad \tilde{p} = \frac{2}{\alpha}\left(\frac{M^2}{2m} - \beta\right),$$

$$\tilde{\gamma} = \sqrt{1 - \frac{2m\beta}{M^2}}, \quad \tilde{e} = \sqrt{1 + \frac{4E}{\alpha^2}\left(\frac{M^2}{2m} - \beta\right)};$$

if $E < 0$, then $\Delta\varphi = 2\pi/\tilde{\gamma}$ and T_r will be the same as in the preceding problem. When $\beta > M^2/(2m)$ we have (using the notations of the problem 2.4)

$$r = \frac{p'}{e'\sinh\gamma'(\varphi - \psi) - 1}, \quad \text{if } E > U_{\max},$$

$$r = \frac{p'}{e''\cosh\gamma'(\varphi - \psi) - 1}, \quad \text{if } E < U_{\max}.$$

2.7. a) A finite orbit is possible if the function $U_{\text{eff}}(r)$ has a minimum. The equation $U'_{\text{eff}}(r) = 0$ can be reduced to the form $f(x) = M^2 x/(\alpha m)$, where $f(x) = x(x + 1)e^{-x}$, $x = \varkappa r$. Using a graph of $f(x)$ one sees easily that this equation has real roots only when $m^2 x/(\alpha m)$ is less than the maximum value of $f(x)$, for $x > 0$. This maximum value is equal to $(2 + \sqrt{5})\exp\left[-\frac{1}{2}(1 + \sqrt{5})\right] \approx 0.84$. A finite orbit is thus possible, provided $M^2 < 0.84\alpha m/\varkappa$.

b) A finite orbit is possible, provided $M^2 < 8mV/(e^2\varkappa^2)$.

2.8. In the equation for the orbit (see (1) of the problem 2.3), we can neglect for small values of r the quantity E, when $n = 2$, and also the term $M^2/(2mr^2)$, when $n > 2$. We

---

[1] The period T_r is the same as in the field $U_0 = -\alpha/r$. To determine T_r it is sufficient to note that adding a term β/r^2 to the field U_0 has the same effect on the radial motion as increasing M. However, the period T_r is independent of M in the Coulomb field U_0.

obtain then (see Fig. 99)

$$\varphi = -\frac{M\ln(r/r_0)}{\sqrt{2m\alpha - M^2}} + \varphi_0, \quad \text{when } n = 2,$$

$$\varphi = \frac{2Mr^{-1+n/2}}{\sqrt{2m\alpha}(n-2)} + C, \quad \text{when } n > 2.$$

It turns out that the number of revolutions is infinite only when $n = 2$.

The time it takes the particle to fall into the center is finite as the radial velocity increases when the center is approached.

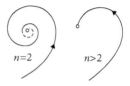

$n=2$ $n>2$

2.9. The number of revolutions of the particle around the centre is infinite only in the case b) $E = 0$, $n = 2$.

Figure 99

2.10. The time it takes the particle to fall into the centre is $\pi \sqrt{mR^3/(8\alpha)}$.

2.11. As it is known, the two-body problem is reduced to two other problems: the motion of the centre of mass and the relative motion of a particle with reduced mass (in this case equal to $m/2$) in the field $U(r)$. We consider only the relative motion. Introducing the relative distance $\mathbf{r} = \mathbf{r}_1 - \mathbf{r}_2$ and taking into account that when time t is close to the initial time $t_\infty \to -\infty$ one can assume that $\mathbf{r}_1 = (vt, 0, 0)$ and $\mathbf{r}_2 = (0, v(t - \tau), 0)$. Using these expressions, we find the initial velocity $\mathbf{v}_\infty = (v, -v, 0)$, the energy $E = \frac{1}{2}(m/2)v_\infty^2 = \frac{1}{2}mv^2$, the angular momentum $\mathbf{M} = (m/2)[\mathbf{r}, \mathbf{v}_\infty] = -(0, 0, E\tau)$, and the effective potential energy

$$U_{\text{eff}}(r) = \frac{\alpha}{r} + \frac{M^2}{mr^2}.$$

After that, the minimum distance which is equal to

$$r_{\min} = \frac{\alpha}{mv^2}\left[1 + \sqrt{1 + \frac{1}{2}\left(mv^3\tau/\alpha\right)^2}\right],$$

can be found from the equation $U_{\text{eff}}(r_{\min}) = E$.

2.12. The problem is reduced to two others: the motion of the centre of mass

$$\mathbf{R} = \frac{m_1\mathbf{r}_1 + m_2\mathbf{r}_2}{m_1 + m_2} \tag{1}$$

with constant velocity $\dot{\mathbf{R}}$ and the relative motion of particles with reduced mass $m = m_1 m_2/(m_1 + m_2)$ in the field $U(r)$, where the relative distance is $\mathbf{r} = \mathbf{r}_1 - \mathbf{r}_2$.

a) Let's take the point A to be the origin of the coordinate system, and the origin of the time will be the moment when particle 1 would fly through this point if there were no particle interaction. Under this condition, the position of the particles would be, respectively, $\mathbf{r}_1 = \mathbf{v}_1 t$ and $\mathbf{r}_2 = \mathbf{v}_2(t - \tau) + \boldsymbol{\rho}$, where $\boldsymbol{\rho} \perp \mathbf{v}_{1,2}$. These expressions can be used to determine all of the following: the law of motion of the centre of mass; the energy $E = \frac{1}{2}mv^2$; the angular momentum $\mathbf{M} = m[\mathbf{r}, \mathbf{v}] = -m([\mathbf{v}_1, \mathbf{v}_2]\tau + [\boldsymbol{\rho}, \mathbf{v}])$; and the effective potential energy

$$U_{\text{eff}}(r) = -\frac{\beta}{r^2} + \frac{M^2}{2mr^2} = -\frac{2m\beta - M^2}{2mr^2}$$

in the relative motion problem (here $\mathbf{v} = \mathbf{v}_1 - \mathbf{v}_2$). Note that from the expressions for E and $\mathbf{M}$, the time t vanishes as expected. It can be seen from the equation

$$E = \frac{1}{2}m\dot{r}^2 + U_{\text{eff}}(r) \tag{2}$$

that the particle falls to the centre under the condition

$$2m\beta > M^2 = m^2 \left([\mathbf{v}_1, \mathbf{v}_2]^2 \tau^2 + \rho^2 v^2\right).$$

This answers the first question.

b) To find the point of impact, we find its time considering relative motion and then determine where at that moment the centre of mass is.

From (2), we find the time of collision

$$t = t_\infty - \int\limits_{r_\infty}^{0} \frac{dr}{\sqrt{2[E - U_{\text{eff}}(r)]/m}}.$$

Notice the we took into account that the radial velocity $\dot{r}$ is negative during the process of particles falling on each other (such process corresponds to the movement from the initial, very large distance r_∞ to the point of incidence $r = 0$). For very large values of r, one can ignore the particle interaction and take the initial time equal to $t_\infty = -r_\infty/v$. In the end, the time of collision of particles is equal to

$$t = \frac{1}{v}\left(\sqrt{r_\infty^2 + a^2} - a - r_\infty\right), \quad a^2 = \frac{2m\beta - M^2}{(mv)^2}.$$

Substituting $\sqrt{r_\infty^2 + a^2} \approx r_\infty + \frac{a^2}{2r_\infty}$ and discarding the term containing $a^2/(2vr_\infty)$, we get $t = -a/v$ and finally find the distance where the particles will collide,

$$\mathbf{R}(t) = -\frac{(m_1\mathbf{v}_1 + m_2\mathbf{v}_2)a + m_2\mathbf{v}_2 v\tau}{(m_1 + m_2)v}.$$

2.13. The relative motion is characterized by the angular momentum $M = mv\rho$ and the energy $E = \frac{1}{2}mv^2$, where $m = m_1m_2/(m_1 + m_2)$ is the reduced mass. The distance to be determined follows from the condition $U_{\text{eff}}(r_{\min}) = E$. Solutions can be obtained easily for $n = 1, 2,$ and 4.

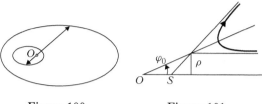

Figure 100   Figure 101

2.14. The particle orbit is:

$$\frac{m}{m_{1,2}} \frac{p}{r_{1,2}} = 1 \pm e \cos\varphi,$$

where

$$m = \frac{m_1 m_2}{m_1 + m_2}, \quad p = \frac{M^2}{m\alpha}, \quad e = \sqrt{1 + \frac{2EM^2}{m\alpha^2}};$$

E and M are the total energy and angular momentum of the system. The particles move in similar conical sections with a common focus, and their radius vectors are at any moment in the opposite directions (Fig. 100).

2.15. One sees easily from Fig. 101 that

$$OS = \rho(\cot\varphi_0 - \cot 2\varphi_0),$$

where

$$\varphi_0 = \rho \int_{r_{\min}}^{\infty} \frac{dr}{r^2 \sqrt{1 - \dfrac{U(r)}{E} - \dfrac{\rho^2}{r^2}}}.$$

At $\rho \to 0$

$$\varphi_0 = \rho \int_{r_{\min}}^{\infty} \frac{dr}{r^2 \sqrt{1 - U/E}} - 2\rho^3 E^{3/2} \frac{\partial}{\partial E} \int_{r_{\min}}^{\infty} \frac{dr}{r^4 \sqrt{E - U}} + \dots$$

(cf. problem 2.27), so that

$$OS = \left(2 \int_{r_{min}}^{\infty} \frac{dr}{r^2 \sqrt{1 - U/E}} \right)^{-1} + \mathcal{O}(\rho^2) \ldots$$

The point of S is the virtual focus of the beam of scattered particles so that up to and including terms of first order in ρ the position of the points where the asymptote of the orbit intersects the axis of the beam is independent on ρ.

2.16. The equation of the orbit is

$$\frac{p}{r} = e \cos(\varphi - \varphi_0) - 1,$$

where $p = \frac{M^2}{m\alpha}$, $e = \sqrt{1 + \frac{2EM^2}{m\alpha^2}}$, while φ_0 is determined from the condition that $\varphi \to \pi$, as $r \to \infty$, so that $\cos \varphi_0 = -1/e$. The region which can not be reached by the particles is bounded by the envelope of the family of orbits.

To find it, we differentiate the equation for the orbit,

$$\frac{M^2}{m\alpha r} + 1 + \cos\varphi - \frac{M}{\alpha} \sqrt{\frac{2E}{m}} \sin\varphi = 0 \qquad (1)$$

with respect to the parameter M

$$\frac{2M}{mr} - \sqrt{\frac{2E}{m}} \sin\varphi = 0, \qquad (2)$$

and eliminate M from (1) and (2). The result is

$$\frac{2\alpha}{Er} = 1 - \cos\varphi.$$

The inaccessible region is thus

$$r < \frac{2\alpha}{E(1 - \cos\varphi)}$$

which is bounded by a paraboloid of rotation (Fig. 102).

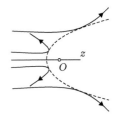

2.17. $\rho > \dfrac{2a\delta}{1 - \delta^2 - (1-\delta)^2 \cos\varphi}$, where $\delta = \dfrac{mav^2}{2\alpha}$, $OA = a$.

Figure 102

2.18. We take the scalar product of the equation

$$[\mathbf{v}, \mathbf{M}] - \frac{\alpha \mathbf{r}}{r} = \mathbf{A}$$

with $\mathbf{r}$. Denoting by φ the angle between $\mathbf{R}$ and $\mathbf{A}$, we get

$$\frac{M^2}{m} - \alpha r = Ar\cos\varphi,$$

or

$$\frac{p}{r} = 1 + e\cos\varphi,$$

where

$$p = \frac{M^2}{m\alpha}, \quad e = \frac{|\mathbf{A}|}{\alpha}.$$

We note that the vector $\mathbf{A}$ is directed from the centre of the field to the point $r = r_{\min}$.

2.19. In a field $U(r) = -\alpha/r$, the body, first as a part of the spacecraft, flew in a circle of radius R at a velocity V. If the separation of the body from the spacecraft occurs at $\mathbf{R} = (0, R)$ and velocity $\mathbf{V} = (V, 0)$, the Laplace vector

$$\mathbf{A}_0 = [\mathbf{V}, \mathbf{M}_0] - \alpha \frac{\mathbf{R}}{R} = mV^2\mathbf{R} - \alpha\frac{\mathbf{R}}{R}$$

is zero and the velocity of the body before separation was $V = \sqrt{\alpha/(mR)}$, the angular momentum $M_0 = mRV = \sqrt{m\alpha R}$.

Obviously, the orbit of the body lies in the same plane as the orbit of the spacecraft, and the velocity of the body at the first moment after separation from the spacecraft is $\mathbf{V} + \mathbf{v}$. The angular momentum of the body has not changed, $m[\mathbf{R}, \mathbf{V} + \mathbf{v}] = \mathbf{M}_0$, since $\mathbf{v} \parallel (-\mathbf{R})$, but the Laplace vector is no longer zero

$$\mathbf{A} = [\mathbf{V} + \mathbf{v}, \mathbf{M}_0] - \alpha \frac{\mathbf{R}}{R} = [\mathbf{v}, \mathbf{M}_0] = mRv\mathbf{V} = \alpha v \frac{\mathbf{V}}{V^2}.$$

As it is known, the Laplace vector is directed towards the pericentre (the point of the orbit with the smallest radius) and is αe, where e is the eccentricity. From here we get the value $e = v/V$. Assuming that $e < 1$, we find that the orbit of the body is an ellipse, whose parameter is $p = M^2/(m\alpha) = R$, and whose pericentre lies on the x-axis. Thus, the equation of the body orbit in polar coordinates r, φ has the form

$$r = \frac{R}{1 + (v/V)\cos\varphi}.$$

2.20. $\delta T = -\dfrac{\partial \delta I}{\partial E}$, where

$$\delta I = T \langle \delta U \rangle = \sqrt{2m} \int\limits_{r_1}^{r_2} \frac{\delta U(r)\, dr}{\sqrt{E - U_{\mathrm{eff}}(r)}}$$

(cf. problem 1.10). Similarly, we can write for the change in angular distances between successive pericentre passages ($r = r_{\min}$) in the form $\delta \Delta \varphi = \dfrac{\partial \delta I}{\partial M}$ (cf. [1], § 15, problem 3; § 49).

2.21. The field $U(r)$ differs little from the Coulomb field $U_0(r) = -\alpha/r$ in the region $r \ll D$. Therefore, a finite orbit which is close to an ellipse with a parameter p and an eccentricity e determined by the integrals of motion E and M will retain its shape, but will change its orientation. The velocity of rotation Ω of the ellipse is determined by the displacement of the pericentre over a period $\Omega = \delta \Delta \varphi / T_0$ which we can evaluate using the equations from the preceding problem with $\delta U = \dfrac{\alpha}{D} - \dfrac{\alpha r}{2D^2}$ while T_0 is the period in the Coulomb field.[1] The result of the calculation is $\Omega = M/(2mD^2)$. We can write the equation of the orbit in the form

$$\frac{p}{r} = 1 + e \cos \gamma \varphi, \quad \gamma = 1 - \frac{\Omega T_0}{2\pi}. \tag{1}$$

The deviation of the curve (1) from the actual orbit is of first order in δU; that is, during one period (1) describes the orbit with the same degree of accuracy as the equation for the fixed ellipse. However, (1) retains the same accuracy during many periods. It is therefore just this equation which can be called the "correct zero approximation".

In other words, only secular first-order effects have been taken into account in (1).

2.22. The field $U(r) = -\alpha/r^{1+\varepsilon}$ differs little from the Coulomb field so that the orbit of the particle in this field will be a slowly precessing ellipse. Expanding $U(r)$ in terms of ε, we can write it in the form

$$U(r) = U_0(r) + \delta U,$$

---

[1] It is convenient to change to an integration over ξ, where (see [1], § 15)

$$r = \frac{\alpha}{2|E|}(1 - e \cos \xi).$$

where

$$U_0(r) = -\frac{\tilde{\alpha}}{r}, \quad \delta U = \frac{\varepsilon\tilde{\alpha}}{r}\ln\frac{r}{R}, \quad \tilde{\alpha} = \alpha R^{-\varepsilon}$$

and R is a constant which characterizes the size of the orbit.

We can evaluate the shift of pericentre for the period

$$\delta\Delta\varphi = \frac{\partial}{\partial M}\int_0^T \delta U\, dt$$

(see problem 2.20) by making the substitution

$$r = -\frac{\tilde{\alpha}}{2E}(1 - e\cos\xi), \quad t = \frac{T}{2\pi}(\xi - e\sin\xi),$$

where

$$e = \sqrt{1 + \frac{2EM^2}{m\tilde{\alpha}^2}}, \quad T = \pi\tilde{\alpha}\sqrt{\frac{m}{2|E|^3}}$$

(see [1], § 15). The result is

$$\Omega = \frac{\delta\Delta\varphi}{T} = \frac{\varepsilon\tilde{\alpha}}{2\pi}\frac{\partial}{\partial M}\int_0^{2\pi}\ln\frac{\tilde{\alpha}(1 - e\cos\xi)}{2|E|R}\,d\xi =$$

$$= -\frac{\varepsilon|E|}{\pi}\frac{\partial e}{\partial M}\int_0^{2\pi}\frac{\cos\xi\, d\xi}{1 - e\cos\xi} = \frac{2\pi}{T}\frac{1 - \sqrt{1 - e^2}}{e^2}\varepsilon.$$

At $\varepsilon > 0$, the direction of the orbit precession coincides with the direction of motion of a particle over orbit, and at $\varepsilon < 0$ the precession is opposite to it.

2.23. $\Omega = -\frac{3}{2}\frac{\beta}{m\omega}\frac{a^2 + b^2}{a^4 b^4}\frac{M}{|M|}$ (see also [7], § 4).

2.24. The Lagrangian function reads

$$L = \frac{m}{2}\left(\dot{x}^2 + \dot{y}^2\right) - \frac{mg}{2l}\left(x^2 + y^2\right) + \frac{m}{2}\frac{x^2 + y^2}{l^2}\left(\dot{x}^2 + \dot{y}^2\right) =$$

$$= \frac{m}{2}\left(\dot{r}^2 + r^2\dot{\varphi}^2\right) - \frac{mg}{2l}r^2 + \frac{mr^2}{2l^2}\dot{r}^2.$$

If we neglect the last term, then it will be convenient to consider the particle motion in the Cartesian coordinates. It leads to harmonic oscillations with frequency $\omega = \sqrt{g/l}$

along the x- and y-axes, that is, to the elliptic orbit. The exact equation of the trajectory is convenient to obtain in the cylindrical coordinates using the integrals of motion E and $\mathbf{M} = (0, 0, M)$:

$$\varphi = \int \frac{M\sqrt{1 + r^2/l^2}\, dr}{r^2\sqrt{2m[E - U_{\text{eff}}(r)]}}, \quad U_{\text{eff}}(r) = \frac{mgr^2}{2l} + \frac{M^2}{2mr^2}.$$

Expansion

$$\sqrt{1 + \frac{r^2}{l^2}} = 1 + \frac{r^2}{2l^2} + \dots$$

leads to the equation

$$\varphi = \int \frac{M\, dr}{r^2\sqrt{2m(E - U_{\text{eff}})}} + \frac{M}{2ml^2}\int \frac{dr}{\sqrt{2m(E - U_{\text{eff}})}} = \varphi_0(r) + \Omega\, t,$$

where $\varphi_0(r)$ corresponds to the motion over an ellipse, and

$$\Omega = \frac{M}{2ml^2} = \frac{ab}{2l^2}\,\omega$$

determines the angular velocity of the precession.

2.25. The Lagrangian function of the Earth–Moon system is

$$L = \tfrac{1}{2}m_1\dot{\mathbf{R}}_1^2 + \tfrac{1}{2}m_2\dot{\mathbf{R}}_2^2 + \frac{Gm_0m_1}{R_1} + \frac{Gm_0m_2}{R_2} + \frac{Gm_1m_2}{|\mathbf{R}_1 - \mathbf{R}_2|},$$

where $\mathbf{R}_1$ and $\mathbf{R}_2$ are radius vectors of the Earth and the Moon in the heliocentric coordinate system, m_1, m_2 are their masses, m_0 is the mass of the Sun, and G is the gravitational constant. We further introduce the coordinates of the centre of mass of the Earth and the Moon (the point O on Fig. 103) and their relative distance

$$\mathbf{R} = \frac{m_1\mathbf{R}_1 + m_2\mathbf{R}_2}{m_1 + m_2}, \quad \mathbf{r} = \mathbf{R}_2 - \mathbf{R}_1,$$

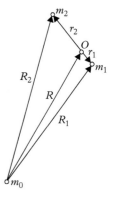

Figure 103

then

$$\mathbf{R}_{1,2} = \mathbf{R} + \mathbf{r}_{1,2}, \quad \mathbf{r}_{1,2} = \mp \frac{m}{m_{1,2}} \mathbf{r},$$

$$m = \frac{m_1 m_2}{m_1 + m_2} \approx m_2.$$

Using the expansion

$$\frac{1}{R_i} = \frac{1}{R} \left(1 + \frac{2\mathbf{R}\mathbf{r}_i}{R^2} + \frac{r_i^2}{R^2} \right)^{-1/2} =$$

$$= \frac{1}{R} - \frac{\mathbf{R}\mathbf{r}_i}{R^3} + \frac{1}{2R^3} \left[3\frac{(\mathbf{R}\mathbf{r}_i)^2}{R^2} - r_i^2 \right] + \ldots$$

and taking into account that

$$m_1\mathbf{r}_1 + m_2\mathbf{r}_2 = 0, \quad m_1 r_1^2 \ll m_2 r_2^2 \approx m_2 r^2,$$

we obtain

$$L = L_1 + L_2 - \delta U, \tag{1}$$

$$L_1 = \frac{m_1 + m_2}{2}\dot{\mathbf{R}}^2 + \frac{Gm_0(m_1 + m_2)}{R},$$

$$L_2 = \frac{m}{2}\dot{\mathbf{r}}^2 + \frac{Gm_1 m_2}{r}, \quad \delta U = -\frac{Gm_0 m}{2R^3}\left[3\frac{(\mathbf{R}\mathbf{r})^2}{R^2} - r^2 \right].$$

If we neglect the term δU, the problems of motion of the centre of mass O and of relative motion are separated and reduced to the Kepler problem (for simplicity, we will talk further about the motions of the Earth around the Sun and the Moon around the Earth).

a) In the problem of the motion of the Earth around the Sun, the smallness of δU is characterized by a ratio

$$\frac{\delta U}{(Gm_0 m_1/R)} \sim \frac{m_2 r^2}{m_1 R^2} \sim 10^{-7}.$$

Considering the orbit of the Moon as a circle of radius r lying in the plane of the Earth's orbit, one has

$$\delta U = -\frac{2\beta}{R^3}\left(3\cos^2 \chi - 1 \right), \quad \beta = \tfrac{1}{4} Gm_0 m_2 r^2, \tag{2}$$

where χ is the angle between vectors $\mathbf{R}$ and $\mathbf{r}$. This angle changes to 2π in 29.5 days (a so-called synodic month which is a time interval between the two sequent new Moons). Averaging expression (2) over that interval leads to the following replacement $\cos^2 \chi \to \frac{1}{2}$. As a result,

$$\delta U = -\frac{\beta}{R^3}.$$

The perihelion displacement per year (see [1], problem 3 to § 15) is

$$\delta\varphi = \frac{6\pi\beta}{\alpha_1 R^2} = \frac{3\pi m_2}{2m_1}\left(\frac{r}{R}\right)^2, \quad \alpha_1 = Gm_0 m_1.$$

The perihelion shift per century equals

$$100\,\delta\varphi = 7.7''.$$

It should be noted that the total displacement of the Earth's perihelion for one century equal to $1158''$ is mainly due to the influence of the Jupiter and the Venus.

It is worth noting that the relativistic correction can be estimated as

$$\sim 100 \cdot 2\pi \left(\frac{V}{c}\right)^2 \sim 1'', \quad V \approx 30 \text{ km/s}$$

(see [2] § 39, § 101 and [7], § 41.2).

b) Studying the motion of the Moon around the Earth, we can also consider δU as a small correction to L_2:

$$\frac{\delta U}{(Gm_1 m_2/r)} \sim \frac{\Omega^2}{\omega^2} \sim 10^{-2}, \quad \Omega^2 = \frac{Gm_0}{R^3}, \quad \omega^2 = \frac{Gm_1}{r^3}.$$

Here the period $2\pi/\omega = 27.3$ days is the *star (or sidereal) month* and the period $2\pi/\Omega$ is 1 year. The correction δU leads to various distortions of the the Moon's orbit as the ripple of eccentricity, the perihelion shift (cf. problem 2.43) and so on. We will consider only one of them, neglecting their mutual influence and assuming the unperturbed orbit of the Moon as a circle; we also take $R = \text{const}$.

We introduce the geocentric coordinate system $Oxyz$ with the x-axis directed along the line of intersection of the plane of the Moon's and the Earth's orbits with (the line of lunar nodes), the y-axis lying in the plane of the Earth's orbit and the z-axis perpendicular to this plane and directed to the northern celestial hemisphere (Fig. 104). The coordinates of the Sun in this system are

$$-\mathbf{R} = R(\cos\varphi, \sin\varphi, 0), \quad \varphi = \Omega t + \varphi_0,$$

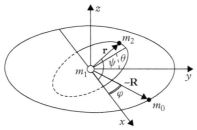

Figure 104

and the coordinates of the Moon are

$$\mathbf{r} = r(\cos\psi, \cos\theta\sin\psi, \sin\theta\sin\psi), \quad \psi = \omega t + \psi_0,$$

where φ_0 and ψ_0 are determined by the moments of passage of the Sun and the Moon through the x-axis.

Averaging quantity

$$(\mathbf{Rr})^2 = R^2 r^2 (\cos\varphi\cos\psi + \cos\theta\sin\varphi\sin\psi)^2$$

over one month leads to the replacements

$$\cos^2\psi \to \tfrac{1}{2}, \quad \sin^2\psi \to \tfrac{1}{2}, \quad \sin\psi\cos\psi \to 0,$$

and to equations

$$\langle(\mathbf{Rr})^2\rangle_{\text{month}} = \tfrac{1}{2} R^2 r^2 \left(\cos^2\varphi + \cos^2\theta\sin^2\varphi\right), \tag{3}$$

while averaging over one year leads to these replacements

$$\cos^2\varphi \to \tfrac{1}{2}, \quad \sin^2\varphi \to \tfrac{1}{2}.$$

It gives

$$\langle(\mathbf{Rr})^2\rangle = \tfrac{1}{4} R^2 r^2 \left(1 + \cos^2\theta\right).$$

As a result,

$$\langle\delta U\rangle = -\frac{Gm_0 m}{8R^3} r^2 \left(3\cos^2\theta - 1\right).$$

The rotation of the orbital plane can be traced by considering the angular momentum $\mathbf{M} = m[\mathbf{r}, \dot{\mathbf{r}}]$ perpendicular to this plane. The equation of its motion is $\dot{\mathbf{M}} = \mathbf{K}$, where

$$\mathbf{K} = -\left[\mathbf{r}, \frac{\partial \delta U}{\partial \mathbf{r}}\right]$$

is the moment of the forces acting upon the Moon. Because the change of the angle θ is the rotation around the x-axis,

$$K_x = -\frac{\partial \delta U}{\partial \theta},$$

that is,

$$\langle \mathbf{K} \rangle = \left(-\frac{3Gm_0 m}{4R^3} r^2 \sin\theta \cos\theta,\, 0,\, 0\right).$$

Since the moment of forces $\langle \mathbf{K} \rangle$ is perpendicular to the vector $\mathbf{M}$, it leads only to its rotation. The x- and y-axes in this case also rotate; therefore, the angular velocity $\mathbf{\Omega}_{\mathrm{pr}}$ of precessions of the vector $\mathbf{M}$ is directed along the z-axis. It can be found from the condition

$$[\mathbf{\Omega}_{\mathrm{pr}}, \mathbf{M}] = \langle \mathbf{K} \rangle,$$

which gives

$$\left(\mathbf{\Omega}_{\mathrm{pr}}\right)_z = -\frac{\langle K_x \rangle}{M_y} = \frac{\langle K_x \rangle}{M\omega r^2 \sin\theta} = -\frac{3\Omega^2}{4\omega}\cos\theta \approx -\frac{\Omega}{17.9}.$$

Thus we found that the precession period of the Moon's orbit is $2\pi/\Omega_{\mathrm{pr}} = 17.9$ years. This motion is called a *retreat of the lunar nodes*. The correct meaning of this period is 18.6 years. Given the crudeness of our approximations, our result could be considered as a good agreement with the correct one.

   This period determines the repetition period of the solar and lunar eclipses cycles. That period (or, more precisely, its triple value containing integer 19 765 days) was already known to the priests of the ancient Babylon.

   Note that by limiting the averaging over a month; that is, using (3), we would discover the unevenness of the precession within a year:

$$\Omega_{\mathrm{pr}} \to \left(1 - \tfrac{1}{2}\cos 2\varphi\right)\Omega_{\mathrm{pr}}.$$

2.26. Similarly to problem 2.21, we have

$$\Omega = \frac{\partial}{\partial M}\langle\delta U\rangle = \frac{a}{2}\frac{\partial e^2}{\partial M}\left[\delta U'(a) + \frac{a}{2}\delta U''(a)\right] =$$

$$= -\sqrt{\frac{a}{m\alpha}}\left[\delta U'(a) + \frac{a}{2}\delta U''(a)\right]\frac{M}{|M|}.$$

If we take into account the next terms of the expansion δU over $(r-a)$, that will give the contribution of $\lesssim e^2\Omega$ into Ω.

2.27. The perihelion displacement for the period

$$\Delta\varphi = \int\limits_{r_1+\delta r_1}^{r_2+\delta r_2}\frac{M\,dr}{r^2\sqrt{2m(E - U_{\text{eff}} - \delta U)}}$$

can be presented as

$$\Delta\varphi = -\frac{4}{3}\sqrt{2m}\frac{\partial^2}{\partial M\partial E}\int\limits_{r_1+\delta r_1}^{r_2+\delta r_2}(E - U_{\text{eff}} - \delta U)^{3/2}dr.$$

Up to the second order in δU, we have

$$\Delta\varphi = 2\pi + \delta_1\varphi + \delta_2\varphi,$$

$$\delta_1\varphi = t\frac{\partial}{\partial M}\langle\delta U\rangle,$$

$$\delta_2\varphi = -\frac{\partial^2}{2\partial M\partial E}[T\langle(\delta U)^2\rangle],$$

where $\langle f\rangle$ is the value of $f(r)$, averaged over the period T of the unperturbed motion (cf. problem 2.20). The angular velocity of the precession is

$$\Omega = \frac{\delta_1\varphi + \delta_2\varphi}{T + \delta T} = \frac{\delta_1\varphi}{T} + \frac{\delta_2\varphi}{T} - \frac{\delta_1\varphi\,\delta T}{T^2},$$

$$\delta T = -\frac{\partial}{\partial E}(T\langle\delta U\rangle).$$

2.28. If we expand the integrand in the equation for the orbit,

$$\varphi = \int \frac{M\,dr}{r^2\sqrt{2m\left(E - \dfrac{M^2}{2mr^2} + \dfrac{\alpha}{r} - \dfrac{\gamma}{r^3}\right)}},$$ (1)

in terms of $\delta U = \gamma/r^3$, we can rewrite this equation in the form

$$\varphi = \varphi_0(r) + \delta\varphi(r),$$ (2)

where

$$\varphi_0(r) = \int \frac{M\,dr}{r^2\sqrt{2m\left(E - \dfrac{M^2}{2mr^2} + \dfrac{\alpha}{r}\right)}},$$ (3)

$$\delta\varphi(r) = \int \frac{\gamma M\,dr}{2r^5\sqrt{2m\left(E - \dfrac{M^2}{2mr^2} + \dfrac{\alpha}{r}\right)^3}}.$$ (4)

If we neglect the correction $\delta\varphi(r)$ in (2), we obviously get the equation of the orbit in a Coulomb field (see [1], § 15)

$$\frac{p}{r} = 1 + e\cos\varphi,$$ (5)

where

$$p = \frac{M^2}{m\alpha}, \quad e = \sqrt{1 + \frac{2EM^2}{m\alpha^2}}.$$

If we take into account the correction $\delta\varphi(r)$ in (2), we get, instead of (5),

$$\frac{p}{r} = 1 + e\cos(\varphi - \delta\varphi(r)).$$ (6)

In the right-hand side of (6) we can expand in terms of $\delta\varphi(r)$ and use in $\delta\varphi(r)$ for r the relation $r = r_0(\varphi)$ which follows from (5). The result is

$$\frac{p}{r} = 1 + e\cos\varphi + e\delta\varphi(r_0(\varphi))\sin\varphi.$$ (7)

Integrating (4), we find[1]

$$\delta\varphi(r_0(\varphi)) =$$

$$= \frac{m^2\alpha\gamma}{M^4}\left\{-3\varphi - \frac{2e^2+1}{e}\sin\varphi - \frac{1+e\cos\varphi}{e^2\sin\varphi}[2e+(e^2+1)\cos\varphi]\right\}. \quad (8)$$

Substituting (8) into (7) gives the result which is accurate up to and including first-order terms in $\zeta = \dfrac{m^2\alpha\gamma}{M^4}$:

$$\frac{p}{r} = 1 + e\cos[(1+3\zeta)\varphi] - \zeta(2e^2+1)\sin^2\varphi-$$

$$-\zeta\frac{1+e\cos\varphi}{e}[2e+(e^2+1)\cos\varphi] \quad (9a)$$

or

$$\frac{p'}{r} = 1 + e'\cos\lambda\varphi + f\cos 2\varphi,$$

$$\lambda = 1+3\zeta,$$

$$p' = p\left[1 + \tfrac{3}{2}\zeta(e^2+2)\right], \quad (9b)$$

$$e' = e\left(1 + \zeta\frac{3e^4-2}{2e^2}\right),$$

$$f = \tfrac{1}{2}\zeta e^2.$$

---

[1] We can rewrite (4) in the form

$$\delta\varphi = \frac{\partial}{\partial M}\int \frac{\sqrt{2m\gamma}\,dr}{2r^3\sqrt{E - \dfrac{M^2}{2mr^2} + \dfrac{\alpha}{r}}},$$

and use (5) to change the integration into one over φ:

$$\delta\varphi = \frac{\partial}{\partial M}\frac{m^2\gamma\alpha}{M^3}\int(1+e\cos\varphi)\,d\varphi =$$

$$= -\frac{3m^2\gamma\alpha}{M^4}(\varphi+e\sin\varphi) + \frac{m^2\gamma\alpha}{M^3}\frac{\partial e}{\partial M}\sin\varphi + \frac{m^2\gamma\alpha}{M^3}\frac{\partial\varphi}{\partial M}(1+e\cos\varphi). \quad (8')$$

From (5) we find

$$\frac{2M}{m\alpha^2 r} = \frac{\partial e}{\partial M}\cos\varphi - e\frac{\partial\varphi}{\partial M}\sin\varphi, \quad \frac{\partial e}{\partial M} = \frac{e^2-1}{eM}$$

and substituting this into (8'), we obtain (8).

In the vicinity $\varphi = 0$ and $\varphi = \pi$, the expression (2) can no longer be applied as $\delta\varphi$ increases infinitely. Equation (9) for the orbit, however, valid also in those regions (cf. problem 1.8).

In the case of an infinite orbit $(E \geqslant 0)$, (9) is the solution of the problem. If $E < 0$, (9) remains the equation for the orbit only during a few revolutions,[1] namely, as long as $3\zeta\varphi \ll 1$. Taking only the secular part $\delta\varphi = -3\zeta\varphi$ in (8) into account, we obtain the equation

$$\frac{p}{r} = 1 + e\cos\lambda\varphi, \quad \lambda = 1 + 3\zeta, \tag{10}$$

which describes the path over a long section of the orbit; that is, a "corrected" zero approximation (as distinct from (5); compare problem 2.21). It is obviously easy to transform (9) also in such a way that it describes the orbit over a long stretch with an accuracy up to and including first-order terms, and we have

$$\frac{p'}{r} = 1 + e'\cos\lambda\varphi + f\cos 2\lambda\varphi. \tag{11}$$

2.29. It is sufficient to prove that the Lagrangian function can be split into two parts when expressed in terms of the centre of mass coordinates

$$\mathbf{R} = \frac{m_1\mathbf{r}_1 + m_2\mathbf{r}_2}{m_1 + m_2},$$

and the relative coordinates, $\mathbf{r} = \mathbf{r}_2 - \mathbf{r}_1$:

$$L = L_1(\mathbf{R}, \dot{\mathbf{R}}) + L_2(\mathbf{r}, \dot{\mathbf{r}}),$$

$$L_1(\mathbf{R}, \dot{\mathbf{R}}) = \tfrac{1}{2}(m_1 + m_2)\dot{\mathbf{R}}^2 + (e_1 + e_2)\mathbf{E}\mathbf{R},$$

$$L_2(\mathbf{r}, \dot{\mathbf{r}}) = \tfrac{1}{2}m\dot{\mathbf{r}}^2 - \frac{e_1 e_2}{r} - m\left(\frac{e_1}{m_1} - \frac{e_2}{m_2}\right)\mathbf{E}\mathbf{r}.$$

The function $L_1(\mathbf{R}, \dot{\mathbf{R}})$ determines the motion of the centre of mass which is the same as the motion of a particle of mass $m_1 + m_2$ and charge $e_1 + e_2$ in uniform field $\mathbf{E}$. The relative motion, determined by $L_2(\mathbf{r}, \dot{\mathbf{r}})$, is the same as the motion of a particle with mass $m = m_1 m_2 / (m_1 + m_2)$ (the reduce mass) in the Coulomb field as well as in uniform field $\mathbf{E}$.

One could, of course, have obtained the same result starting from the equations of motion of a particles.

2.30. The Lagrangian function

---

[1] In particular, the radius vector $\mathbf{r}$ must be a periodic function of φ.

$$L = \tfrac{1}{2}\left(m_1\dot{\mathbf{r}}_1^2 + m_2\dot{\mathbf{r}}_2^2\right) - \frac{e_1 e_2}{|\mathbf{r}_1 - \mathbf{r}_2|} + \frac{e_1}{c}\mathbf{A}(\mathbf{r}_1)\dot{\mathbf{r}}_1 + \frac{e_2}{c}\mathbf{A}(\mathbf{r}_2)\dot{\mathbf{r}}_2$$

can be split into two parts depending only on $\mathbf{R}$, $\dot{\mathbf{R}}$ and $\mathbf{r}$, $\dot{\mathbf{r}}$ (using the notations of problem 2.29, c is the velocity of light), if $e_1/m_1 = e_2/m_2$. In this case we get:

$$L = \tfrac{1}{2}(m_1 + m_2)\dot{\mathbf{R}}^2 + \frac{e_1 + e_2}{c}\mathbf{A}(\mathbf{R})\dot{\mathbf{R}} + \tfrac{1}{2}m\dot{\mathbf{r}}^2 - \frac{e_1 e_2}{r} + \frac{e_1 m_2^2 + e_2 m_1^2}{c(m_1 + m_2)^2}\mathbf{A}(\mathbf{r})\dot{\mathbf{r}}.$$

2.31. $T = \tfrac{1}{2}\sum_{n=1}^{N}\mu_n\dot{\boldsymbol{\xi}}_n^2$, $\quad \mathbf{p} = \mu_N\dot{\boldsymbol{\xi}}_N$, $\quad \mathbf{M} = \sum_{n=1}^{N}\mu_n[\boldsymbol{\xi}_n, \dot{\boldsymbol{\xi}}_n]$, where

$$\frac{1}{\mu_n} = \frac{1}{\sum_{k=1}^{n}m_k} + \frac{1}{m_{n+1}} \quad (n = 1, \ldots, N-1), \quad \mu_N = \sum_{k=1}^{N}m_k.$$

2.32. Let the coordinates of the incident particle and of the particle initially at rest be x_1 and x_2, respectively, and let at $t = 0$, $x_1 = -R$, $x_2 = 0$. The centre of mass of the system moves accordingly to the equation $X = -\tfrac{1}{2}R + \tfrac{1}{2}vt$. The relative motion ($x = x_2 - x_1$) follows from the equation law

$$t = -\sqrt{\frac{m}{4}}\int_{R}^{x}\frac{dx}{\sqrt{\dfrac{mv^2}{4} - \dfrac{\alpha}{x^n}}}.$$

The first particle thus has the following position:

$$x_1 = X - \tfrac{1}{2}x = -\tfrac{1}{2}R + \tfrac{1}{2}\int_{x}^{R}\frac{dx}{\sqrt{1 - 4\alpha/(mv^2 x^n)}} - \tfrac{1}{2}x.$$

The distance between the particles decreases until it becomes $x_{\min} = \left(\dfrac{4\alpha}{mv^2}\right)^{1/n}$ and then increases again. When it is again equal to R, the first particle has come to the point

$$x_{1f} = x_{\min}\left[\int_{1}^{R/x_{\min}}\left(\frac{1}{\sqrt{1 - z^{-n}}} - 1\right)dz - 1\right]. \tag{1}$$

The point where the first particle comes to rest is the limit of x_{1f}, as $R \to \infty$.

If $n \leqslant 1$, then $x_{1f} \to \infty$, as $R \to \infty$; that is, both particles go to infinity after the collision.

2.34. The equation of motion (c is the velocity of light)

$$m\dot{\mathbf{v}} = \frac{eg}{cr^3}[\mathbf{v}, \mathbf{r}] \tag{1}$$

has the integrals of motion

$$\tfrac{1}{2} mv^2 = E, \tag{2}$$

$$m[\mathbf{r}, \mathbf{v}] - \frac{eg}{c}\frac{\mathbf{r}}{r} = \mathbf{J}. \tag{3}$$

Multiplying (1) by $\mathbf{r}$, we obtain $\mathbf{r}\dot{\mathbf{v}} = 0$ or $\dfrac{d^2}{dt^2}\dfrac{r^2}{2} - v^2 = 0$, where

$$r^2 = r_0^2 + v^2(t - t_0)^2. \tag{4}$$

Multiplying (3) by $\mathbf{r}/r$, we obtain

$$-\frac{eg}{c} = \mathcal{J}\cos\theta, \tag{5}$$

where θ is the polar angle in the spherical coordinates with the z-axis parallel to the vector $\mathbf{J}$. Projection (3) onto the z-axis

$$mr^2\dot{\varphi}\sin^2\theta - \frac{eg}{c}\cos\theta = \mathcal{J}$$

together with (5) gives

$$\dot{\varphi} = \frac{\mathcal{J}}{mr^2}. \tag{6}$$

At $t = t_0$, we have $v = r_0\dot{\varphi}\sin\theta$, besides, from (6) we get $\mathcal{J} = \dfrac{mr_0 v}{\sin\theta}$, and the result is (taking into account (5))

$$\tan\theta = -\frac{mr_0 vc}{eg}. \tag{7}$$

Integrating the equation $\dot{\varphi} = \dfrac{r_0 v}{r^2(t)\sin\theta}$, we obtain

$$\varphi = \varphi_0 + \frac{1}{\sin\theta}\arctan\frac{v(t - t_0)}{r_0}. \tag{8}$$

The particle thus moves with a constant velocity v on the surface of the cone $\theta = $ const. By introducing

$$\psi = (\varphi - \varphi_0)\sin\theta, \tag{9}$$

we rewrite (8) as

$$\tan \psi = \frac{v(t - t_0)}{r_0}.$$ (10)

Equations (4), (9), and (10) can be interpreted as follows: the motion of a point on the cone involute is uniform and straightforward, while r and ψ are the polar coordinates in the plane of the involute.

2.35. The motion of a charged particle in an electromagnetic field is determined by the Lagrangian function

$$L = \tfrac{1}{2} m \mathbf{v}^2 - e\varphi + \frac{e}{c} \mathbf{v} \mathbf{A},$$ (1)

where φ and $\mathbf{A}$ are the scalar and vector potentials, and c is the velocity of light (see [7], §10 and [2], §16–17).

Using cylindrical coordinates, we have

$$L = \frac{m}{2}(\dot{r}^2 + r^2\dot{\varphi}^2 + \dot{z}^2) + \frac{e}{c}\frac{\mu r^2 \dot{\varphi}}{(r^2 + z^2)^{3/2}}.$$ (2)

We see from the equation of motion for z

$$m\ddot{z} + \frac{3e}{c}\frac{\mu r^2 \dot{\varphi} z}{(r^2 + z^2)^{5/2}} = 0$$ (3)

that for $z = 0$ the component of the force along the z-axis vanishes. The orbit thus lies in the plane $z = 0$ when $z(0) = \dot{z}(0) = 0$.

As φ is a cyclic coordinate, we have

$$\frac{\partial L}{\partial \dot{\varphi}} = mr^2\dot{\varphi} + \frac{e\mu}{cr} = p_\varphi = \text{const.}$$ (4)

From this equation it follows that the projection of the momentum $M_z = m[\mathbf{r}, \mathbf{v}]_z = mr^2\dot{\varphi}$ is not saved, but the sum of $M_z + \frac{e\mu}{cr} = p_\varphi$ is retained. For the case of the infinite motion, p_φ is the value of M_z at $r \to \infty$. Moreover, the energy conservation law is satisfied (since $\partial L/\partial t = 0$):

$$\tfrac{1}{2}m(\dot{r}^2 + r^2\dot{\varphi}^2) = E.$$ (5)

Using (4) to eliminate $\dot{\varphi}$ from (5) we get

$$\tfrac{1}{2}m\dot{r}^2 + U_{\text{eff}}(r) = E,$$ (6)

where

$$U_{\text{eff}}(r) = \frac{1}{2mr^2}\left(p_\varphi - \frac{e\mu}{cr}\right)^2. \tag{7}$$

The radial motion thus takes place as in one-dimensional field $U_{\text{eff}}(r)$.

We have drawn $U_{\text{eff}}(r)$ given by (7) for the cases $p_\varphi < 0$ and for the case $p_\varphi > 0$ in Figs. 105a and 105b, respectively.

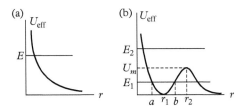

Figure 105

In the case when $p_\varphi < 0$ the orbit is always infinite. To give a qualitative description of the orbit, it is useful to use (4) to write

$$\dot\varphi = -\frac{|p_\varphi|}{mr^2} - \frac{e\mu}{mcr^3}. \tag{8}$$

The velocity with which the radius vector of the particle turns round has the same direction all the time and increases when the particle approaches the dipole. Curve 1 in Fig. 106 shows such an orbit. The orbit is symmetric with respect to straight line connecting the centre of the field with the point $r = r_{\text{min}}$.

When $p_\varphi > 0$, scattering can occur for any energy E, but if

$$E < U_m = \frac{c^2 p_\varphi^4}{32me^2\mu^2}$$

(see the line $E = E_1$ in Fig. 105b), there is also the possibility of finite orbit. From the equation

$$\dot\varphi = \frac{p_\varphi}{mr^2} - \frac{e\mu}{mcr^3}$$

it follows that $\dot\varphi > 0$ when $r > r_1 = \frac{e\mu}{cp_\varphi}$ and $\dot\varphi < 0$ when $r < r_1$. At $r = r_1$ there is a "sticking point" in φ.

A particle with an energy $E > U_m$ (see the line $E = E_2$ in Fig. 93b) is scattered while in two points (where $r = r_1$) its velocity is parallel to its radius vector (curve 2 in Fig. 106).

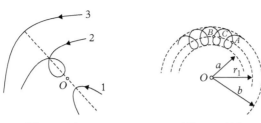

Figure 106   Figure 107

At $E < U_m$ there can occur scattering without sticking points for φ (curve 3 of Fig. 106) or a finite orbit in an annulus $a \leqslant r \leqslant b$ (Fig. 107). In the latter case the particle can perform both a direct (section AB) and "counter" (section BC) motion as far as φ is concerned.

2.36. a) It is convenient to use cylindrical coordinates and to take the vector potential in the form

$$A_\varphi = \tfrac{1}{2} rB, \ A_z = A_r = 0.$$

In the z-direction the motion is uniform, while in a plane perpendicular to the z-axis we have a finite motion. In Fig. 108 we show the projection of the orbit on this plane. The orbits a, b, and c correspond, respectively, to the cases[1] $p_\varphi > 0$ and energy values $E_\perp = E - \tfrac{1}{2} mv_z^2$ in the regions

$$U_1 < E_\perp < U_2, \ E_\perp = U_2, \ E_\perp > U_2,$$

where

$$U_1 = (\tilde{\Omega} - \Omega)p_\varphi, \ U_2 = \frac{\lambda p_\varphi}{2\Omega}, \ \Omega = \frac{eB}{2mc}, \ \tilde{\Omega} = \sqrt{\Omega^2 + \lambda}.$$

An orbit for the case when $p_\varphi < 0$ is given in Fig. 108d, while Fig. 108e depicts the orbit for the case where $p_\varphi = 0$.

The orbit for the particle in the xy-plane can easily be determined once we know the motion of a free isotropic oscillator of frequency $\tilde{\Omega}$ (see [1], §23, problem 3 and [7], §4)[2]

$$r^2 = a^2 \cos^2 \tilde{\Omega}t + b^2 \sin^2 \tilde{\Omega}t, \quad \varphi = -\Omega t + \arctan\left(\frac{b}{a} \tan \tilde{\Omega}t\right).$$

---

[1] To fix our ideas we assume $B > 0$; p_φ is the generalized momentum corresponding to the coordinate φ.
[2] The branches of $\arctan\left(\frac{b}{a} \tan \tilde{\Omega}t\right)$ should be chosen such that the angle φ is a continuous function of t.

Here the minimum (b) and maximum (a) radii are determined by the energy $E = \frac{1}{2} m\tilde{\Omega}^2 (a^2 + b^2) - p_\varphi \Omega$ and the angular momentum $p_\varphi = m\tilde{\Omega} ab$; the origins for t and φ are chosen that $\varphi(0) = 0$, $r(0) = a$.

Note that the period of radial oscillations, $T = \pi/\tilde{\Omega}$ is independent of E and p_φ. The angle over which the radius vector turns during this period is

$$\Delta\varphi = \pi(\pm 1 - \Omega/\tilde{\Omega}) \text{ for } p_\varphi \gtrless 0,$$

where the sign is the same as that for p_φ, and

$$\Delta\varphi = -\pi\Omega/\tilde{\Omega} \text{ for } p_\varphi = 0.$$

The angle $\Delta\varphi$ does not depend on E.[1]

What will be the motion when $\lambda < 0$?

It is interesting to compare the motion of the particle in this problem with the motion of the charged particle in crossed electric and magnetic fields (see [2], § 22).

2.37. The equation of the orbit is

$$\varphi = \sqrt{\frac{m}{2}} \int \frac{\dfrac{p_\varphi}{mr^2} - \Omega}{\sqrt{E - U_{\text{eff}}(r)}} \, dr, \tag{1}$$

where $\Omega = \dfrac{eB}{2mc}$, c is the velocity of light and

$$U_{\text{eff}}(r) = -\frac{\alpha}{r} + \frac{1}{2} mr^2 \left(\Omega - \frac{p_\varphi}{mr^2}\right)^2.$$

By looking at a graphs of $U_{\text{eff}}(r)$ one can study the character of the motion qualitatively. One must then pay attention to the fact that $\dot\varphi$ changes sign when r passes through the value $r_0 = \sqrt{\dfrac{p_\varphi}{m\Omega}}$. The result is orbits such as the ones in Fig. 108a–e.[2] The different orbits occur when

a) $p_\varphi > 0$, $U_{\text{min}} < E < U_0$, where U_{min} is the minimum value of $U_{\text{eff}}(r)$ and $U_0 = U_{\text{eff}}(r_0)$;

b) $p_\varphi > 0$, $E = U_0$;

c) $p_\varphi > 0$, $E > U_0$;

d) $p_\varphi < 0$;

e) $p_\varphi = 0$.

---

[1] Another way to solve this problem is given in problem 6.37.

[2] As we are only interested in a qualitative study, using the shape of $U_{\text{eff}}(r)$, we can use the same approximate treatment as in problem 2.36. Of course, the exact form of the orbits is different in the two problems.

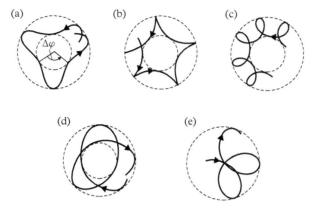

Figure 108

In the latter case, the particle falls into the centre at the first loop.

Let us consider in somewhat more detail two limiting cases.
Equation (1) can be written in the form

$$\varphi = \frac{p_\varphi}{\sqrt{2m}} \int \frac{dr}{r^2 \sqrt{E + p_\varphi \Omega + \frac{\alpha}{r} - \frac{p_\varphi^2}{2mr^2} - \frac{1}{2} m\Omega^2 r^2}} - \Omega t. \tag{2}$$

We must thus describe the effect of the magnetic field as a change in the energy $E' = E + p_\varphi \Omega$ and an extra term in the field $U(r) = -\frac{\alpha}{r}$ which is $\delta U(r) = \frac{1}{2} m\Omega^2 r^2$, each of which leads to a precession of the orbit and an additional precession with an angular velocity $-\Omega$. If the magnetic field B is sufficiently small, the term $\delta U(r)$ can be considered to be a small correction to

$$U_0(r) = \frac{p_\varphi^2}{2mr^2} - \frac{\alpha}{R},$$

provided the following condition is satisfied,

$$\delta U(r) \ll |U_0(r)|, \tag{3}$$

for the whole range of the motion.

The precession velocity caused by δU can be found from

$$\Omega' = \frac{\delta \Delta \varphi}{T} = \frac{1}{T} \frac{\partial}{\partial p_\varphi} (T \langle \delta U \rangle), \tag{4}$$

where the averaging of δU is over the motion of the particle in field U_0 with an energy E' and an angular momentum p_φ, while T is the period of the motion (cf. problems 2.20

and 2.21). If we perform the calculation,[1] we get

$$\Omega' = -\frac{3\Omega^2 p_\varphi}{4|E'|};$$

(5)

we can assume δU to be a small correction if, apart from (3), the condition $\delta\Delta\varphi \ll 2\pi$ is also satisfied, or

$$\Omega^2 p_\varphi \alpha \sqrt{m} |E'|^{-5/2} \ll 1.$$

(6)

It is, of course, impossible to consider δU to be a small correction when $E' \geqslant 0$, as in that case neglecting δU may introduce a qualitatively change the character of the motion.

The quantity Ω' may turn out to be either small or large compared to Ω. The sign Ω' is the opposite to that of p_φ; that is, the direction of this velocity is the opposite of that of the motion of the particle in its orbit. The direction of the velocity Ω is determined by the magnetic field.

The orbit is thus an ellipse precessing with the angular velocity

$$\Omega_{\text{prec}} = -\Omega + \Omega'.$$

(7)

To be more exact, the orbit is a fixed ellipse if the system of reference rotates with an angular velocity Ω_{prec}, because it is possible that $\Omega T \gtrsim 1$.

It is interesting to compare the results with the Larmor theorem (see [7], § 17.3, [2], § 45, and problem 9.26).

Is there a case when although E is positive we can consider the field $\delta U(r)$ to be a small correction?

Let us now consider the case when the field $U(r) = -\alpha/r$ can be considered to be a small correction. If $U(r)$ were not present, the motion would be along a circle. Its radius a and the distance b from of its centre from the centre of the field can be express in terms of the maximum and minimum distances of the particle from the centre of the field

$$r_{1,2}^2 = \frac{E + p_\varphi\Omega \pm \sqrt{(E + 2p_\varphi\Omega)E}}{m\Omega^2}.$$

(8)

There are now two possibilities (see Fig. 109), depending on how the circle is placed with respect to the origin. If $p_\varphi\Omega < 0$, case 109a occurs, and if $p_\varphi\Omega > 0$, case 109b occurs. In both cases, we have

$$b^2 = \frac{E + 2p_\varphi\Omega}{2m\Omega^2}, \quad a^2 = \frac{E}{2m\Omega^2}.$$

(9)

---

[1] To evaluate $\langle \delta U \rangle$, it is convenient to use the variables used in problems 2.21 and 2.22; since the period in the field U_0 is independent of p_φ, we can take T from under the differentiation sign in (4).

Taking $U(r)$ into account leads to a systematic displacement of this circle (the so-called drift), while its radius and the distance from the origin of the fields which are determined by a and b do not change; that is, its centre moves along a circle of radius b. The angular velocity of the displacement of the centre of the circle is

$$\gamma = \frac{\partial}{\partial p_\varphi} \langle U \rangle,$$

where the averaging of $U(r)$ is over the uniform motion along the circle. Let us restrict ourselves to the case when $a \ll b$. In that case, we may simple assume that

$$\langle U(r) \rangle = -\frac{\alpha}{b}, \tag{10}$$

so that

$$\gamma = \frac{\alpha}{2m\Omega b^3}.$$

We note that in this case the linear drift velocity is equal to $c|\mathbf{E}|/B$, where $e|\mathbf{E}| = \alpha/b^2$ is the force acting upon the particle at a distance b (cf. [2], §22).

2.38. The problem of the motion of two identical charged particles in a uniform magnetic field can be reduced to the problem on the motion of the centre of mass and the problem of the relative motion (see problem 2.30).

For the centre of mass coordinates, we have

$$X = R\cos\omega t, \quad Y = -R\sin\omega t, \tag{1}$$

where $\omega = eB/(mc)$.

The relative motion is the same as that of a motion of a particle of mass $m/2$ and charge $e/2$ in the field $U(r) = e^2/r$ and in a uniform magnetic field $\mathbf{B}$. This motion is similar to the one considered in the preceding problem (we need only change m to $m/2$, e to $e/2$ and α to $-e^2$ in formulae). Let us restrict ourselves to the case where the radius a of the orbit is small compared to the distance b from the centre of the fields (Fig. 109b). We can easily find the frequency of the radial oscillations to a higher degree of accuracy than we was done in preceding problem, by expanding

$$U_{\text{eff}}(r) = \frac{e^2}{r} + \frac{p_\varphi^2}{mr^2} + \tfrac{1}{16}2m\omega^2 r^2 - \tfrac{1}{2}p_\varphi\omega$$

in a series in the vicinity of the minimum where $r = b$ (see [1], §21). From the condition

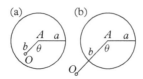

Figure 109

$U'_{\text{eff}}(b) = 0$, we find

$$p_\varphi = \sqrt{\frac{m^2}{16}\omega^2 b^4 - \frac{m}{2}e^2 b} \approx \frac{m}{2}b^2\left(\frac{\omega}{2} - \gamma\right), \quad \gamma = \frac{2e^2}{m\omega b^3}, \tag{2}$$

and we thus get finally for $\omega_r = \sqrt{\frac{2U''_{\text{eff}}(b)}{m}}$ the result $\omega_r = \omega - \frac{1}{2}\gamma$, and for the separation of the particles

$$r = b + a\cos(\omega_r t + \alpha). \tag{3}$$

To find $\varphi(t)$, we apply the principle of conservation of generalized momentum to $p_\varphi = \frac{1}{2}mr^2\left(\dot\varphi + \frac{1}{2}\omega\right)$. Using (2) and (3), we get

$$\varphi(t) = -\gamma t - \frac{a}{b}\sin(\omega_r t + \alpha) + \varphi_0. \tag{4}$$

Using (3) and (4), we get for the relative coordinates

$$x = r\cos\varphi = b\cos(\gamma t - \varphi_0) + a\cos\left(\omega t + \frac{1}{2}\gamma t + \beta\right),$$
$$y = r\sin\varphi = -b\sin(\gamma t - \varphi_0) - a\sin\left(\omega t + \frac{1}{2}\gamma t + \beta\right), \tag{5}$$

where $\beta = \alpha - \varphi_0$. The first terms here correspond to the motion of the centre of the circle with a drift velocity $b\gamma$, and the second terms to the motion along this circle with an angular velocity $\omega + \frac{1}{2}\gamma$.

Figure 110

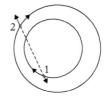

Figure 111

Coordinates of the particles, $x_{1,2} = X \pm \frac{1}{2}x$, $y_{1,2} = Y \pm \frac{1}{2}y$ can be written in the form

$$x_{1,2} = \pm \tfrac{1}{2} b \cos(\gamma t - \varphi_0) + \rho_{1,2} \cos(\omega t + \psi_{1,2}),$$
$$y_{1,2} = \mp \tfrac{1}{2} b \sin(\gamma t - \varphi_0) - \rho_{1,2} \sin(\omega t + \psi_{1,2}), \tag{6}$$

where

$$\rho_{1,2} = \sqrt{R^2 + \tfrac{1}{4} a^2 \pm a R \cos\left(\tfrac{1}{2}\gamma t + \beta\right)},$$

$$\tan\psi_{1,2} = \frac{\pm a \sin\left(\tfrac{1}{2}\gamma t + \beta\right)}{2R \pm a \cos\left(\tfrac{1}{2}\gamma t + \beta\right)}.$$

The centres of the circles along which the particles move thus rotate around the origin with an angular velocity γ (a drift velocity $b\gamma/2$) while their radii oscillate with a frequency $\gamma/2$ (Fig. 110).

Another limiting case when $a \gg b$ (distance of the centre of the orbits to the origin small compared to the radii of the orbits (Fig. 111)) may give a clear insight into the mechanism of energy "exchange". The work carried out by the the force of the interaction on the second particle is clearly positive, while the work carried on the first particle is negative, when taken over many periods.

2.39. One can easily prove that the given quantity is a constant by using the equations of motion and writing this quantity in the form

$$\mathbf{AF} + \tfrac{1}{2}[\mathbf{F}, \mathbf{r}]^2,$$

where

$$\mathbf{A} = [\mathbf{v}, \mathbf{M}] - \frac{\alpha \mathbf{r}}{r}$$

is the Laplace vector (see [1], §15 and [7], §14). For small values of F the orbit will be close to an ellipse, with its semi-major axis along the direction of the vector $\mathbf{A}$ and with an eccentricity $e = |\mathbf{A}|/\alpha$. In this case $\mathbf{AF} \approx$ const, or $e\cos\psi \approx$ const, where ψ is the angle between $\mathbf{A}$ and $\mathbf{F}$.

2.40. When there is a small extra term $\delta U(\mathbf{r})$ in the potential energy, the quantities characterizing the motion of the particle, such as the angular momentum, pericentrum position and so on, change, although they do not change their values appreciable over a shot time interval (a few periods of the unperturbed motion). However, these changes may add up over an extended time, so that some of the quantities may happen to change by large amounts.

In particular, the orbit remains elliptical for a shot time interval. Its semi-major axis, $a = \frac{\alpha}{2|E|}$, which is determined by the energy, does not change over a long time, while the eccentricity $e = \sqrt{1 - \frac{M^2}{m\alpha a}}$ and the orientation of the orbit are both liable to secular changes.

a) The change in the angular momentum is determined by the equation

$$\dot{\mathbf{M}} = [\mathbf{r}, \mathbf{F}]. \tag{1}$$

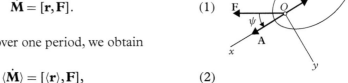

Figure 112

Average this equation over one period, we obtain

$$\langle \dot{\mathbf{M}} \rangle = [\langle \mathbf{r} \rangle, \mathbf{F}], \tag{2}$$

where

$$\langle \mathbf{r} \rangle = \frac{1}{T} \int_0^T \mathbf{r}(t)\, dt. \tag{3}$$

For averaging, we use a coordinate system with the z-axis parallel to $\mathbf{M}$ and the x-axis parallel to $\mathbf{A}$ (Fig. 112). Here

$$\mathbf{A} = [\mathbf{v}, \mathbf{M}] - \frac{\alpha \mathbf{r}}{r}$$

is an additional integral of motion in the Kepler problem. Recall that the vector $\mathbf{A}$ is directed from the centre of the field to the pericentre, and $|\mathbf{A}| = \alpha e$. Clearly, the vector $-\langle \mathbf{r} \rangle$ is parallel to the x-axis.

Making the substitution

$$x = a(\cos\xi - e), \quad t = \frac{T}{2\pi}(\xi - e\sin\xi),$$

we get

$$\langle x \rangle = \frac{a}{2\pi} \int_0^{2\pi} (\cos\xi - e)(1 - \cos\xi)\, d\xi = -\frac{3ae}{2}. \tag{4}$$

Therefore

$$\langle \mathbf{r} \rangle = -\frac{3ae}{2} \frac{\mathbf{A}}{|A|} = -\frac{3a}{2\alpha} \mathbf{A}. \tag{5}$$

b) If the force **F** is at right angles to **M**, it is obvious from symmetry consideration that the orbit lies in a plane and the vector **M** retains its direction—apart possibly from the sign.

Omitting the averaging sign, we can write (2) and (5) in the form

$$\dot{M} = \tfrac{3}{2}\, aeF \sin\psi, \tag{6}$$

where ψ is the angle between **A** and **F**. Using the fact that

$$e \cos\psi = e_0 = \text{const} \tag{7}$$

(see problem 2.39), and eliminating e and ψ from (6) we find

$$\dot{M} = \tfrac{3}{2}\, aF\sqrt{1 - e_0^2 - \frac{M^2}{ma\alpha}}. \tag{8}$$

Integrating (8), we get

$$M = M_0 \cos(\Omega t + \beta), \tag{9}$$

where

$$\Omega = \frac{3F}{2}\sqrt{\frac{a}{m\alpha}}, \quad M_0 = \sqrt{ma\alpha(1 - e_0^2)},$$

as well as

$$e = \sqrt{1 - (1 - e_0^2)\cos^2(\Omega t + \beta)}.$$

The orbit is thus an ellipse which oscillates about the direction of **F** and an eccentricity which changes in tune with the oscillations (Fig. 113). The direction of motion along the ellipse also changes, together with the sign of **M**. The period $2\pi/\Omega$ of the oscillation of the ellipse is much longer than the period T of the rotation of the particle along the ellipse.

c) In the general case, we consider also the change in the vector **A**. Using the equations of motion, we easily get

$$\dot{\mathbf{A}} = \frac{1}{m}[\mathbf{F}, \mathbf{M}] + [\mathbf{v}, [\mathbf{r}, \mathbf{F}]]. \tag{10}$$

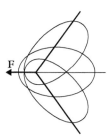

Figure 113

For the averaging in (10), we use the equations

$$\langle x\dot{x} \rangle = \left\langle \frac{d}{dt}\frac{x^2}{2} \right\rangle = 0,$$

$$\langle y\dot{y} \rangle = \left\langle \frac{d}{dt}\frac{y^2}{2} \right\rangle = 0,$$

$$\langle x\dot{y} \rangle + \langle y\dot{x} \rangle = \left\langle \frac{d}{dt}xy \right\rangle = 0, \tag{11}$$

$$\langle x\dot{y} \rangle - \langle y\dot{x} \rangle = \frac{M}{m}.$$

As a result, we get

$$\langle \dot{\mathbf{A}} \rangle = \frac{3}{2m}[\mathbf{F},\mathbf{M}]. \tag{12}$$

We have thus for $\mathbf{M}$ and $\mathbf{A}$, averaged over one period (in the following we omit the averaging sign), the following set of equations

$$\dot{\mathbf{A}} = \frac{3}{2m}[\mathbf{F},\mathbf{M}],$$
$$\dot{\mathbf{M}} = \frac{3a}{2\alpha}[\mathbf{F},\mathbf{A}]. \tag{13}$$

The components of these vectors along the direction of $\mathbf{F}$ are conserved:

$$\mathbf{MF} = \text{const}, \quad \mathbf{AF} = \text{const}. \tag{14}$$

(The same result could also easily be obtained from other considerations). For the transverse component $\mathbf{M}$,

$$\mathbf{M}_\perp = \mathbf{M} - \frac{\mathbf{F}(\mathbf{MF})}{F^2}, \tag{15}$$

we obtain from (13)

$$\ddot{\mathbf{M}}_\perp + \Omega^2 \mathbf{M}_\perp = 0. \tag{16}$$

In a coordinate system $OX_1X_2X_3$ with the X_3-axis parallel to $\mathbf{F}$ we have for the solution of (13):

$$M_1 = B_1 \cos \Omega t + C_1 \sin \Omega t,$$
$$M_2 = B_2 \cos \Omega t + C_2 \sin \Omega t. \tag{17}$$

We then obtain from (13):

$$A_1 = -\frac{3F}{2m\Omega}(B_2 \sin \Omega t - C_2 \cos \Omega t),$$

$$A_2 = \frac{3F}{2m\Omega}(B_1 \sin \Omega t - C_1 \cos \Omega t). \tag{18}$$

As we should expect, the constants $B_{1,2}$ and $C_{1,2}$ are determined by the initial values of the vectors $\mathbf{M}$ and $\mathbf{A}$.

The end point of the vector $\mathbf{M}$ describes an ellipse with its centre on the X_3-axis and lying in a plane μ which is parallel to the OX_1X_2-plane (Fig. 114). The end of the vector $\mathbf{A}$ also describes an ellipse with centre on the X_3-axis and lying in a plane σ which is parallel to μ; this second ellipse is similar to the first one, but rotated over an angle $\pi/2$: $\mathbf{A}$ remains thus at right angles to $\mathbf{M}$ all the time. The plane of the orbit is perpendicular to $\mathbf{M}$, and the vector $\mathbf{A}$ determines the direction to the pericentre of the orbit.

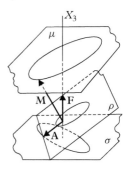

Figure 114

The plane of the orbit thus rotates (precesses) around $\mathbf{F}$. The angle between the plane of the orbit ρ and $\mathbf{F}$ oscillates about its mean value. The eccentricity and the angle between the projection of $\mathbf{F}$ upon the plane ρ and the direction to the pericentre also oscillate about their mean values. All these motions occur with a frequency Ω.

We should bear in mind that we have neglected those corrections in $\mathbf{F}$ which are of the first order but which do not lead to secular effects. Our solution is valid for a time interval of the order of several periods of the orbital precession.

The reader should checked whether the next approximations will lead to a qualitative change in the character of the motion (e.g., to a possible departure of the particle to infinity). The exact solution of the problem of the motion of a particle in the field $U = -\frac{\alpha}{r} - \mathbf{Fr}$, which can be given in parabolic coordinates (see problem 12.12b), shows that such effects do not take place if $E < 0$ and if $\mathbf{F}$ is sufficiently small.

It should be emphasized that the appearance of secular changes of the orbit under the action of infinitesimal constant perturbation is connected with a degeneracy of the unperturbed motion.

In [8], §7.3 one can find a solution of this problem using the canonical perturbation theory.

2.41. According to the Larmore's theorem (see [7], §17.3 and [2], §45), the orbit of a particle in a uniform magnetic field $\mathbf{B}$ rotates around the centre of the Coulomb field with the angular velocity $\mathbf{\Omega} = -\frac{q\mathbf{B}}{2mc}$, where q and m are the charge and mass of a particle and c is the velocity of light. In this case, the angular momentum $\mathbf{M}$ and the Laplace vector $\mathbf{A}$ ratate with the same angular velocity:

$$\dot{\mathbf{M}}_1 = [\boldsymbol{\Omega}, \mathbf{M}], \quad \dot{\mathbf{A}}_1 = [\boldsymbol{\Omega}, \mathbf{A}]. \tag{1}$$

In the preceding problem (see (13)), the averaged velocity change of the vectors $\mathbf{M}$ and $\mathbf{A}$ under the influence of the constant force $\mathbf{F} = q\mathbf{E}$ has been determined:

$$\dot{\mathbf{M}}_2 = \frac{3a}{2\alpha}[\mathbf{F}, \mathbf{A}], \quad \dot{\mathbf{A}}_2 = \frac{3}{2m}[\mathbf{F}, \mathbf{M}]. \tag{2}$$

The averaged velocity change of the vectors $\mathbf{M}$ and $\mathbf{A}$ under the influence of both fields is the following sum:

$$\dot{\mathbf{M}} = \dot{\mathbf{M}}_1 + \dot{\mathbf{M}}_2, \quad \dot{\mathbf{A}} = \dot{\mathbf{A}}_1 + \dot{\mathbf{A}}_2. \tag{3}$$

a) Let us direct the x-axis along the electric field and the z-axis along the magnetic one. Then the equations (3) take the form

$$\dot{A}_x = -\Omega A_y, \quad \dot{A}_y = \Omega A_x - \frac{3F}{2m}M, \quad \dot{M} = \frac{3aF}{2\alpha}A_y.$$

The solution of this system is

$$A_x = \frac{\Omega}{\omega}C\cos(\omega t + \beta) + C_1,$$
$$A_y = C\sin(\omega t + \beta),$$
$$M = -\frac{3aF}{2\alpha\omega}C\cos(\omega t + \beta) + \frac{2m\Omega}{3F}C_1,$$

where $\omega = \sqrt{\Omega^2 + 9aF^2/(4m\alpha)}$ and constants C, β, and C_1 are determined by the initial values of $\mathbf{A}$ and $\mathbf{M}$.

Thus, the end of the vector $\mathbf{A}$ moves along an ellipse with the axes which are parallel to the axes x and y (Fig. 115) and with the centre on the x-axis. The orbit wiggles (or rotates with $\Omega C > \omega C_1$), and its eccentricity changes periodically. When $9aF^2C > 4m\alpha\Omega\omega C_1$, the elliptical orbit periodically turns into the straight-line segment.

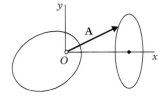

Figure 115

b) Let us denote

$$\Omega_F = \frac{3}{2} \mathbf{F} \sqrt{\frac{a}{ma}}, \quad \mathbf{N} = \mathbf{A} \sqrt{\frac{ma}{\alpha}},$$

then we can write (3) as

$$\dot{\mathbf{M}} = [\boldsymbol{\Omega}, \mathbf{M}] + [\boldsymbol{\Omega}_F, \mathbf{N}], \quad \dot{\mathbf{N}} = [\boldsymbol{\Omega}, \mathbf{N}] + [\boldsymbol{\Omega}_F, \mathbf{M}].$$

Adding and subtracting these equations, we find

$$\dot{\mathbf{J}}_{1,2} = [\boldsymbol{\omega}_{1,2}, \mathbf{J}_{1,2}],$$

where

$$\mathbf{J}_{1,2} = \tfrac{1}{2}(\mathbf{M} \pm \mathbf{N}), \quad \boldsymbol{\omega}_{1,2} = \boldsymbol{\Omega} \pm \boldsymbol{\Omega}_F.$$

Thus, the vectors $\mathbf{J}_i$, $i = 1$, 2 rotate with constant angular velocities $\boldsymbol{\omega}_i$

$$\mathbf{J}_i(t) = \mathbf{J}_i(0) \cos \omega_i t + \left[\frac{\boldsymbol{\omega}_i}{\omega_i}, \mathbf{J}_i(0) \right] \sin \omega_i t + \boldsymbol{\omega}_i \frac{\mathbf{J}_i(0)\boldsymbol{\omega}_i}{\omega_i^2} (1 - \cos \omega_i t),$$

and vectors

$$\mathbf{M} = \mathbf{J}_1 + \mathbf{J}_2, \quad \mathbf{A} = \sqrt{\frac{\alpha}{ma}} (\mathbf{J}_1 - \mathbf{J}_2)$$

are completely determined by their initial values.

We will not analyse this answer in detail, only note that $\omega_1 = \omega_2$ when the electric and magnetic fields are mutually orthogonal.

2.42. Let the xz-plane be where the particle moves. The equations for the Laplace vector $\mathbf{A}$ (see (10) in problem 2.40 where one must substitute $\mathbf{F} = -\partial \delta U(\mathbf{r})/\partial \mathbf{r}$) can be presented in the form

$$\dot{A}_x = 2\beta(5xz\dot{z} - 2\dot{x}z^2) = -\frac{6\beta}{m} zM + 2\beta \frac{d}{dt}(xz^2),$$

$$\dot{A}_z = -2\beta(4\dot{x}z - \dot{z}x^2) = -\frac{4\beta}{m} xM - 2\beta \frac{d}{dt}(x^2 z),$$

where $M = m(z\dot{x} - x\dot{z})$ is a slowly changing function of time.

After averaging these equations, we obtain

$$\dot{A}_x = -6\beta \frac{M}{m}\langle z\rangle = 9\frac{\beta a}{m\alpha}A_z M,$$

$$\dot{A}_z = -4\beta \frac{M}{m}\langle x\rangle = 6\frac{\beta a}{m\alpha}A_x M,$$

where a is a large semi-axis of an ellipse. Hence

$$\frac{\dot{A}_x}{\dot{A}_z} = \frac{3A_z}{2A_x};$$

that is, in contrast to problem 2.40b, the end of the vector $\mathbf{A}$ varies not along a straight line but along a hyperbola

$$A_x^2 = \tfrac{3}{2}A_z^2 + \text{const.}$$

The time dependence of $\mathbf{A}$ can be found from the equation

$$t = \frac{m\alpha}{9\beta a}\int \frac{dA_x}{MA_z},$$

in which M and A_z must be expressed as the functions of A_x.

For example, for the case when at the initial moment $A_z(0) = 0, A_x(0) = \alpha e_0$ (here e_0 is the initial eccentricity value), we have

$$A_z = \alpha\sqrt{\tfrac{2}{3}\left(u^2 - e_0^2\right)}, \quad M = \sqrt{\tfrac{5}{3}m\alpha a(c^2 - u^2)},$$

where

$$u = \frac{A_x}{\alpha}, \quad c = \sqrt{\frac{3 + 2e_0^2}{5}}.$$

Over time, the orbit slowly rotates in the xz-plane and turns into a segment whose angle with the x-axis is

$$\psi = -\arctan\left(\sqrt{2/3}\,k\right), \quad k = \sqrt{\frac{3 - 3e_0^2}{3 + 2e_0^2}},$$

during the time

$$\tau = \frac{1}{3\beta}\sqrt{\frac{m\alpha}{10a^3}} \int_{e_0}^{c} \frac{du}{\sqrt{(u^2 - e_0^2)(c^2 - u^2)}}.$$

The integral in the previous expression is reducible to the complete elliptic integral of the first kind (see footnote in the solution of problem 1.7)

$$\tau = \frac{T\alpha}{6\pi\sqrt{10}\,c\beta a^3}\mathrm{K}(k),$$

where T is the period of unperturbed motion.

Let us use our model to estimate the lifetime of a satellite whose initial orbit has a small eccentricity $e_0 \sim 0.1$ and is close to a circle of radius $a \sim 40\,000$ km. In this case, the period of the satellite T is about one day, and the time of the fall on the Earth (i.e. the time at which its minimum distance becomes equal to the radius of the Earth) is of the order of τ. The estimate is valid for the relation: $\alpha/\beta \sim (m_1/m_2)\,R^3$, where m_1 and m_2 are the respective masses of the Earth and the Moon, and R is the distance from the Earth to the Moon. From here we get the estimate

$$\tau \sim \frac{T_2^2}{T}, \quad T_2 = \frac{2\pi}{\sqrt{Gm_1/R^3}},$$

where T_2 is the period of the Moon's rotation and G is the gravitational constant. Thus, the time during which such a satellite will fall to the Earth, is about a few years. However, the correct fall time is much less due to the presence of the air resistance even at high altitudes.

2.43. Here we will use the geocentric coordinate system. Let's assume the strait line between the Earth and the Sun as the z-axis, and let the x-axis be in the plane of the Earth's orbit. This coordinate system rotates around the y-axis with an angular velocity Ω. In this frame of reference, δU is evidently time-independent, so that the integral of motion is the energy

$$E = \tfrac{1}{2}mv^2 - \frac{\alpha}{r} + \delta U - \tfrac{1}{2}m\Omega^2 r^2,$$

where $\alpha = Gm_1 m$, G is the gravitational constant and m_1 is the Earth's mass. The force

$$\delta\mathbf{F} = -\frac{\partial\delta U}{\partial\mathbf{r}}$$

and the force of inertia

$$\mathbf{F}_i = m\Omega^2 \mathbf{r} + 2m[\mathbf{v}, \mathbf{\Omega}]$$

leads to the distortion of the elliptical orbit of the Moon, while the major semi-axis of the ellipse $a = \alpha/(2|E|)$ stays almost unchanged.

The rate of change of the Laplace vector $\mathbf{A}$ consists of two terms: $\dot{\mathbf{A}}_1$ and $\dot{\mathbf{A}}_2$ that correspond to $\delta\mathbf{F}$ and $\mathbf{F}_i$ (cf. problem 2.41). The term $\dot{\mathbf{A}}_1$ has already been found in problem 2.41a:

$$\dot{A}_{1x} = \tfrac{9}{2}\Omega\zeta A_z, \quad \dot{A}_{1z} = 3\Omega\zeta A_x,$$

where $\zeta = M\Omega a/\alpha$. Since the eccentricity of the Moon's orbit is small $e = 0.055$, we have

$$M \approx ma^2\omega, \quad \alpha \approx m\omega^2 a^3, \quad \zeta \approx \frac{\Omega}{\omega}\omega \approx \frac{29.5}{365} \approx \frac{1}{12},$$

where ω is the angular velocity of the Moon around the Earth in the given (rotating) coordinate system.

The force of inertia leads to rotation of the vector $\mathbf{A}$ with an angular velocity $-\mathbf{\Omega}$ (note that the orbit would be fixed in $0x_0yz_0$ frame of reference in the absence of the term δU):

$$\dot{A}_{2x} = -\Omega A_z, \quad \dot{A}_{2z} = \Omega A_x.$$

Thus[1]

$$\dot{A}_x = -\Omega\left(1 - \tfrac{9}{2}\zeta\right)A_z, \quad \dot{A}_z = \Omega(1 + 3\zeta)A_x. \tag{1}$$

Integration of (1) gives

$$A_x = B\cos(\Omega't + \phi), \quad A_z = B\left(1 + \tfrac{15}{4}\zeta\right)\sin(\Omega't + \phi), \tag{2}$$

where $\Omega' = \Omega(1 - \tfrac{3}{4}\zeta)$, B, and ϕ are constants. In the $0xyz$-frame of reference, the vector $\mathbf{A}$ rotates around the y-axis with an average angular velocity $-\mathbf{\Omega}'$. In the system $0x_0yz_0$, it rotates with the angular velocity (so-called precession)

$$\mathbf{\Omega}_{\text{prec}} = \mathbf{\Omega} - \mathbf{\Omega}' = \tfrac{3}{4}\zeta\mathbf{\Omega}.$$

---

[1] Note that according to (1)
$$A\dot{A} = A_x\dot{A}_x + A_z\dot{A}_z = 7.5\,\Omega\zeta A_x A_z.$$
Using the relations $A = \alpha e$, $e^2 = 1 - M^2/(ma\alpha)$, we can estimate $\dot{\zeta} = \Omega a\dot{M}/\alpha \sim 7.5\,e^2\zeta^2\Omega \ll \Omega$. The value of ζ in (1) can be considered constant.

The small changes of $|\mathbf{A}|$ defined by (2) correspond to the small ripples of the orbit's eccentricity.

2.44. The Lagrangian function of the system (q is the particle charge and c is the velocity of light)

$$L = \tfrac{1}{2} mv^2 + \frac{\alpha}{r} + \frac{q}{c} \frac{[\boldsymbol{\mu}, \mathbf{r}]}{r^3} \mathbf{v}$$

is the same—apart from the notations—as the one considered in [2] (problem 2 to § 105).

The equations of motion for the angular momentum $\mathbf{M}$ and the Laplace vector $\mathbf{A}$ are the following

$$\dot{\mathbf{M}} = \frac{q}{mcr^3} [\mathbf{M}, \boldsymbol{\mu}],$$

$$\dot{\mathbf{A}} = \frac{q}{mcr^3} [\mathbf{A}, \boldsymbol{\mu}] + \frac{3q(\mathbf{M}, \boldsymbol{\mu})}{m^2 cr^5} [\mathbf{M}, \mathbf{r}]\dot{}$$

When averaging over the period T of unperturbed motion, these equations describe the systematic change of vectors $\mathbf{M}$ and $\mathbf{A}$:

$$\langle \dot{\mathbf{M}} \rangle = [\boldsymbol{\Omega}', \mathbf{M}], \quad \boldsymbol{\Omega}' = -\frac{q}{cma^3(1-e^2)^{3/2}} \boldsymbol{\mu}, \tag{1}$$

$$\langle \dot{\mathbf{A}} \rangle = [\boldsymbol{\Omega}, \mathbf{A}], \quad \boldsymbol{\Omega} = \boldsymbol{\Omega}' - 3\mathbf{M} \frac{(\boldsymbol{\Omega}'\mathbf{M})}{M^2}, \tag{2}$$

where a and e are again the semi-major axis and the eccentricity of the unperturbed ellipse. Equation (1) can also be rewritten as

$$\langle \dot{\mathbf{M}} \rangle = [\boldsymbol{\Omega}, \mathbf{M}], \tag{3}$$

since the vector $\boldsymbol{\Omega} - \boldsymbol{\Omega}'$ is parallel to $\mathbf{M}$.

From (2) and (3), it is seen that the ellipse, along which the particle moves, precesses "as a whole" with an angular frequency $\boldsymbol{\Omega}$. Other interpretation can also be given based on (1) and (2): in the coordinate system, rotating with the frequency $\boldsymbol{\Omega}'$, the vector $\mathbf{M}$ as well as the plane of motion of a particle are stationary, while the vector $\mathbf{A}$ and the perihelion of the orbit revolve with the constant frequency $\boldsymbol{\Omega} - \boldsymbol{\Omega}'$ around the direction $\mathbf{M}$.

We also indicate that it is helpful to perform the averaging of $1/r^3$ and $\mathbf{r}/r^5$ using the angle φ instead of the variable t:

$$\left\langle \frac{1}{r^3} \right\rangle = \frac{1}{T} \int_0^T \frac{dt}{r^3(t)} = \frac{m}{TM} \int_0^{2\pi} \frac{d\varphi}{r(\varphi)} = \frac{m}{TMp} \int_0^{2\pi} (1 + e\cos\varphi)\, d\varphi = \frac{2\pi m}{TMp},$$

$$\left\langle \frac{\mathbf{r}}{r^5} \right\rangle = \beta \mathbf{A}, \quad \beta = \frac{1}{A^2 T} \int\limits_0^T \frac{\mathbf{Ar}}{r^5}\, dt = \frac{m}{ATM} \int\limits_0^{2\pi} \frac{\cos\varphi}{r^2(\varphi)}\, d\varphi = \frac{2\pi m e}{TMp^2 A}.$$

2.45. When averaging the equation for the Laplace vector (see (10) of problem 2.40)

$$\dot{\mathbf{A}} = \frac{1}{m}[\mathbf{F},\mathbf{M}] + [\mathbf{v}, [\mathbf{r},\mathbf{F}]]$$

we take into account that $\langle \mathbf{F} \rangle = 0$ and that, according to the unperturbed equations of motion,

$$m\ddot{\mathbf{v}} = \frac{d}{dt}m\dot{\mathbf{v}} = \frac{d}{dt}\left(-\alpha\frac{\mathbf{r}}{r^3}\right) = -\frac{\alpha\mathbf{v}}{r^3} + \frac{3\alpha\mathbf{r}(\mathbf{rv})}{R^5}.$$

It gives (see the preceding problem)

$$\langle \dot{\mathbf{A}} \rangle = -\frac{\alpha\beta}{m^2}\left\langle \frac{[\mathbf{v},\mathbf{M}]}{r^3} \right\rangle = \frac{\alpha\beta}{m^2}\left\langle \frac{\mathbf{A}}{r^3} + \frac{\alpha\mathbf{r}}{r^4} \right\rangle = -\frac{3\alpha\beta}{2m^2 a^3(1-e^2)^{3/2}}\mathbf{A}.$$

The rate of change of the vector $\mathbf{A}$ is directed in the direction opposite to the vector $\mathbf{A}$ itself. Vector $\mathbf{A}$ is directed to pericentre of the orbit and its magnitude is $A = \alpha e$. Thus, the additional force does not cause precession of the orbit, but leads to the reduction of the eccentricity.

It can also be shown (see [2], §75, problem 1) that a particle will fall on the centre over a finite time due to the energy loss.

§3

Scattering in a given field. Collision between particles

3.1. a) Fig. 116 shows that the angle of deflection of a particle θ is twice the angle of the slope of the tangent to the surface of revolution at the point of collisions. We have therefore

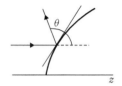

Figure 116

$$\tan(\theta/2) = \frac{d\rho}{dz} = \frac{b}{a}\cos\frac{z}{a}.$$

Hence we have

$$\rho^2 = b^2 - a^2\tan^2(\theta/2)$$

and thus

$$d\sigma = \pi|d\rho^2| = \pi a^2\tan(\theta/2)\frac{d\theta}{\cos^2(\theta/2)} = \frac{a^2\,d\Omega}{4\cos^4\frac{\theta}{2}}.$$

The possible deflection of the particle lie within the range of angles from zero, as $\rho \to b$, to $\theta_m = 2\arctan(b/a)$, as $\rho \to 0$.

As a result,

$$\frac{d\sigma}{d\Omega} = \begin{cases} \dfrac{a^2}{4\cos^4(\theta/2)}, & \text{when } 0 < \theta < \theta_m, \\ 0, & \text{when } \theta_m < \theta. \end{cases}$$

b) $\dfrac{d\sigma}{d\Omega} = \dfrac{A^{2/(1-n)}}{2(1-n)\sin\theta\,\cos^2(\theta/2)}\,[n\cot(\theta/2)]^{(1+n)/(1-n)}.$

Exploring Classical Mechanics: A Collection of 350+ Solved Problems for Students, Lecturers, and Researchers. First Edition.
Gleb L. Kotkin and Valeriy G. Serbo, Oxford University Press (2020). © Gleb L. Kotkin and Valeriy G. Serbo 2020.
DOI: 10.1093/oso/9780198853787.001.0001

c)

$$\frac{d\sigma}{d\Omega} = \begin{cases} \dfrac{\sqrt{2}a\left(b - a\sqrt{\operatorname{tg}(\theta/2)}\right)}{4\sin^{3/2}\theta\sqrt{\cos(\theta/2)}}, & \text{when } 0 < \theta < \theta_m \\ 0, & \text{when } \theta > \theta_m \end{cases}$$

where $\theta_m = 2\arctan\left(b^2/a^2\right)$.

3.2. It is the paraboloid of revolution $\rho^2 = \alpha z/E$. The reader should check whether the trajectories of the particles scattered by the field $U(r) = -\alpha/r$ and by the paraboloid approach one another as $r \to \infty$.

3.3. At $E > V$

$$\frac{d\sigma}{d\Omega} = \begin{cases} \dfrac{a^2 n^2}{4\cos\frac{\theta}{2}} \dfrac{\left(n\cos\frac{\theta}{2} - 1\right)\left(n - \cos\frac{\theta}{2}\right)}{\left(1 + n^2 - 2n\cos\frac{\theta}{2}\right)^2} + \dfrac{a^2}{4}, & \text{when } 0 < \theta < \theta_m, \\ 0, & \text{when } \theta_m < \theta < \pi, \end{cases}$$

where

$$n = \sqrt{1 - (V/E)}, \quad \theta_m = 2\arccos n.$$

Unlike the case of scattering on a potential well (see [1], § 19, problem 2), here it is necessary to take into account the possibility of the elastic reflection at impact parameters $an < \rho < a$.

3.4. a)

$$\sigma = \begin{cases} \pi\left(\dfrac{\beta}{E} - \dfrac{\alpha^2}{4E^2}\right), & \text{when } E > \dfrac{\alpha^2}{4\beta}, \\ 0, & \text{when } E < \dfrac{\alpha^2}{4\beta}. \end{cases}$$

How does the cross-section change when α changes sign?

b)

$$\sigma = \begin{cases} \pi\left(2\sqrt{\dfrac{\gamma}{E}} - \dfrac{\beta}{E}\right), & \text{when } E > \dfrac{\beta^2}{4\gamma}, \\ 0, & \text{when } E < \dfrac{\beta^2}{4\gamma}. \end{cases}$$

3.5. a) We first consider the motion of a particle in the field $U(r) = -\alpha/r^n$. The behaviour of

$$U_{\text{eff}}(r) = \frac{E\rho^2}{r^2} - \frac{\alpha}{r^n}$$

for different values of the impact parameters ρ is shown in Fig. 117.

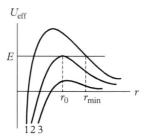

Figure 117

For large values of ρ (curve 1), the particles are scattered, and the minimum distance from the centre of the field $r_{\min}(\rho)$, which is determined by the equation $U_{\text{eff}}(r_{\min}) = E$, decreases with the decreasing ρ up to the value r_0, achieved when $\rho = \rho_0$ (curve 2). At even smaller ρ, the particle falls into the centre (curve 3).

The values r_0 and ρ_0 are determined by the following equations

$$U_{\text{eff}}(r_0) = E, \quad U'_{\text{eff}}(r_0) = 0$$

and equal to

$$r_0 = \left[\frac{(n-2)\alpha}{2E}\right]^{1/n}, \quad \rho_0 = \sqrt{\frac{n}{n-2}}\left[\frac{(n-2)\alpha}{2E}\right]^{1/n}.$$

If $R > r_0$, then all particles for which $r_{\min} \leqslant R$ will hit the sphere and the cross-section will be

$$\sigma = \pi\rho^2(r_{\min} = R) = \pi R^2\left(1 + \frac{\alpha}{ER^n}\right).$$

If $R < r_0$, only those particles which fall into the centre of the field will hit the sphere, and in that case

$$\sigma = \pi\rho_0^2 = \frac{\pi n}{n-2}\left[\frac{(n-2)\alpha}{2E}\right]^{2/n}.$$

b) If $\gamma > ER^4$ the cross-section is

$$\sigma = \pi\left(2\sqrt{\frac{\gamma}{E}} - \frac{\beta}{E}\right)$$

provided $2\sqrt{\gamma E} > \beta$; if $\gamma < ER^4$ then

$$\sigma = \pi R^2 \left(1 + \frac{\gamma}{ER^4} - \frac{\beta}{ER^2}\right)$$

provided $2\sqrt{\gamma E} > \beta$ or provided both $2\sqrt{\gamma E} < \beta$ and $R < R_1$ or $R > R_2$, otherwise $\sigma = 0$; here

$$R_{1,2}^2 = \frac{\beta}{2E} \mp \sqrt{\frac{\beta^2}{4E^2} - \frac{\gamma}{E}}.$$

3.6. Let us take the origin of the coordinate system in the centre of the planet direct the x-axis parallel to $\mathbf{v}_\infty$ and the y-axis parallel to the impact parameter $\boldsymbol{\rho}$. The motion of meteorites occurs in the field

$$U(r) = -\frac{\alpha}{r} = -\frac{Gm_0 m}{r},$$

where G is the gravitational constant. The orbit of the meteorite in the polar coordinates is

$$r(\varphi) = \frac{p}{1 + e\cos(\varphi - \varphi_A)}, \quad p = \frac{M^2}{m\alpha} = \frac{\rho^2}{\xi R}, \quad e = \sqrt{1 + \left(\frac{\rho}{\xi R}\right)^2}, \quad \xi = \frac{Gm_0}{Rv_\infty^2},$$

where φ_A is the polar angle of point of the meteorites' smallest distance from the origin, coinciding with the direction of the Laplace vector

$$\mathbf{A} = [\mathbf{v}, \mathbf{M}] - \alpha\frac{\mathbf{r}}{r} = \alpha e\left(1, \frac{\rho}{\xi R}\right) = \alpha e(\cos\varphi_A, \sin\varphi_A).$$

The polar coordinates of the meteorites falling on the planet satisfy the equation $r(\varphi) = R$ at $\varphi_A < |\varphi| < \pi$. In this case, the impact parameter of such meteorites is less than the maximum ρ_{max} as defined by the equation $r_{min} = r(\varphi_a) = R$, from which we obtain $\rho_{max} = R\sqrt{1 + 2\xi}$. The meteorite falling on the invisible part of the planet satisfies the equation $r(\varphi) = R$ at $\varphi_A < |\varphi| < \pi/2$ with impact parameters from the interval $\rho_{min} < \rho < \rho_{max}$. The corresponding impact parameter ρ_{min} is determined by the equation $r(\varphi = \pi/2) = R$, from which we obtain the equation

$$\frac{\rho_{min}^2/(\xi R)}{1 + \rho_{min}/(\xi R)} = R.$$

Hence

$$\rho_{min} = \tfrac{1}{2}R\left(1 + \sqrt{1 + 4\xi}\right).$$

Thus, the fraction of particles falling on the invisible part of the planet is equal to

$$\delta = \frac{\pi\rho_{max}^2 - \pi\rho_{min}^2}{\pi\rho_{max}^2} = \frac{1}{2} - \frac{\sqrt{1+4\xi}}{2(1+2\xi)}.$$

At high energy of the flying particles, $\xi \ll 1$, this fraction is small (ξ^2). With decreasing the energy of flying particles, the value of δ increases as well, reaching the value of $\delta = 1/2$ at $\xi \gg 1$.

3.7. a) $\dfrac{d\sigma}{d\Omega} = \dfrac{R^2(1+\lambda)}{4\left(1+\lambda\sin^2\dfrac{\theta}{2}\right)^2}$, where $\lambda = \dfrac{4RE(RE+\alpha)}{\alpha^2}$.

How can one explain the result which one obtains when $\alpha + 2RE = 0$?

b) Let us introduce the Cartesian coordinates in the plane of the orbit of the particle with the x-axis directed along the axis of the beam and the y-axis directed along the impact parameter. The motion of the particle in the region $r \leqslant R$ is a harmonic oscillation along each of the coordinates, and in the initial moment of this oscillation (for $t = 0$) $y_0 = \rho, \dot{y}_0 = 0, x_0 = -\sqrt{R^2 - \rho^2}, \dot{x}_0 = v$, thus

$$x = -\sqrt{R^2 - \rho^2}\cos\omega t + \frac{v}{\omega}\sin\omega t, \quad y = \rho\cos\omega t. \tag{1}$$

The moment of release of the particle from action of the forces is determined by the equation

$$x^2 + y^2 = R^2, \tag{2}$$

and the angle θ between the particle velocity at this moment and the x-axis by the equation

$$\sin\theta = \frac{\dot{y}}{v} = -\frac{\rho\omega}{v}\sin\omega t. \tag{3}$$

Substituting (1) and (3) into (2), we obtain

$$\rho^4 - \rho^2 R^2(1+\lambda\sin^2\theta) + \tfrac{1}{4}R^4(1+\lambda)^2\sin^2\theta = 0,$$

where $\lambda = v^2/(R^2\omega^2)$. As a result,

$$\rho_{1,2}^2 = \frac{R^2}{2}(1+\lambda\sin^2\theta \mp \sqrt{\cos^2\theta - \lambda^2\sin^2\theta\cos^2\theta}),$$

and the cross-section

$$d\sigma = \pi(|d\rho_1^2| + |d\rho_2^2|) = \pi d(\rho_1^2 - \rho_2^2) = \frac{R^2(1 + \lambda^2 \cos 2\theta)d\Omega}{2\sqrt{1 - \lambda^2 \sin^2 \theta}}.$$

If $\lambda > 1$, the scattering is only possible at angles smaller than

$$\theta_m = \arcsin \frac{R^2 \omega^2}{v^2},$$

and at $\theta \to \theta_m$, the differential cross-section $d\sigma/d\Omega$ increases indefinitely. Such behaviour of the cross-section is called *rainbow* scattering (see [12], chapter 5, § 5). A similar type of cross-section behaviour leads to formation rainbows in the scattering of light by drops of water.

See also problems 3.9 and 3.11 as examples of the rainbow scattering.

3.8. a)
$$\frac{d\sigma}{d\Omega} = \begin{cases} \dfrac{a^2}{\theta_m^2}, & \text{when } \theta \le \theta_m = \pi\dfrac{V}{2E}, \\ 0, & \text{when } \theta > \theta_m. \end{cases}$$

b)
$$\frac{d\sigma}{d\Omega} = \begin{cases} \dfrac{\pi a^2}{4\theta_m \theta} \sin(\pi\theta/\theta_m), & \text{when } \theta \le \theta_m = \pi\dfrac{V}{2E}, \\ 0, & \text{when } \theta > \theta_m. \end{cases}$$

c)
$$\frac{d\sigma}{d\Omega} = \begin{cases} \dfrac{a^2\theta_m^2(\theta_m - \theta)}{\theta^3(2\theta_m - \theta)^2}, & \text{when } \theta \le \theta_m = \pi\dfrac{V}{E}, \\ 0, & \text{when } \theta > \theta_m. \end{cases}$$

d) When calculating the scattering angle (see [1], § 20 and [7], § 6.2)

$$\theta = \int_{-\infty}^{\infty} \frac{F_y dx}{2E} = 2\frac{V}{E} \frac{\rho\sqrt{R^2 - \rho^2}}{R^2}$$

we take into account that the force is

$$F_y = \begin{cases} \dfrac{2Vy}{R^2} = \dfrac{2V\rho}{R^2}, & \text{when } -\sqrt{R^2 - \rho^2} < x < \sqrt{R^2 - \rho^2}, \\ 0, & \text{when } |x| > \sqrt{R^2 - \rho^2}. \end{cases}$$

From here we find

$$\rho_{1,2}^2 = \frac{1}{2}R^2(1 \mp \sqrt{1 - \theta^2/\theta_m^2}),$$

where $\theta_m = V/E$, so thus

$$d\sigma = \pi(|d\rho_1^2| + |d\rho_2^2|) = \pi d(\rho_1^2 - \rho_2^2).$$

Finally,

$$\frac{d\sigma}{d\Omega} = \begin{cases} \dfrac{R^2}{2\theta_m\sqrt{\theta_m^2 - \theta^2}}, & \text{when } 0 < \theta < \theta_m, \\ 0, & \text{when } \theta > \theta_m. \end{cases}$$

Compare this result with the result of problem 3.7b, which was obtained for the field that is different from the given by a sign.

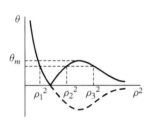

Figure 118

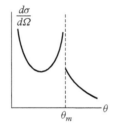

Figure 119

3.9. One can easily evaluate the angle of deflection of a particle, using a general formula,[1] and we have

$$\theta = \left| \frac{3\pi\beta}{4E\rho^4} - \frac{\pi\alpha}{2E\rho^2} \right|. \tag{1}$$

The function $\theta(\rho^2)$ is shown in Fig. 118. From (1) we find

$$\rho_1^2 = \frac{\pi\alpha}{4E\theta}\left(\sqrt{1 + \frac{\theta}{\theta_m}} - 1\right), \qquad \rho_{2,3}^2 = \frac{\pi\alpha}{4E\theta}\left(1 \mp \sqrt{1 - \frac{\theta}{\theta_m}}\right),$$

where $\theta_m = \dfrac{\pi\alpha^2}{12E\beta}$.

---

[1] It is simplest to use both terms of (1) from [1], § 20, problem 2.

For the cross-section, we have

$$d\sigma = \pi(|d\rho_1^2| + |d\rho_2^2| + |d\rho_3^2|) = \pi d(-\rho_1^2 + \rho_2^2 - \rho_3^2) =$$
$$= \frac{\pi\alpha}{8E\theta^3}\left(\frac{1+\theta/(2\theta_m)}{\sqrt{1+\theta/\theta_m}} + \frac{2-\theta/\theta_m}{\sqrt{1-\theta/\theta_m}} - 1\right)d\Omega. \tag{2}$$

This result is valid provided each of the terms in (1) is much less than unity. An estimate shows that the condition $\theta \ll 1$ is sufficient for this. Equation (2) is obtained when $\theta < \theta_m$. If $\theta_m \ll 1$ and $\theta_m < \theta \ll 1$, we have for the cross-section

$$d\sigma = \pi|d\rho_1^2| = \frac{\pi\alpha}{8E\theta^3}\left(\frac{1+\theta/(2\theta_m)}{\sqrt{1+\theta/\theta_m}} - 1\right)d\Omega.$$

Fig. 119 shows how $d\sigma/d\Omega$ depends on θ. The differential cross-section becomes infinite both as $\theta \to 0$ and as $\theta \to \theta_m$. The total cross-section for scattering into a range of angles adjoining $\theta = 0$ is infinite as the small-angle scattering corresponds to large impact parameters.

The total cross-section for scattering of particles into a range of angles $\theta_m - \delta < \theta < \theta_m$,

$$\int_{\theta_m-\delta}^{\theta_m} 2\pi \frac{d\sigma}{d\Omega}\theta\, d\theta = \frac{\pi^2\alpha\delta^{1/2}}{2E\theta_m^{3/2}},$$

is finite and tends to zero as $\delta \to 0$.

What is the relation between the number of scattered particles reaching a counter and its size, if the counter is located at the angle θ_m?

3.10. The velocity of the particle after scattering is at an angle

$$\theta = \pi - \frac{\pi}{\sqrt{1-a^2/\rho^2}}, \quad a^2 = \frac{\alpha}{E}. \tag{1}$$

to its original direction. A counter registers particles scattered over an angle $|\theta| < \pi$ together with those particles which have made several revolution around the scattering centre (Fig. 120a). The observed angle of deflection χ lies in the range $0 < \chi < \pi$ and satisfies the relation

$$-\theta = 2\pi l \pm \chi, \tag{2}$$

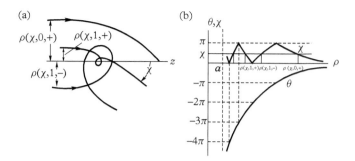

Figure 120

where $l = 0, 1, \ldots$ corresponds to the upper sign in (2) and $l = 1, 2, \ldots$ to the lower sign in (2) (Fig. 120b). From (1) and (2), we have

$$\rho^2(\chi, l, \pm) = a^2 + \frac{\pi a^2}{2} \left(\frac{1}{2\pi l \pm \chi} - \frac{1}{2\pi l + 2\pi \pm \chi} \right).$$

The cross-section is

$$d\sigma = \pi \sum_{l=0}^{\infty} |d\rho^2(\chi, l, +)| + \pi \sum_{l=1}^{\infty} |d\rho^2(\chi, l, -)|.$$

Using the fact that

$$\frac{d\rho^2(\chi, l, +)}{d\chi} < 0, \quad \frac{d\rho^2(\chi, l, -)}{d\chi} > 0,$$

we find that

$$d\sigma = \pi d \left[-\sum_{l=0}^{\infty} \rho^2(\chi, l, +) + \sum_{l=1}^{\infty} \rho^2(\chi, l, -) \right] =$$

$$= \tfrac{1}{2} \pi^2 a^2 d \left(\frac{1}{2\pi - \chi} - \frac{1}{\chi} \right) = \frac{\pi a^2 (2\pi^2 - 2\pi \chi + \chi^2)}{2\chi^2 (2\pi - \chi)^2 \sin \chi} d\Omega.$$

The cross-section $d\sigma/d\Omega$ turns out to infinity as $\chi \to \pi$. This leads to the fact that the cross-section for scattering in a small finite solid angle

$$\Delta\Omega = \int_{\pi - \chi_0}^{\pi} 2\pi \sin \chi \, d\chi = \pi \chi_0^2$$

is equal to

$$\Delta\sigma = \frac{a^2}{2\pi} \int_{\pi-\chi_0}^{\pi} \frac{d\Omega}{\chi} = a^2 \chi_0 = a^2 \sqrt{\frac{\Delta\Omega}{\pi}}.$$

The appearance of such features of the differential cross-section is because the angle of deviation equal to π is reached at the values of the impact parameter $\rho(\pi, l, \pm)$, which are nonzero. In a solid angle $\Delta\Omega$, proportional to the square of a small value $\Delta\chi \equiv \chi_0$, flying particles fall before scattering through sites $2\pi\rho\Delta\rho$, areas which are proportional to the first degree $\Delta\rho \propto \chi_0$. This feature of scattering is called *radiance* (see [12], Ch. 5, § 5).

The same feature is present in scattering by an angle where $\chi = 0$, but in this case it is masked by an infinite cross-section due to scattering of particles with the arbitrarily large impact parameters. Radiance in the forward scattering could occur, for example, under a constraint the diameter of the projectile to the centre of the beam of particles.

3.11. a) One sees easily that the condition $E \gg V$ means that the angle over which the particle is deflected during the scattering is small. The change in momentum is (see [7], § 6.2)

$$\Delta p = -\frac{\partial}{\partial\rho} \int_{-\infty}^{\infty} U(|\boldsymbol{\rho} + \mathbf{v}t|)\, dt = \frac{2\sqrt{\pi}\, V}{v} xe^{-x^2},$$

where $x = \varkappa\rho$. The angle of deflection is

$$\theta = \frac{\Delta p}{p} = \sqrt{\pi}\frac{V}{E} xe^{-x^2}. \tag{1}$$

We cannot solve this equation for x in analytic form. However, from the graph of the function xe^{-x^2} (Fig. 121), we see that (1) has two roots when $\theta < \theta_m = \frac{V}{E}\sqrt{\frac{\pi}{2e}}$.

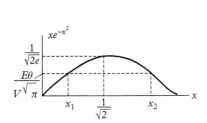
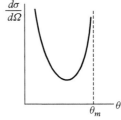

Figure 121   Figure 122

Using (1) and the relation

$$d\theta = \sqrt{\pi}\frac{V}{E}(1 - 2x^2)e^{-x^2}\, dx,$$

we can write the expression for the cross-section

$$d\sigma = \pi(|d\rho_1^2| + |d\rho_2^2|) = \frac{2\pi}{\varkappa^2}(x_1\,dx_1 - x_2\,dx_2)$$

in the form

$$\frac{d\sigma}{d\Omega} = \frac{1}{\varkappa^2\theta^2}\left(\frac{x_2^2}{2x_2^2 - 1} + \frac{x_1^2}{1 - 2x_1^2}\right).$$

When $\theta \ll \theta_m$, then $x_1 \ll 1$ and $x_2 \gg 1$, so that

$$\frac{d\sigma}{d\Omega} = \frac{1}{2\varkappa^2\theta^2}.$$

When $\theta_m - \theta \ll \theta_m$, we can solve (1) by expanding xe^{-x^2} in a series near the maximum. We then get

$$x_{1,2} = \frac{1 \mp \sqrt{1 - \theta/\theta_m}}{\sqrt{2}}, \quad \frac{d\sigma}{d\Omega} = \frac{1}{2\varkappa^2\theta_m^2\sqrt{1 - \theta/\theta_m}}.$$

Fig. 122 shows $d\sigma/d\Omega$ as function of θ. The singularity at $\theta = \theta_m$ is integrable (cf. problem 3.9).

Discuss whether the presence of the singularity in the cross-section at $\theta = \theta_m$ is connected with the approximations made in the solution of the problem?

b) $\frac{d\sigma}{d\Omega} = \frac{1}{\varkappa^2\theta^2}\left(\frac{x_1 + x_1^2}{1 - 2x_1} + \frac{x_2 + x_2^2}{2x_2 - 1}\right)$, where $\frac{x_{1,2}}{(1 + x_{1,2})^3} = \left(\frac{2E\theta}{\pi V}\right)^2$.

When $\theta \ll \theta_m = \frac{\pi}{3\sqrt{3}}\frac{V}{E}$ we have

$$\frac{d\sigma}{d\Omega} = \frac{\pi V}{4\varkappa^2 E\theta^3}.$$

When $\theta_m - \theta \ll \theta_m$ we have

$$\frac{d\sigma}{d\Omega} = \frac{\sqrt{3}}{2\sqrt{2}\varkappa^2\theta_m^2\sqrt{1 - \theta/\theta_m}}.$$

3.12. a) A particle with a velocity $\mathbf{v}$ before the collision will have a velocity $\mathbf{v'} = \mathbf{v} - 2\mathbf{n}(\mathbf{nv})$ after the collision, where $\mathbf{n}$ is a unit vector which is normal to the surface of the ellipsoid. Using the relations[1]

---

[1] We know from differential geometry that

$$\mathbf{n} \propto \mathrm{grad}\left(\frac{x^2}{a^2} + \frac{y^2}{b^2} + \frac{z^2}{c^2} - 1\right),$$

while the value of N is determined by the relation $\mathbf{n}^2 = 1$.

$$\mathbf{v} = v(0, 0, 1), \quad \mathbf{n} = \frac{1}{N}\left(\frac{x}{a^2}, \frac{y}{b^2}, \frac{z}{c^2}\right),$$

we get

$$\mathbf{v}' = v\left(-\frac{2xz}{N^2 a^2 c^2}, -\frac{2yz}{N^2 b^2 c^2}, 1 - \frac{2z^2}{N^2 c^4}\right). \tag{1}$$

Introducing spherical coordinates with the z-axis along $\mathbf{v}$ we can write

$$\mathbf{v}' = v(\sin\theta\cos\varphi, \sin\theta\sin\varphi, \cos\theta),$$

and from (1) we then have

$$\tan\varphi = \frac{a^2 y}{b^2 x}, \quad \cos\theta = 1 - \frac{2z^2}{N^2 c^4}, \quad \sin^2\theta = \left(\frac{2z}{N^2 c^2}\right)^2\left(\frac{x^2}{a^4} + \frac{y^2}{b^4}\right). \tag{2}$$

For the cross section, we have

$$d\sigma = dx\,dy = \left|\frac{\partial(x, y)}{\partial(\theta, \varphi)}\right| d\theta\, d\varphi,$$

where the dependence of x and y on θ and φ follows from (2) and from the equation for the ellipsoid.

To evaluate the Jacobian, it is helpful to introduce an auxiliary variable u such that

$$x = a^2 u\cos\varphi, \quad y = b^2 u\sin\varphi.$$

From (1) and (2), it then follows that

$$\sin\theta = \frac{2zu}{N^2 c^2}, \quad 1 - \cos\theta = \frac{2z^2}{N^2 c^4}, \quad \tan(\theta/2) = \frac{z}{uc^2}$$

and from the equation of the ellipsoid, we find

$$u^{-2} = a^2\cos^2\varphi + b^2\sin^2\varphi + c^2\tan^2(\theta/2).$$

Moreover, we have

$$\frac{\partial(x, y)}{\partial(\theta, \varphi)} = \frac{\partial(x, y)}{\partial(u, \varphi)}\frac{\partial(u, \varphi)}{\partial(\theta, \varphi)} = \frac{a^2 b^2}{2}\frac{\partial u^2}{\partial\theta}.$$

We thus get, finally,

$$\frac{d\sigma}{d\Omega} = \frac{a^2 b^2 c^2}{4\cos^4\frac{\theta}{2}\left(a^2\cos^2\varphi + b^2\sin^2\varphi + c^2\tan^2\frac{\theta}{2}\right)^2}.$$

What is the limit which we must take to obtain from this result the cross-section for scattering by a paraboloid?

b) $\frac{d\sigma}{d\Omega} = \dfrac{a^2 b^2 c^2}{\cos^3\theta(a^2\cos^2\varphi + b^2\sin^2\varphi + c^2\tan^2\theta)^2}.$

c) $\frac{d\sigma}{d\Omega} = \dfrac{a^2 b^2 c^2\cos\theta}{\sin^4\theta\left(a^2\cos^2\varphi + b^2\sin^2\varphi + c^2\cot^2\theta\right)^2}.$

3.13. a) The change in the scattering momentum is (see [7], §6.2)

$$\Delta\mathbf{p} = -\frac{\partial}{\partial\boldsymbol{\rho}}\int_{-\infty}^{\infty} U(\boldsymbol{\rho} + \mathbf{v}t)dt = -\frac{\partial}{\partial\boldsymbol{\rho}}\frac{\pi(\mathbf{a}\boldsymbol{\rho})}{v\rho}. \tag{1}$$

Let the z-axis be parallel to $\mathbf{v}$ and the y-axis be perpendicular to $\mathbf{a}$. We then have

$$\Delta p_x = -\frac{\pi a_x}{v}\frac{\rho_y^2}{\rho^3}, \qquad \Delta p_y = \frac{\pi a_x}{v}\frac{\rho_x\rho_y}{\rho^3}. \tag{2}$$

The direction of the velocity after scattering can be characterized by two spherical angles

$$\tan\varphi = \frac{\Delta p_y}{\Delta p_x}, \qquad \theta = \frac{\Delta p}{p}. \tag{3}$$

It is clear from (2) that scattering only occurs when

$$\tfrac{1}{2}\pi < \varphi < \tfrac{3}{2}\pi.$$

From (2) and (3), we find

$$\rho_x = \pm\frac{\pi a_x}{2E}\frac{\sin\varphi\cos\varphi}{\theta}, \qquad \rho_y = \mp\frac{\pi a_x}{2E}\frac{\cos^2\varphi}{\theta}. \tag{4}$$

For the cross-section, we find

$$d\sigma = \sum d\rho_x\, d\rho_y = \sum\left|\frac{\partial(\rho_x, \rho_y)}{\partial(\theta, \varphi)}\right| d\theta\, d\varphi = \left(\frac{\pi a_x}{E}\right)^2\frac{\cos^2\varphi}{2\theta^4}\, d\Omega. \tag{5}$$

The summation in (5) addresses the two possible values of ρ (see (4)).

b) $\frac{d\sigma}{d\Omega} = \frac{|\mathbf{a}_\perp|}{2E\theta^3}$, where $\mathbf{a}_\perp$ is the component of $\mathbf{a}$ perpendicular to $\mathbf{v}_\infty$. The cross-section turns out to be symmetrical with respect to $\mathbf{v}_\infty$ (although the field is by no means symmetric with respect to this directions.)

3.14. The change in the angle of deflection of the particle is given by the equation (see problem 2.20)

$$\delta\theta(\rho) = -\frac{1}{E}\frac{\partial}{\partial\rho}\int_{r_{min}}^{\infty}\frac{\delta U(r)\,dr}{\sqrt{1-\frac{\rho^2}{r^2}-\frac{U(r)}{E}}}. \tag{1}$$

We then find from the equation $\theta = \theta_0(\rho) + \delta\theta(\rho)$:

$$\rho = \rho_0(\theta) - \delta\theta(\rho_0(\theta))\frac{d\rho_0(\theta)}{d\theta}$$

(cf. problem 1.8). The function $\theta_0(\rho)$ and hence the function $\rho_0(\theta)$ are the expressions obtained when $\delta U(r) = 0$. We then get for the cross-section

$$\frac{d\sigma}{d\theta} = \pi\left|\frac{d\rho^2}{d\theta}\right| = \pi\left|\frac{d\rho_0^2(\theta)}{d(\theta)} - \frac{d}{d\theta}[2\rho_0(\theta)\delta\theta(\rho_0(\theta))]\frac{d\rho_0(\theta)}{d\theta}\right| =$$

$$= \frac{d\sigma_0}{d\theta} - \frac{d}{d\theta}\left[\delta\theta(\rho_0(\theta))\frac{d\sigma_0}{d\theta}\right].$$

a) $\delta\frac{d\sigma}{d\theta} = -\frac{\pi\beta}{E}\frac{d}{d\theta}\left\{\frac{\pi-\theta-\sin\theta}{\sin\theta}\right\};$

b) $\delta\frac{d\sigma}{d\theta} = -\frac{2\pi\gamma}{\alpha}\frac{d}{d\theta}\left\{\frac{3}{2}(\pi-\theta)[1+\tan^2(\theta/2)] - 3\tan(\theta/2) - \sin\theta\right\};$

c) $\delta\frac{d\sigma}{d\theta} = -\frac{4\gamma}{\pi\sqrt{\beta E}}\frac{d}{d\theta}\left\{\frac{(\pi-\theta)^2}{\sqrt{\theta(2\pi-\theta)}}\right\}.$

3.15. The energy acquired by the particle,

$$\varepsilon = \frac{(\mathbf{p}+\Delta\mathbf{p})^2}{2m} - \frac{p^2}{2m} \approx \mathbf{v}\Delta\mathbf{p},$$

is to the first order determined solely by the change of the longitudinal component of the momentum. As we assume that the deflection of the particle is small, we can (after differentiating) put $\mathbf{r} = \boldsymbol{\rho} + \mathbf{v}(t-\tau)$ in the expression for the force

$$\mathbf{F} = -\frac{\partial U(r, t)}{\partial \mathbf{r}},$$

acting upon the particle. Here ρ is the impact parameter and τ the time when the particle is at its smallest distance from the centre. Therefore, we have

$$\varepsilon = \mathbf{v} \int_{-\infty}^{\infty} \mathbf{F}(t)dt = \varepsilon_m e^{-\varkappa^2 \rho^2 \cos\varphi},$$

$$\varepsilon_m = \sqrt{\pi}\, V_2 \frac{\omega}{\varkappa v} e^{-\omega^2/(4\varkappa^2 v^2)}, \quad \varphi = \omega\tau,$$

and the scattering cross-section for particles with a given value of τ when $\cos\varphi > 0$ ($\cos\varphi < 0$) is

$$\frac{d\sigma}{d\varepsilon} = \begin{cases} \dfrac{\pi}{\varkappa^2 |\varepsilon|}, & \text{when } 0 < \varepsilon < \varepsilon_m \cos\varphi \quad (0 > \varepsilon > \varepsilon_m \cos\varphi), \\ 0, & \text{when } |\varepsilon| > \varepsilon_m |\cos\varphi|. \end{cases}$$

In the incident beam, there are particles with different values of τ. If we average the cross-section over the phase φ, for $\varepsilon > 0$ using, for example, the formula

$$\left\langle \frac{d\sigma}{d\varepsilon} \right\rangle = \frac{1}{2\pi} \int_{-\alpha}^{\alpha} \frac{d\sigma}{d\varepsilon} d\varphi, \quad \alpha = \arccos\frac{\varepsilon}{\varepsilon_m},$$

we get

$$\left\langle \frac{d\sigma}{d\varepsilon} \right\rangle = \begin{cases} \dfrac{1}{\varkappa^2 |\varepsilon|} \arccos\dfrac{|\varepsilon|}{\varepsilon_m}, & \text{when } |\varepsilon| < \varepsilon_m, \\ 0, & \text{when } |\varepsilon| > \varepsilon_m. \end{cases}$$

3.16. The distribution of decay particles is

$$\frac{dN}{N} = \frac{\lambda^2 \sin\theta\, d\theta}{\cos^3\theta \sqrt{1 - \lambda^2 \tan^2\theta}}, \quad \lambda = \frac{V^2 - v_0^2}{2Vv_0}$$

in the angle interval $0 \leqslant \theta \leqslant \arctan(1/\lambda)$ if $V > v_0$ and in the angle interval $(\pi - \arctan(1/|\lambda|) \leqslant \theta \leqslant \pi$ if $V < v_0$.

3.17. The distribution of decay particles by energies $T = \frac{1}{2} mv^2$ has the form

$$\frac{dN}{N} = \frac{6(T_{\max} - T)(T - T_{\min})}{(T_{\max} - T_{\min})^3} dT, \quad T_{\min} \leqslant T \leqslant T_{\max},$$

where

$$T_{\min} = \tfrac{1}{2} m(v_0 - V)^2, \quad T_{\max} = \tfrac{1}{2} m(v_0 + V)^2.$$

3.18.

$$\tan\theta_1 = \cot\theta_2 = \alpha/(E\rho), \qquad E = \tfrac{1}{2} mv^2,$$

$$v_1 = \frac{E\rho v}{\sqrt{\alpha^2 + E^2\rho^2}}, \qquad v_2 = \frac{\alpha v}{\sqrt{\alpha^2 + E^2\rho^2}}.$$

3.19.

$$\tfrac{1}{2}\pi \leqslant \theta \leqslant \pi, \text{ when } m_1 < m_2,$$

$$\theta = \tfrac{1}{2}\pi, \text{ when } m_1 = m_2,$$

$$0 \leqslant \theta \leqslant \tfrac{1}{2}\pi, \text{ when } m_1 > m_2.$$

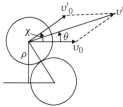

Figure 123

3.20. The velocity component normal to the sphere's surface at the point of impact becomes zero in the centre-of-mass system while the tangential component v_0' is conserved (see Fig. 123). The scattering cross-section as function of the angle of deflection χ of the particle in the centre-of-mass system is

$$d\sigma = \pi |d\rho^2| = 4a^2\pi |d\cos^2\chi| = 4a^2 \cos\chi \, d\Omega.$$

To transform to the laboratory system, we use the equation

$$\text{tg}\,\theta = \frac{v_0' \sin\chi}{v_0' \cos\chi + v_0} = \frac{\sin\chi \cos\chi}{1 + \cos^2\chi}$$

to find the equation

$$\cos^2\chi_{1,2} = \tfrac{3}{2}\cos^2\theta - 1 \pm \tfrac{1}{2}\cos\theta\sqrt{9\cos^2\theta - 8}.$$

Taking into account that there are two possible connections between χ and θ, we obtain

$$d\sigma = 4\pi a^2(|d\cos^2\chi_1| + |d\cos^2\chi_2|) =$$

$$= 4\pi a^2 d(\cos^2\chi_2 - \cos^2\chi_1) = 4a^2\frac{5 - 9\sin^2\theta}{\sqrt{1 - 9\sin^2\theta}} d\Omega,$$

where $0 < \theta < \arcsin(1/3)$.

If the spheres impinging upon those at rest are identical so that there are no means of distinguishing between the two after the collisions, we must add to the cross-section which we have just obtained the cross-section

$$d\sigma' = 4a^2 \cos\theta \, d\Omega \quad (0 < \theta < \tfrac{1}{2}\pi),$$

which refers to the spheres which were originally at rest flying off at an angle θ

3.21. When passing the distance x intensity is

$$I(x) = I(0)e^{-n\sigma x}.$$

3.22. $dN = \sigma n_1 n_2 |\mathbf{v}_1 - \mathbf{v}_2| \, dV \, dt.$

3.23. a) $F_{\mathrm{fr}} = 2\pi m v^2 n \displaystyle\int_0^\pi f(\theta)(1 - \cos\theta)\sin\theta \, d\theta;$[1]

b) $\langle \Theta^2 \rangle = 2\pi \left(\dfrac{m}{M}\right)^2 nl \displaystyle\int_0^\pi f(\theta)\sin^3\theta \, d\theta;$

where l is the path travelled by a particle of mass M; v, its velocity; and n, the concentration of light particles.

---

[1] The quantity $\displaystyle\int (1 - \cos\theta)\frac{d\sigma}{d\Omega} \, d\Omega$ is called the *transport cross-section*—as distinct from the total cross-section $\displaystyle\int \frac{d\sigma}{d\Omega} \, d\Omega.$

§4

Lagrangian equations of motion. Conservation laws

4.1. Assuming that $t = 0$ when the particle is at $x = 0$, we find $C = 0$, and from the condition that $x = a$ when $t = \tau$, we find that $B = \frac{a}{\tau} - a\tau$. Substituting the function

$$x(t) = At^2 + \left(\frac{a}{\tau} - A\tau\right)t$$

into the action, we find

$$S = \int_0^\tau L(x, \dot{x})\, dt = \int_0^\tau \left(\tfrac{1}{2} m\dot{x}^2 + Fx\right) dt = \frac{mA^2\tau^3}{6} + \frac{ma^2}{2\tau} - \frac{FA\tau^3}{6} + \frac{Fa\tau}{2}.$$

From the condition that the action be a minimum, $\frac{\partial S}{\partial A} = 0$, it follows that $A = \frac{F}{2m}$. It is clear that in the present case the law of motion

$$x(t) = \frac{Ft^2}{2m} + \left(\frac{a}{\tau} - \frac{F\tau}{2m}\right)t$$

is exact. However, the only thing the solution given here allows us to state is that this law is in some sense the best one among all possible laws of this kind.

In order that we can be certain that this law of motion gives a smaller value for S than any other $x(t)$, that is, that it is the true law, we must verify that it satisfies the Lagrangian equation of motion.

Exploring Classical Mechanics: A Collection of 350+ Solved Problems for Students, Lecturers, and Researchers. First Edition.
Gleb L. Kotkin and Valeriy G. Serbo, Oxford University Press (2020). © Gleb L. Kotkin and Valeriy G. Serbo 2020.
DOI: 10.1093/oso/9780198853787.001.0001

4.2.

$$x(t) = \begin{cases} v_x t - a, & \text{when } 0 < t < t_0, \\ \sqrt{v_x^2 - 2V/m}\,(t - \tau) + a, & \text{when } t_0 < t < \tau, \end{cases}$$

$$y(t) = at/\tau,$$

where $v_x = [2aV/(m\tau)]^{1/3}$ and $t_0 = a/v_x$.

4.3. From the relation

$$\tilde{L}(Q, \dot{Q}, t) = L(q(Q, t), \dot{q}(Q, \dot{Q}, t), t) = L\left(q, \frac{\partial q}{\partial Q}\dot{Q} + \frac{\partial q}{\partial t}, t\right)$$

we get

$$\frac{d}{dt}\frac{\partial \tilde{L}}{\partial \dot{Q}} = \frac{\partial q}{\partial Q}\frac{d}{dt}\frac{\partial L}{\partial \dot{q}} + \frac{\partial L}{\partial \dot{q}}\frac{d}{dt}\frac{\partial q}{\partial Q},$$

$$\frac{\partial \tilde{L}}{\partial Q} = \frac{\partial L}{\partial q}\frac{\partial q}{\partial Q} + \frac{\partial L}{\partial \dot{q}}\frac{\partial \dot{q}}{\partial Q}.$$

Bearing in mind that $\dfrac{\partial \dot{q}}{\partial Q} = \dfrac{d}{dt}\dfrac{\partial q}{\partial Q}$, we obtain

$$\frac{d}{dt}\frac{\partial \tilde{L}}{\partial \dot{Q}} - \frac{\partial \tilde{L}}{\partial Q} = \frac{\partial q}{\partial Q}\left(\frac{d}{dt}\frac{\partial L}{\partial \dot{q}} - \frac{\partial L}{\partial q}\right). \tag{1}$$

The validity of the equation $\dfrac{d}{dt}\dfrac{\partial L}{\partial \dot{q}} - \dfrac{\partial L}{\partial q} = 0$ thus leads to the validity of the analogous equation

$$\frac{d}{dt}\frac{\partial \tilde{L}}{\partial \dot{Q}} - \frac{\partial \tilde{L}}{\partial Q} = 0.$$

If there are several degrees of freedom, instead of (1) we get the following equations:

$$\frac{d}{dt}\frac{\partial \tilde{L}}{\partial \dot{Q}_i} - \frac{\partial \tilde{L}}{\partial Q_i} = \sum_k \frac{\partial q_k}{\partial Q_i}\left(\frac{d}{dt}\frac{\partial L}{\partial \dot{q}_k} - \frac{\partial L}{\partial q_k}\right).$$

4.4. $\tilde{L}\left(Q, \dfrac{dQ}{d\tau}, \tau\right) = L\left(q, \dfrac{dq}{dt}, t\right)\dfrac{dt}{d\tau}.$

Here

$$q = q(Q, \tau), \quad \frac{dq}{dt} = \frac{dq}{d\tau}\frac{d\tau}{dt},$$

$$\frac{dq}{d\tau} = \frac{\partial q}{\partial Q}\frac{dQ}{d\tau} + \frac{\partial q}{\partial \tau}, \quad \frac{dt}{d\tau} = \frac{\partial t}{\partial Q}\frac{dQ}{d\tau} + \frac{\partial t}{\partial \tau}.$$

4.5. $\tilde{L} = \frac{1}{2}m\frac{\dot{x}^2}{1+\lambda x} - (1+\lambda\dot{x})\,U(x), \quad \dot{x} = \frac{dx}{d\tau}.$

4.6. $\tilde{L} = -\sqrt{1 - \left(\frac{dq}{d\tau}\right)^2}.$

The problem is a purely formal one. However, both this Lagrangian function and the transformation consider ("improved" by the introduction of dimensional factors) have a simple physical meaning in the theory of relativity (e.g. see [2], § 4 and § 8).

4.7. The generalized momentum $P_l = \dfrac{\partial L}{\partial \dot{Q}_l}$ and the energy

$$E' = \sum_l P_l \dot{Q}_l - L$$

transform as

$$P_l = \sum_k \frac{\partial f_k}{\partial Q_l}p_k, \quad E' = E - \sum_k p_k \frac{\partial f_k}{\partial t}.$$

4.8. Applying the formulae of the preceding problem, we obtain

a) $p'_r = m\dot{r}' = p_r, \; p'_\varphi = mr'^2(\dot{\varphi}' + \Omega) = p_\varphi,$
   $E' = E - \Omega p_\varphi;$

b)
$$\begin{cases} p'_x = p_x\cos\Omega t + p_y\sin\Omega t, \\ p'_y = -p_x\sin\Omega t + p_y\cos\Omega t. \end{cases} \tag{1}$$

It follows from (1) that $\mathbf{p}' = \mathbf{p}$, while (1) also gives the rules for the law of transformation of vector components when we are changing to a coordinate system which is rotated over an angle Ωt. We emphasize that $\mathbf{p}' \neq m\mathbf{v}'$ (cf. [1], § 39).

4.9. a) $E' = E - \mathbf{V}\mathbf{p}, \; \mathbf{p}' = \mathbf{p};$
   b) $E' = E - \mathbf{V}\mathbf{p} + \frac{1}{2}mV^2, \quad \mathbf{p}' = \mathbf{p} - m\mathbf{V}.$

The two expressions for the energy differ by a constant. Usually, we employ the second formula, as it agrees with the definition of the energy in the theory of relativity.

4.10. In the frame of reference rotating together with the rod, the law of conservation of energy is correct (taking into account the centrifugal energy):

$$-\tfrac{1}{2}m\Omega^2 a^2 = \tfrac{1}{2}mv_r^2 - \tfrac{1}{2}m\Omega^2 l^2,$$

where v_r is the velocity directed along the rod at the moment of the bead slipping off. From here we get $v_r^2 = \Omega^2(l^2 - a^2)$. In the laboratory reference of frame, the velocity $\mathbf{v}_t$ directed across the rod should be added ($v_t = \Omega l$) to this velocity $\mathbf{v}_r$. Thus, the value of the total velocity is

$$v = \sqrt{v_r^2 + v_t^2} = \Omega\sqrt{2l^2 - a^2}.$$

The velocity of the slipping bead can be changed significantly if the end of the rod is bent in the direction towards or against the motion.

4.11. In a rotating frame of reference $x'y'$ (Fig. 7), the Lagrangian function is equal to (see [7], §17.2)

$$L(\mathbf{r'}, \mathbf{v'}, t) = \tfrac{1}{2}m\left(v'^2 + \Omega^2 r'^2\right) + m\mathbf{v'}[\mathbf{\Omega}, \mathbf{r'}].$$

We direct the x'-axis along rod a and introduce the angle φ, which determines the direction of rod b with respect to the direction of rod a (see Fig. 7). Substituting the vector $\mathbf{r'} = (a + b\cos\varphi, b\sin\varphi)$ in the Lagrangian function, we obtain

$$L(\varphi, \dot{\varphi}, t) = \tfrac{1}{2}mb^2\dot{\varphi}^2 + mab\Omega^2\cos\varphi$$

(after removing $\tfrac{1}{2}m\Omega^2\left(a^2 + b^2\right) + m\Omega\dot{\varphi}b\left(a\cos\varphi + b\right)$, representing the total derivative of the function of φ with respect to the time). The Lagrangian equation

$$b\ddot{\varphi} = -a\Omega^2\sin\varphi$$

is identical to that of the pendulum of length b in the field of gravity $g = a\Omega^2$, so the frequency of small oscillations is $\omega = \Omega\sqrt{a/b}$.

If we add a uniform constant magnetic field $\mathbf{B}$, perpendicular to the plane of rotation, then the addition to the Lagrangian function will be

$$\Delta L(\varphi, \dot{\varphi}, t) = mab\Omega\omega_B\cos\varphi, \quad \omega_B = \frac{eB}{mc},$$

where e is the particle charge and c is the velocity of light. (This result occurs if we choose the vector potential $\mathbf{A} = \frac{1}{2}[\mathbf{B}, \mathbf{r}]$ and remove the terms representing the total derivative of the function of φ with respect to time.)

Thus, the effect of the magnetic field is reduced to a simple substitution of Ω^2 for $\Omega(\Omega + \omega_B)$ in the Lagrangian function without magnetic field. In particular, for $\Omega + \omega_B = 0$, the oscillations of the angle φ will be completely "turned off".

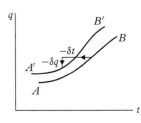

Figure 124

4.12. Let $q_i = g_i(t)$ describe the motion of the system (trajectory AB in Fig. 124). As the form of the action is invariant when we change to the variables q_i', t', the motion is also describes by the equations $q_i' = g_i(t')$. If we express these equations in terms of the variables q_i, t, we have up to first order in ε:

$$q_i(t - \delta t) = g_i(t) - \delta q_i,$$

with

$$\delta q_i = \varepsilon f_i(g(t), t), \quad \delta t = \varepsilon h(g(t), t)$$

(trajectory $A'B'$ in Fig. 124).

Small changes in the coordinates and time of the beginning and the end of the motion when we change from the trajectory AB to the trajectory $A'B'$ lead to the following change in the action:

$$S_{A'B'} - S_{AB} = \left[\frac{\partial S}{\partial t}(-\delta t) + \sum_i \frac{\partial S}{\partial q_i}(-\delta q_i) \right]_A^B.$$

Here we have (see [1], § 43)

$$\frac{\partial S}{\partial t} = L - \sum_i \frac{\partial L}{\partial \dot{q}_i} \dot{q}_i = -E(t), \quad \frac{\partial S}{\partial q_i} = \frac{\partial L}{\partial \dot{q}_i} = p_i(t).$$

On the other hand, according to the conditions of the problem $S_{AB} = S_{A'B'}$, so that

$$E(t_A)\varepsilon h(q_A, t_A) - \sum_i p_i(t_A)\varepsilon f_i(q_A, t_A) =$$

$$= E(t_B)\varepsilon h(q_B, t_B) - \sum_i p_i(t_B)\varepsilon f_i(q_B, t_B),$$

or

$$Eh - \sum_i p_i f_i = \text{const.}$$

The proved theorem is, in essence, a united derivation of the various conservation laws. The importance of this theorem increases if we take into account the fact that the same theorem takes also place in the field theory (Noether's theorem, see [13] and [14]).

4.13. $\sum_i \dfrac{\partial L}{\partial \dot{q}_i}(\dot{q}_i h - f_i) - Lh - F = \text{const.}$

4.14. a) The linear omentum;

   b) the angular momentum;

   c) the energy;

   d) $M_z + \dfrac{h}{2\pi}\, p_z = \text{const}$, where h is the pitch of the screw;

   e) $Ex - p_x t = \text{const}$: the integral of motion of the centre of inertia of the system (see [2], §14).

4.15. a) The potential energy is $U(\mathbf{r}) = -\mathbf{Fr}$, and this energy and at the same time the action remain unchanged under transformation in direction at right angles to $\mathbf{F}$ or under rotating around axis parallel to $\mathbf{F}$. Integral of motion are thus the linear momentum components at right angles to $\mathbf{F}$ and the angular momentum component parallel to $\mathbf{F}$. As the Lagrangian function is independent of the time, the energy is an integral of the motion.

The statement that different points in some region are "equal" means that the value of the potential energy (but not of the force!) is the same in all those points.

b) The similarity transformation $\mathbf{r}' = \alpha\mathbf{r}$ leaves the form of the action invariant if the time is transformed as $t' = \beta t$ simultaneously. The contribution of the kinetic energy to the action

$$\int \frac{m}{2}v'^2\, dt' = \frac{\alpha^2}{\beta} \int \frac{m}{2}v^2\, dt$$

remains unchanged at $\beta = \alpha^2$, and the contribution of the potential energy

$$-\int U(\mathbf{r}')dt' = -\alpha^n\beta \int U(\mathbf{r})dt = -\alpha^{n+2} \int U(\mathbf{r})dt$$

remains unchanged when $n = -2$. To use the theorem, formulated in problem 4.12, we write the infinitesimal similarity transformation as $\alpha = 1 + \varepsilon$:

$$\mathbf{r}' = (1+\varepsilon)\mathbf{r}, \quad t' = (1+2\varepsilon)t, \quad \varepsilon \to 0,$$

so thus $\mathbf{f} = \mathbf{R}$, $h = 2t$ and the integral of motion reads

$$\frac{\partial L}{\partial \mathbf{v}} (\mathbf{v}h - \mathbf{f}) - Lh = 2Et - m\mathbf{v}\mathbf{r} = C. \tag{1}$$

From (1) we can find $r(t)$ taken into account that $\mathbf{r}\mathbf{v} = \frac{1}{2}\frac{dr^2}{dt}$:

$$r^2 = \frac{2E}{m}t^2 - \frac{2}{m}Ct + C_1. \tag{2}$$

If $E < 0$, the particle falls into the centre (in this case, $\dot{r} \to \infty$). If $E > 0$, it is helpful to introduce other constants τ, B instead of C, C_1 and write (2) as

$$r^2 = \frac{2E}{m}(t - \tau)^2 + B.$$

At $B > 0$ the dependence $r(t)$ is the same as for the free motion of particles with the velocity $v_0 = \sqrt{2E/m}$ and with the impact parameter $\rho = \sqrt{B}$. At $B < 0$, the particle falls into the centre.

The fields for which the conditions of this problem are satisfied are given, for example, in problems 12.6, 12.7 and in [1], problem 2 to § 15.

 c) $E - \mathbf{V}\mathbf{p} = \text{const}$;
 d) $\mathbf{r}\mathbf{p} - 2Et = \text{const}$, where $\mathbf{p} = m\mathbf{v} + \frac{e}{c}\mathbf{A}(\mathbf{r})$, if $\mathbf{A}(\alpha\mathbf{r}) = \alpha^{-1}\mathbf{A}(\mathbf{r})$;
 e) $E - p_\varphi \Omega = \text{const}$.

4.16. One can use the results of problem 4.15b (see (2), taking into account the initial conditions)

$$r^2(t) = \frac{2E}{m}t^2 + 2\mathbf{r}_0\mathbf{v}_0 t + r_0^2. \tag{1}$$

The fall time is determined from the quadratic equation $r^2(t) = 0$, which gives two roots

$$t_{1,2} = \frac{1}{2E}\left(-m\mathbf{r}_0\mathbf{v}_0 \pm \sqrt{\beta}\right), \quad \beta = (m\mathbf{r}_0\mathbf{v}_0)^2 - 2mEr_0^2.$$

From (1) we can see that the fall must happen at $E < 0$; its time is determined by the positive root t_2. As for the case when $E > 0$, the fall is possible only at $\mathbf{r}_0\mathbf{v}_0 < 0$ and $\beta > 0$. In this case, one should choose the smallest t_2 among the two positive roots.

Please note that in this problem the value $M^2 - 2m(\mathbf{ar}/r)^2$ is an integral of motion which coincides with the $(-\beta)$ (see problem 10.26). Therefore, the equation for the radial motion

$$E = \tfrac{1}{2} m \dot{r}^2 - \frac{\beta}{r^2}$$

leads to the following fall time

$$t = -m \int_{r_0}^{0} \frac{dr}{\sqrt{2mE - \beta/r^2}} = \frac{1}{2E} \left(\sqrt{\beta + 2mEr_0^2} - \sqrt{\beta} \right),$$

same as t_2.

4.17. $m\mathbf{r} - \mathbf{p}t = \text{const}$ (cf. problem 4.14e).

Is this integral of motion the eighth independent integral for a closed system (apart from E, $\mathbf{M}$, and $\mathbf{p}$)?

4.18. a) Let the z-axis be parallel to $\mathbf{B}$. A translation along the z-axis or a rotation around it lives the form of $\mathbf{A}$, and hence also the form of the action, invariant. We have thus the following integrals of motion:

$$p_z = \frac{\partial L}{\partial \dot{z}} = m\dot{z} \quad \text{and} \quad M_z = xp_y - yp_x = m(x\dot{y} - y\dot{x}) + \frac{eB}{2c}(x^2 + y^2).$$

Moreover, the energy

$$E = \tfrac{1}{2} m(\dot{x}^2 + \dot{y}^2 + \dot{z}^2)$$

is an integral of motion.

b) $E = \tfrac{1}{2} m(\dot{x}^2 + \dot{y}^2 + \dot{z}^2)$, $p_y' = m\dot{y} + eBx/c$, $p_z' = m\dot{z}$ (cf. problem 10.8).

Symmetry considerations allow us to determine various integrals of motion depending on the choice of the vector potential for a given field $\mathbf{B}$. However, the quantities $E, p_z = p_z'$, M_z, and p_y' are all integrals of motion, independent of the choice of $\mathbf{A}$.

4.19. a) $E = \tfrac{1}{2} m(\dot{r}^2 + r^2\dot{\varphi}^2 + \dot{z}^2)$, $p_\varphi = mr^2\dot{\varphi} + \frac{e\mu}{c} \frac{r^2}{(r^2 + z^2)^{3/2}}$, where we use the cylindrical coordinates and the z-axis is taken parallel to the vector $\boldsymbol{\mu}$ (cf. problem 2.35).

b) From the symmetry properties of the given field we can obtain the following integral of motion:

$$p_z = m\dot{z}, \quad M_z \equiv p_\varphi = mr^2\dot{\varphi} + \frac{e\mu}{c}, \quad E = \tfrac{1}{2} m(\dot{r}^2 + r^2\dot{\varphi}^2 + \dot{z}^2).$$

However, the motion in this "field" is free-particle motion. Indeed, the Lagrangian function,

$$L = \tfrac{1}{2} m v^2 + \frac{e\mu}{c} \dot{\varphi},$$

differs from the Lagrangian function of a free particle only by the total derivative with respect to time of the function $e\mu\varphi/c$ (of course, in this case we have $\mathbf{B} = \operatorname{rot} \mathbf{A} = 0$).

We note that in the case when μ is a function of time, p_z and M_z remain integrals of motion.

4.20. a) $\ddot{x} + x = 0$. The same equation could have obtained from the Lagrangian function $L_1(x, \dot{x}) = \dot{x}^2 - x^2$. It is well known that if two Lagrangian functions differ by a total derivative of a function of the coordinates and the time, they lead to the same Lagrangian equations of motion. The reverse statement is incorrect.

b) $\ddot{x} + \alpha \dot{x} + \omega^2 x = 0$.

4.21. a) In the spherical coordinates the Lagrangian equations of motion for a particle moving in a field U are

$$m(\ddot{r} - r\dot{\varphi}^2 \sin^2\theta - r\dot{\theta}^2) + \frac{\partial U}{\partial r} = 0,$$

$$m(r^2\ddot{\theta} + 2r\dot{r}\dot{\theta} - r^2\dot{\varphi}^2 \sin\theta\cos\theta) + \frac{\partial U}{\partial \theta} = 0,$$

$$m(r^2\ddot{\varphi}\sin^2\theta + 2r\dot{r}\dot{\varphi}\sin^2\theta + 2r^2\dot{\theta}\dot{\varphi}\sin\theta\cos\theta) + \frac{\partial U}{\partial \varphi} = 0$$

We can easily write them in the form

$$m(\dot{\mathbf{v}})_i = (\mathbf{F})_i,$$

where the components of the force are the components of $(-\operatorname{grad} U)$:

$$F_r = -\frac{\partial U}{\partial r}, \quad F_\theta = -\frac{1}{r}\frac{\partial U}{\partial \theta}, \quad F_\varphi = -\frac{1}{r\sin\theta}\frac{\partial U}{\partial \varphi}.$$

Hence we have

$$(\dot{\mathbf{v}})_r = \ddot{r} - r\dot{\varphi}^2 \sin^2\theta - r\dot{\theta}^2,$$
$$(\dot{\mathbf{v}})_\theta = r\ddot{\theta} + 2\dot{r}\dot{\theta} - r\dot{\varphi}^2 \cos\theta\sin\theta,$$
$$(\dot{\mathbf{v}})_\varphi = r\ddot{\varphi}\sin\theta + 2\dot{r}\dot{\varphi}\sin\theta + 2r\dot{\theta}\dot{\varphi}\cos\theta.$$

b) $(\dot{\mathbf{v}})_i = \dfrac{1}{2h_i}\left(\dfrac{d}{dt}\dfrac{\partial}{\partial \dot{q}_i} - \dfrac{\partial}{\partial q_i}\right)\dfrac{ds^2}{dt^2} = h_i\ddot{q}_i + \displaystyle\sum_{k=1}^{3}\left(2\dot{q}_i\dot{q}_k\dfrac{\partial h_i}{\partial q_k} - \dot{q}_k^2\dfrac{h_k}{h_i}\dfrac{\partial h_k}{\partial q_i}\right).$

4.22. a) The Lagrangian function

$$L = \tfrac{1}{2} m \sum_{i,k=1}^{3} g_{ik} \dot{q}_i \dot{q}_k - U(q_1, q_2, q_3),$$

with

$$g_{ik} = \sum_{l=1}^{3} \frac{\partial x_l}{\partial q_i} \frac{\partial x_l}{\partial q_k},$$

leads to the equations

$$m \sum_{k=1}^{3} g_{sk} \ddot{q}_k + m \sum_{k,l=1}^{3} \Gamma_{s,kl} \dot{q}_k \dot{q}_l = -\frac{\partial U}{\partial q_s} \quad (s = 1, 2, 3), \tag{1}$$

where the

$$\Gamma_{s,kl} = \tfrac{1}{2} \left(\frac{\partial g_{sk}}{\partial q_l} + \frac{\partial g_{ls}}{\partial q_k} - \frac{\partial g_{kl}}{\partial q_s} \right).$$

are the so-called Christoffel symbols of the first kind.

b) Using the notation $q_4(x, t) = t$, we can proceed as under 4.22a and obtain the same formulae, merely replacing $\sum_1^3$ by $\sum_1^4$.

What is the meaning of the terms in (1) which contain $\Gamma_{1,k4}$ $(k = 1, 2, 3, 4)$ if the q_l are Cartesian coordinates in a rotating frame of reference (see problem 4.8)?

4.23. Since the proposed Lagrangian function differs from the Lagrangian function of the free-moving particles only by the term

$$L_1 = -\frac{eg}{c} \dot{\varphi} \cos\theta$$

the components of the force in the spherical coordinates (see problem 4.21a) have the form

$$F_r = 0,$$

$$F_\theta = \frac{1}{r} \frac{\partial L_1}{\partial \theta} = \frac{eg\dot{\varphi} \sin\theta}{cr},$$

$$F_\varphi = -\frac{1}{r\sin\theta} \frac{d}{dt} \frac{\partial L_1}{\partial \dot{\varphi}} = \frac{eg\dot{\theta}}{cr}$$

and coincide with the components of the Lorentz force

$$\frac{e}{c}[\mathbf{v},\mathbf{B}] = \frac{eg}{cr^3}[\mathbf{v},\mathbf{r}].$$

Since $\partial L/\partial t = 0$, the integral of motion is the energy $E = \frac{1}{2}mv^2$. Similarly, since $\partial L/\partial\varphi = 0$, the integral of motion is the generalized momentum

$$p_\varphi = mr^2\dot\varphi\sin^2\theta - \frac{eg}{c}\cos\theta = \mathcal{J}_z.$$

In the given Lagrangian function, the direction of the z-axis can be chosen arbitrary; this leads to the conservation of the vector

$$\mathbf{J} = m[\mathbf{r},\mathbf{v}] - \frac{eg}{c}\frac{\mathbf{r}}{r}.$$

Besides, the action corresponding to this Lagrangian function is invariant under the similarity transformation. Therefore, the integral of motion is

$$p_r r - 2Et = m\dot r r - 2Et$$

(cf. problem 4.15b).

4.24. The equations of motion are

$$-\mathscr{L}\dot I = \varphi_A - \varphi_B, \quad \frac{q_2}{C} = \varphi_B - \varphi_A.$$

We shall assume that the potential source is a capacitor with a very large capacitance C_0 and that its charge at the moment when $q_1 = 0$ is Q. The energy of the system including the potential source and the inductance is

$$E_0 = \frac{(Q+q_1)^2}{2C_0} + \frac{1}{2}\mathscr{L}\dot q_1^2.$$

Shifting the zero of the energy and considering the limit $Q/C_0 \to U$ as $C_0 \to \infty$, we get

$$E = E_0 - \frac{Q^2}{2C_0} = Uq_1 + \frac{1}{2}\mathscr{L}\dot q_1^2.$$

This is the form of the energy which leads to the Lagrangian function given in the problem.

Similar to this, the energy of a particle of mass m in a uniform force field $-F(t)$ is $\frac{1}{2}m\dot x^2 + Fx$.

The same Lagrangian function may be obtained from that of an electromagnetic field including interactions between an electromagnetic field and charges (see [3], §27 and §28):

$$L = \frac{1}{8\pi} \int (\mathbf{E}^2 - \mathbf{B}^2) \, dV + \frac{1}{c} \int \mathbf{A} \mathbf{j} \, dV - \int \varphi \rho \, dV \qquad (1)$$

(using Gaussian units).

Generally speaking, the electromagnetic field is a system with an infinite number of degrees of freedom. However, the fields inside the capacitor and inductance are specified by the charge q_2 or the current $\dot{q}_1$. Using equations

$$c \operatorname{rot} \mathbf{B} = 4\pi \mathbf{j}, \quad \operatorname{div} \mathbf{E} = 4\pi \rho$$

(and given that the fields are concentrated inside a limited volume), we obtain

$$\int \varphi \rho \, dV = \frac{1}{4\pi} \int \varphi \operatorname{div} \mathbf{E} \, dV =$$

$$= \frac{1}{4\pi} \int \operatorname{div}(\varphi \mathbf{E}) \, dV - \frac{1}{4\pi} \int \mathbf{E} \operatorname{grad} \varphi \, dV = \frac{1}{4\pi} \int \mathbf{E}^2 \, dV,$$

and similarly

$$\frac{1}{c} \int \mathbf{A} \mathbf{j} \, dV = \frac{1}{4\pi} \int \mathbf{B}^2 \, dV,$$

so that

$$L = \frac{1}{8\pi} \int \left(\mathbf{B}^2 - \mathbf{E}^2 \right) dV.$$

Therefore, the Lagrangian function can be expressed in terms of the electric field energy inside a capacitor as

$$\frac{1}{8\pi} \int \mathbf{E}^2 \, dV = \frac{q_2^2}{2C}$$

and in term of the magnetic field energy inside the inductance as

$$\frac{1}{8\pi} \int \mathbf{B}^2 \, dV = \tfrac{1}{2} \mathcal{L} \dot{q}_1^2$$

(see [3], § 2 and § 33).

4.25.   a) $L = \tfrac{1}{2} \mathcal{L} \dot{q}_1^2 - \dfrac{q_2^2}{2C} + U(q_2 - q_1);$

b) $L = \frac{1}{2}\mathscr{L}\dot{q}^2 - \frac{q^2}{2C}$;

c) $L = \frac{1}{2}\mathscr{L}_1\dot{q}_1^2 + \frac{1}{2}\mathscr{L}_2\dot{q}_2^2 - \frac{q_1^2}{2C_1} - \frac{q_2^2}{2C_2} - \frac{(q_1+q_2)^2}{2C}$.

4.26.   a) $L = \frac{1}{2}ml^2\dot{\varphi}^2 + \frac{1}{2}\mathscr{L}\dot{q}^2 + mgl\cos\varphi - \frac{q^2}{2C(\varphi)}$;

b) $L = \frac{1}{2}m\dot{x}^2 + \frac{1}{2}\mathscr{L}(x)\dot{q}^2 - \frac{1}{2}kx^2 + mgx - \frac{q^2}{2C}$.

4.27. Let φ be the angle of rotation of the frame around the AB-axis, such that $\varphi = 0$ gives the direction of the magnetic field; let $\dot{q}$ be the current in the frame (the positive direction is the one from A to D). The Lagrangian function for the system is

$$L = \frac{1}{2}ma^2\dot{\varphi}^2 + \frac{1}{2}\mathscr{L}\dot{q}^2 + Ba^2\dot{q}\sin vfi.$$

The integrals of motion are the energy,

$$E = \frac{1}{2}ma^2\dot{\varphi}^2 + \frac{1}{2}\mathscr{L}\dot{q}^2, \tag{1}$$

and the momentum conjugate to the cyclic coordinate q which is associated with the total magnetic flux through the frame,

$$\frac{\partial L}{\partial \dot{q}} = \mathscr{L}\dot{q} + Ba^2\sin\varphi = \Phi_0.$$

The current through the frame is thus uniquely determined by its position:

$$\dot{q} = \frac{\Phi_0 - Ba^2\sin\varphi}{\mathscr{L}}.$$

Substituting this value $\dot{q}$ in (1) we obtain

$$E = \frac{1}{2}ma^2\dot{\varphi}^2 + U_{\text{eff}}(\varphi), \quad U_{\text{eff}}(\varphi) = \frac{\left(\Phi_0 - Ba^2\sin\varphi\right)^2}{2\mathscr{L}}. \tag{2}$$

The problem of system's motion is thus reduced to a one-dimensional one.

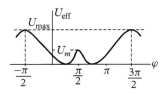

Figure 125

Let us consider the case $0 < \Phi_0 < Ba^2$ in more detail. The function $U_{\mathrm{eff}}(\varphi)$ for this case is given in Fig. 125. One sees that when $E > U_{\max} = (\Phi_0 + Ba^2)^2/(2\mathscr{L})$ the frame rotates and $\dot{\varphi}$ is a periodic time function with period

$$T = \sqrt{2ma} \int\limits_{-\pi/2}^{\pi/2} \frac{d\varphi}{\sqrt{E - U_{\mathrm{eff}}(\varphi)}}.$$

When $U_{\max} > E > (\Phi_0 - Ba^2)^2/(2\mathscr{L}) = U_m$, the frame performs periodic oscillations within the angular interval $\varphi_1 < \varphi < \pi - \varphi_1$, where

$$\varphi_1 = \arcsin \frac{\Phi_0 - \sqrt{2\mathscr{L}E}}{Ba^2};$$

the period tends to infinity as $E \to U_{\max}$ (see problem 1.5). When $0 < E < U_m$ one can have oscillations either in the interval $\varphi_1 < \varphi < \varphi_2$ or in the interval $\pi - \varphi_2 < \varphi < \pi - \varphi_1$, where

$$\varphi_2 = \arcsin \frac{\Phi_0 + \sqrt{2\mathscr{L}E}}{Ba^2}.$$

How will the character of the frame's rotation change if one assume it to have a small resistance?

4.28. a) The equations of motion for the system can be obtained using a Lagrangian function with an additional term responsible for the constraints (see [5], § 2.4)

$$L^* = \tfrac{1}{2}m(\dot{x}^2 + \dot{z}^2) - mgz + \lambda(z - ax^2),$$

where λ is a time-dependent Lagrangian multiplier. The equation of motion

$$m\ddot{x} = -2\,lmax, \tag{1}$$
$$m\ddot{z} + mg = \lambda \tag{2}$$

together with the equation of constraints $z = ax^2$ completely determine the motion of the particle.

On the right-hand side of (1) and (2) there are the components of the reaction forces along the two axes: $R_x = -2\lambda ax$ and $R_z = \lambda$. They can be rewritten in terms of the coordinate and velocity of the particle with the equation of constraints as

$$R_x = -2axR_z, \quad R_z = \frac{(2a\dot{x}^2 + g)m}{1 + 4a^2x^2}.$$

b)
$$m\ddot{r} - mg\cos\varphi - mr\dot{\varphi}^2 = \lambda,$$
$$mr^2\ddot{\varphi} + 2mr\dot{r}\dot{\varphi} + mgr\sin\varphi = 0,$$
$$r = l.$$

The reaction force,

$$\lambda = -mg\cos\varphi - ml\dot{\varphi}^2,$$

lies along **r**.

4.29.

$$L^* = \tfrac{1}{2}m(r^2\dot{\varphi}^2 + \dot{r}^2) + mgr\cos\varphi + \lambda(\varphi - \Omega t);$$

$\lambda = 2mr\dot{r}\Omega + mgr\sin\Omega t$ is the generalized force corresponding to the coordinate φ (the moment of the torque force).

4.30. a) $\dfrac{dE}{dt} = -\dfrac{\partial L}{\partial t} + \displaystyle\sum_{i=1}^{s}\dot{q}_i R_i.$

b) The law of the transformation of the left-hand sides of the equations of motion is given in problem 4.3:

$$\frac{d}{dt}\frac{\partial\widetilde{L}}{\partial\dot{Q}_i} - \frac{\partial\widetilde{L}}{\partial Q_i} = \sum_k \frac{\partial q_k}{\partial Q_i}\left(\frac{d}{dt}\frac{\partial L}{\partial\dot{q}_k} - \frac{\partial L}{\partial q_k}\right).$$

The same law of transformation must be for the right-hand sides:

$$\widetilde{R}_i = \sum_k \frac{\partial q_k}{\partial Q_i}R_k. \tag{1}$$

If the law of transformation of coordinates is not included evidently the time, the velocities are transformed according to the law

$$\dot{q}_i = \sum_k \frac{\partial q_i}{\partial Q_k}\dot{Q}_k,$$

which is reverse to (1).

In other words, the components of the force R_k form a covariant vector, while the velocity components form a contravariant vector in s-dimensional space (see [2], § 83).

Therefore one can find the reaction forces R_i in terms of any generalized coordinates if the constraint and friction forces are known in Cartesian coordinates. In particular, if the friction forces are given in terms of a dissipative function, $R_i = -\dfrac{\partial F}{\partial\dot{q}_i}$, the transformation of F is reduced to a change of variables.

4.31. The equations stated in this problem can be obtained by eliminating the λ_β from the following equations:

$$\frac{d}{dt}\frac{\partial L}{\partial \dot{q}_\beta} - \frac{\partial L}{\partial q_\beta} = \lambda_\beta, \quad \beta = 1, \ldots, r,$$

$$\frac{d}{dt}\frac{\partial L}{\partial \dot{q}_n} - \frac{\partial L}{\partial q_n} = -\sum_{\beta=1}^{r} \lambda_\beta b_{\beta n}, \quad n = r+1, \ldots, ., s.$$

The following relations must be taken into account:

$$\frac{\partial \tilde{L}}{\partial \dot{q}_n} = \frac{\partial L}{\partial \dot{q}_n} + \sum_{\beta=1}^{r} \frac{\partial L}{\partial \dot{q}_\beta} b_{\beta n},$$

$$\frac{\partial \tilde{L}}{\partial q_n} = \frac{\partial L}{\partial q_n} + \sum_{\beta=1}^{r} \sum_{m=r+1}^{s} \frac{\partial L}{\partial \dot{q}_\beta} \frac{\partial b_{\beta m}}{\partial q_n} \dot{q}_m.$$

Thus, the equations of motion for a system with non-holonomous constraints differ from the Lagrangian equations of motion although the constraints equations permit one to eliminate certain coordinates and velocities from the Lagrangian function.

4.32. a) Taking into account that q_n is occur both in L_n and in L_{n+1}, we obtain the Lagrangian equations of motion

$$\frac{d}{dt}\frac{\partial L_n}{\partial q_n} = \frac{\partial L_n}{\partial q_n} + \frac{\partial L_n}{\partial(\Delta q_n)} - \frac{\partial L_{n+1}}{\partial(\Delta q_{n+1})} \quad (\Delta q_n = q_n - q_{n-1}), \tag{1}$$

whence we find as $a \to 0$

$$\frac{1}{a}\frac{\partial L_n}{\partial \dot{q}_n} \to \frac{\partial \mathscr{L}}{\partial(\partial q/\partial t)}, \quad \frac{1}{a}\frac{\partial L_n}{\partial q_n} \to \frac{\partial \mathscr{L}}{\partial q},$$

$$\frac{1}{a}\left[\frac{\partial L_n}{\partial(\Delta q_n)} - \frac{\partial L_{n+1}}{\partial(\Delta q_{n+1})}\right] \to -\frac{\partial}{\partial x}\frac{\partial \mathscr{L}}{\partial(\partial q/\partial x)},$$

so that the equations (1) are reduced to the following equations:

$$\frac{\partial}{\partial t}\frac{\partial \mathscr{L}}{\partial(\partial q/\partial t)} + \frac{\partial}{\partial x}\frac{\partial \mathscr{L}}{\partial(\partial q/\partial x)} = \frac{\partial \mathscr{L}}{\partial q}. \tag{2}$$

Here the derivatives $\partial/\partial t$ and $\partial/\partial x$ should be applied to the functions $q(x, t)$ and its derivatives.

The system of N ordinary differential equations (1) thus transforms to one equation in the partial derivatives (2). For a continuous system the variable x indicates a fixed point on the string.

We do not consider the physical consequences of the (2) since systems with an infinite number of degrees of freedom are the subject of field theory, but not of mechanics (see [5], Ch. 11; [2], § 26 and § 32).

b) $E = \int \left\{ \dfrac{\partial \mathcal{L}}{\partial(\partial q/\partial t)} \dfrac{\partial q}{\partial t} - \mathcal{L} \right\} dx.$

4.33. The Lagrangian function is

$$L = \tfrac{1}{2} m v^2 - U(r) + \frac{e}{c} \mathbf{A}(\mathbf{r}) \mathbf{v},$$

where $\mathbf{A}(\mathbf{r})$ is the vector potential of the magnetic field, $\mathbf{B} = \operatorname{rot} \mathbf{A}$ (of course, $\mathbf{A}(\mathbf{r})$ can always be taken as a homogeneous function of the coordinates of degree $n + 1$). If as a result of the similarity transformation,

$$\mathbf{r} \to \alpha \mathbf{r}, \quad t \to \alpha^{1-k/2} t,$$

the transformation of the vector potential is the same as that for the velocity, that is, $n = -1 + k/2$, we have $L \to \alpha^k L$. Therefore, the equations of motion remain invariant after this transformation and the principle of mechanical similarity holds (see [1], § 10).

It is clear that the principle of mechanical similarity remains valid for a magnetic field if it is constant in space and if its value changes by a factor $\alpha^{-1+k/2}$ under a similarity transformation (e.g. see, problems 2.34– 2.37 and 6.37).

4.34. The kinetic energy of the system is $T = \sum_a \tfrac{1}{2} m_a v_a^2$, so that

$$2T = \sum_a \frac{\partial T}{\partial \mathbf{v}_a} \mathbf{v}_a = \frac{d}{dt} \left(\sum_a m_a \mathbf{v}_a \mathbf{r}_a \right) - \sum_a \mathbf{r}_a m_a \dot{\mathbf{v}}_a.$$

The first term $\frac{d}{dt} \left(\sum_a m_a \mathbf{v}_a \mathbf{r}_a \right)$, which is the total time derivative with respect of time of a bounded function becomes zero after averaging over large time interval (see [1], § 10 and [7], § 7). Substituting

$$m_a \dot{\mathbf{v}}_a = -\frac{\partial U}{\partial \mathbf{r}_a} + \frac{e_a}{c} [\mathbf{v}_a, \mathbf{B}]$$

into the second term and averaging over the time we obtain

$$\left\langle 2T + \mathbf{B} \sum_a \frac{e_a}{c} [\mathbf{r}_a, \mathbf{v}_a] \right\rangle = k \langle U \rangle,$$

where the pointed brackets $\langle\ \rangle$ indicate the time averaging. In particular, if the magnetic field **B** is homogeneous and constant, then

$$2\langle T\rangle = k\langle U\rangle - 2\langle\mu\rangle\mathbf{B},$$

where

$$\langle\mu\rangle = \left\langle\frac{1}{2c}\sum_a e_a[\mathbf{r}_a,\mathbf{v}_a]\right\rangle$$

is the average magnetic moment of the particle system. If $\frac{e_a}{m_a} = \frac{e}{m}$, then $\mu = \frac{e}{2mc}\mathbf{M}$, where **M** is the angular momentum of the system.

4.35. a) Write $\frac{dA}{dt}$ in the form of two terms

$$\frac{dA}{dt} = [m(\ddot{x}_1\dot{x}_2\dot{x}_3 + \dot{x}_1\ddot{x}_2\dot{x}_3 + \dot{x}_1\dot{x}_2\ddot{x}_3) - \dot{x}_1\dot{U}_{23} - \dot{x}_2\dot{U}_{13} - \dot{x}_3\dot{U}_{12}] -$$
$$- (\ddot{x}_1 U_{23} + \ddot{x}_2 U_{13} + \ddot{x}_3 U_{12}). \tag{1}$$

Using the equations of motion

$$m\ddot{x}_1 = F_{12} + F_{13}, \quad m\ddot{x}_2 = F_{21} + F_{23}, \quad m\ddot{x}_3 = F_{31} + F_{32}, \tag{2}$$

$$F_{ik} = -F_{ki} = -\frac{\partial U_{ik}}{\partial x_i} = \frac{2g^2}{(x_i - x_k)^3} \tag{3}$$

and introducing the relative distances

$$x_1 - x_2 = x, \quad x_2 - x_3 = y, \quad x_1 - x_3 = z,$$

write the second term of (1) as

$$-\frac{2g^4}{m}\left[\left(\frac{1}{x^3} + \frac{1}{z^3}\right)\frac{1}{y^2} + \left(-\frac{1}{x^3} + \frac{1}{y^3}\right)\frac{1}{z^2} - \left(\frac{1}{z^3} + \frac{1}{y^3}\right)\frac{1}{x^2}\right]. \tag{4}$$

Combine the terms with the same powers of z, we rewrite (4) as

$$\frac{2g^4(x-y)}{m}\left(\frac{1}{x^3y^3} - \frac{x^2 + xy + y^2}{z^2x^3y^3} - \frac{x+y}{z^3x^2y^2}\right).$$

Substituting $z = x + y$, it is easy to prove that this expression becomes zero.

The first term of (1) is zero for arbitrary forces. To show this, it is sufficient to use the equations of motion in form (2) and substitute

$$\dot{U}_{ik} = \frac{\partial U_{ik}}{\partial(x_i - x_k)}(\dot{x}_i - \dot{x}_k) = -(\dot{x}_i - \dot{x}_k)F_{ik}.$$

Finally, note that in this field the similarity transformation does not change an action, and, as a consequence, the is the fourth (additional) integral of motion (see problem 4.15b)

$$m(x_1\dot{x}_1 + x_2\dot{x}_2 + x_3\dot{x}_3) - 2Et. \tag{5}$$

4.36. When two particles approache each other, their interaction energy becomes infinite. As a result, the particles can not pass "one through the other" and the order of their location on the line is conserved.

When two particles of equal mass but with arbitrary interaction energy (only allowing the expulsion of particles) collide, these particles simply exchange velocities. (This is follows from the laws of conservation of energy and momentum.) If the collisions of three particles occur one after another, so that during the convergence of two particles the third particle is far away from them, it will simply lead to the exchange of velocities. Thus the collisions will end when the fastest particle is ahead and the slowest one is behind, in other words, in this case

$$v_1' = v_3, \quad v_2' = v_2, \quad v_3' = v_1. \tag{1}$$

In general, when all three particles simultaneously converge, the values of the velocities will not be conserved.

The more surprising is that for the forces specified in the preceding problem, the result (1) is conserved. This can be shown using all three integrals of motion: P, E, and A. Taking into account that the functions $U_{ik} \to 0$ as $t \to \pm\infty$ and comparing P, E, and A at $t \to +\infty$ and at $t \to -\infty$, we obtain three equations:

$$\begin{aligned} v_1' + v_2' + v_3' &= v_1 + v_2 + v_3, \\ (v_1')^2 + (v_2')^2 + (v_3')^2 &= v_1^2 + v_2^2 + v_3^2, \\ v_1'v_2'v_3' &= v_1v_2v_3. \end{aligned} \tag{2}$$

Solving this system with respect to v_i', we get, generally speaking, six different solutions. However, all these solutions can be guessed.

It is easy to verify that the solution (1) satisfies the system (2). Furthermore, since the equations (2) are obviously symmetric with respect to all six possible permutations of the particles, it is clear that the remaining roots of the set (2) can be obtained by simple permutations from (1).

After that, it is easily to prove that only the result (1) can be realized at $t \to +\infty$, since any other options imply the possibility of the further collisions of the particles (due to inequalities $v_3 > v_2 > v_1$).

§5

Small oscillations of systems with one degree of freedom

5.1. a) $\omega^2 = \frac{V\alpha^2}{m}\sqrt{1-\left(\frac{F}{V\alpha}\right)^2}$; a minimum in $U(x)$ occurs when $F < V\alpha$;

b) $\omega^2 = \frac{8\pi}{3}\frac{V\alpha^4}{m}x_0^2\left(\frac{\Gamma(3/4)}{\Gamma(1/4)}\right)^2$, where the amplitude x_0 is determined from the equality

$$E = \tfrac{1}{2}m\dot{x}^2 + \tfrac{1}{3}V\alpha^4 x^4 = \tfrac{1}{3}V\alpha^4 x_0^4.$$

5.2. The Lagrangian function of the system is (see [1], § 5, problem 4)

$$L = ma^2[\dot{\theta}^2(1+2\sin^2\theta) + \Omega^2\sin^2\theta + 2\Omega_0^2\cos\theta],$$

where we have introduce $\Omega_0^2 = 2g/a$.

When $\Omega > \Omega_0$ the potential energy of the system,

$$U(\theta) = -ma^2(\Omega^2\sin^2\theta + 2\Omega_0^2\cos\theta),$$

has a minimum when $\cos\theta_0 = \Omega_0^2/\Omega^2$. Expanding $U(\theta)$ in the neighbourhood of θ_0, and in the kinetic energy putting

$$1 + 2\sin^2\theta = 1 + 2\sin^2\theta_0 = 3 - 2\left(\frac{\Omega_0}{\Omega}\right)^4 \equiv \frac{M}{2ma^2},$$

we get

$$L = \tfrac{1}{2}m\dot{x}^2 - \tfrac{1}{2}kx^2,$$

Exploring Classical Mechanics: A Collection of 350+ Solved Problems for Students, Lecturers, and Researchers. First Edition.
Gleb L. Kotkin and Valeriy G. Serbo, Oxford University Press (2020). © Gleb L. Kotkin and Valeriy G. Serbo 2020.
DOI: 10.1093/oso/9780198853787.001.0001

where $k = U''(\theta_0)$, $x = \theta - \theta_0$. Hence we get

$$\omega^2 = \frac{k}{M} = \Omega^2 \frac{\Omega^4 - \Omega_0^4}{3\Omega^4 - 2\Omega_0^4} \quad (\Omega > \Omega_0).$$

When $\Omega \gg \Omega_0$, the oscillation frequency is proportional to the angular velocity of rotation, $\omega = \Omega/\sqrt{3}$ and $\theta_0 = \pi/2$; when $\Omega \to \Omega_0$, small oscillations occur with a frequency $\omega \to 0$ and $\theta_0 \to 0$.

If $\Omega < \Omega_0$, we can consider oscillations near $\theta_0 = 0$ for elastic collisions of the lateral particles

$$\omega^2 = \Omega_0^2 - \Omega^2 \quad (\Omega < \Omega_0).$$

If $\Omega = \Omega_0$, the potential energy U has a minimum at $\theta_0 = 0$ and in the neighbourhood of it, we can write

$$U = ma^2 \Omega_0^2 \left(-2 + \tfrac{1}{4}\theta^4 \right),$$

that is, the oscillations are non-linear in an essential way. Retaining also in the kinetic energy terms up to the fourth order, we get

$$\frac{2\pi}{\omega} = T = 4 \int_0^{\theta_m} \frac{\sqrt{1 + 2\theta^2}\, d\theta}{\sqrt{\Omega_0^2(\theta_m^4 - \theta^4)}}.$$

Here θ_m is the amplitude of the oscillations (see [1], § 11, problem 2a).

5.3. The Lagrangian function of the system is

$$L = \tfrac{1}{2} mR^2 \dot{\varphi}^2 - U(\varphi), \quad U(\varphi) = 2mgR \left[\sin^2 \frac{\varphi}{2} + \frac{2s^3}{\sin(\varphi/2)} \right],$$

where $s = \left(\dfrac{q^2}{8mgR^2} \right)^{1/3}$.

When $s < 1$, the stable equilibrium position φ_0 is determined by the condition $\sin(\varphi_0/2) = s$, and $\omega^2 = \dfrac{3g}{R}(1 - s^2)$.

When $s > 1$, the point A is a position of stable equilibrium and $\omega^2 = \dfrac{g}{R}(s^3 - 1)$.

When $s = 1$, the potential energy $U(\varphi)$ has a minimum at $\varphi_0 = \pi$ and in the neighbourhood of it

$$U(\varphi) = \frac{3}{32} mgR(\varphi - \pi)^4 + 6mgR,$$

that is, the oscillations are non-linear in an essential way. As a result, we get

$$\frac{2\pi}{\omega} = T = 16\sqrt{\frac{R}{3g}} \int_0^{\varphi_m} \frac{d\alpha}{\sqrt{\varphi_m^4 - \alpha^4}},$$

where φ_m is the amplitude of the oscillations (see [1], § 11, problem 2a).

5.4. $r = r_0 + a\cos\omega(t - t_0),$

$$\varphi = \varphi_0 + \Omega(t - t_0) - \frac{2a\Omega}{r_0\omega} \sin[\omega(t - t_0)],$$

where r_0, φ_0, a, and t_0 are integration constants $(a \ll r_0)$, and

$$\Omega = \sqrt{\frac{n\alpha}{m}}\, r_0^{-(n+2)/2}, \quad \omega = \Omega\sqrt{2 - n}.$$

5.5. At the point $\theta = \theta_0$ the effective potential energy

$$U_{\mathrm{eff}}(\theta) = \frac{M_z^2}{2ml^2 \sin^2\theta} - mgl\cos\theta$$

(see [1], § 14, problem 1) has a minimum, so that $U'_{\mathrm{eff}}(\theta_0) = 0$. We thus obtain

$$M_z^2 = \frac{m^2 l^3 g \sin^4\theta_0}{\cos\theta_0},$$

and the frequency of small oscillations is

$$\omega(\theta_0) = \sqrt{\frac{U''_{\mathrm{eff}}(\theta_0)}{ml^2}} = \sqrt{\frac{g}{l} \frac{1 + 3\cos^2\theta_0}{\cos\theta_0}}.$$

For this calculation to be the applicable the condition

$$\tfrac{1}{2} U''_{\mathrm{eff}}(\theta_0)(\Delta\theta)^2 \gg \tfrac{1}{6} |U'''_{\mathrm{eff}}(\theta_0)|(\Delta\theta)^3,$$

with $\Delta\theta$ the oscillation amplitude, must be satisfied. If $\theta_0 \sim 1$, it is satisfied for $\Delta\theta \ll 1$. If, however, $\theta_0 \ll 1$, we have $U'''_{\mathrm{eff}}(\theta_0) \propto 1/\theta_0$, and the oscillations in θ can be considered to be small only when $\Delta\theta \ll \theta_0$. The result obtained, $\omega = 2\sqrt{g/l}$, is nevertheless valid also for $\Delta\theta \sim \theta_0$ when the oscillations in θ are no longer harmonic. Indeed, in this case small harmonic oscillations with frequency $\sqrt{g/l}$ occur along the x- and y-axes, that is, the pendulum moves along an ellipse executing two oscillations in the angle θ for each revolution (see [1], § 23, problem 3 and [7], § 4).

5.6. The effective potential energy for radial oscillations of the molecule is

$$U_{\text{eff}}(r) = \tfrac{1}{2} m\omega_0^2 (r - r_0)^2 + \frac{M^2}{2mr^2},$$

where r is the distance between the atoms and m the reduced mass. The extra term which is assumed to be a small correction leads to a small shift in the equilibrium position

$$\delta r_0 = \frac{M^2}{m^2 \omega_0^2 r_0^3}.$$

We determine the shift in the frequency by expanding $U_{\text{eff}}(r)$ in a series near the point $r_0 + \delta r_0$:

$$U_{\text{eff}}(r) = \tfrac{1}{2} m\omega_0^2 (r - r_0 - \delta r_0)^2 + \frac{M^2}{2mr_0^2} + \frac{3M^2}{2mr_0^4}(r - r_0 - \delta r_0)^2.$$

We get from this a correction to the frequency

$$\delta\omega = \frac{3M^2}{2m^2 \omega_0 r_0^4} = \frac{3\Omega^2}{2\omega_0},$$

where $\Omega = M/(mr_0^2)$ is the average angular velocity of the molecule rotation.

5.7. a) We get for the displacement from the equilibrium position:

$$x = x_0 \cos\omega t + \frac{\dot{x}_0}{\omega} \sin\omega t = \sqrt{x_0^2 + \frac{\dot{x}_0^2}{\omega^2}} \cos(\omega t + \varphi),$$

$$\tan\varphi = -\frac{\dot{x}_0}{\omega x_0}, \qquad \omega^2 = \frac{2k}{m}.$$

b) Let the tension in each spring be f. For small displacements $|y| \ll \sqrt{fl/k}$, where l is the separation between the points where the springs are fixed, the oscillations are harmonic $y = A\cos(\omega t + \varphi)$, and $\omega^2 = 2f/(ml)$.

When $f = kl$, the oscillation frequency is equal to that in part 5.7a. In the case of springs in which there is no tension ($f = 0$) the oscillations are non-linear, the restoring force is $F = -ky^3/l^2$, and the frequency (cf. problem 5.1b) is

$$\omega = \frac{\sqrt{\pi}\,\Gamma(3/4)}{\Gamma(1/4)} \sqrt{\frac{2k}{m}} \frac{y_m}{l},$$

where y_m is the amplitude of the oscillations.

If the particle can move in the xy-plane, its motion – for the case when $f \neq 0$ and x and y are small – consists of harmonic oscillations along the x- and y-axes with frequencies $\omega_x^2 = 2k/m$ and $\omega_y^2 = 2f/(ml)$, respectively (see problem 6.3).

5.8. Let y be the coordinate of the particle reckoned from the upper suspension point, and $2l$ the distance between the two suspension points. The Lagrangian function of the system,

$$L = \tfrac{1}{2} m\dot{y}^2 - k(y-l)^2 + mgy = \tfrac{1}{2} m\dot{y}^2 - k\left(y-l-\frac{mg}{2k}\right)^2 + \text{const},$$

describes a harmonic oscillator with frequency $\omega^2 = 2k/m$ and equilibrium position $y_0 = l + mg/(2k)$, so that $y = y_0 + A\cos(\omega t + \varphi)$.

If we take the displacement from the equilibrium position as coordinate, we can eliminate the gravity field from the Lagrangian function.

5.9. For the angle between the pendulum and the vertical we have

$$\varphi = \frac{a\Omega^2}{g - l\Omega^2}\sin\Omega t, \qquad \left|\frac{a\Omega^2}{g - l\Omega^2}\right| \ll 1$$

(see also problem 8.3).

It is also possible the oscillations of the pendulum near the direction of the radius vector

$$\varphi = \Omega t + \frac{g}{a\Omega^2}\sin\Omega t, \qquad \Omega^2 \gg \frac{g}{a}.$$

5.10. The result for the current in circuit

$$I = \frac{dq}{dt} = \frac{U_0\sin(\omega t - \varphi)}{\sqrt{R^2 + (\omega\mathcal{L} - 1/(\omega C))^2}}, \qquad \tan\varphi = \frac{\omega\mathcal{L} - 1/(\omega C)}{R}$$

can be obtained by solving the Lagrangian equations of motion for q. The Lagrangian function of the system is

$$L = \tfrac{1}{2}\mathcal{L}\dot{q}^2 - \frac{q^2}{2C} + qU$$

(see problem 4.25), and the dissipative function is equal to $\tfrac{1}{2}R\dot{q}^2$ (see [3], §48).

5.11. The general solution of the equation of motion (see [1], §26)

$$\ddot{x} + 2\lambda\dot{x} + \omega_0^2 x = \frac{F}{m}\cos\gamma t$$

under such conditions that $\omega^2 = \omega_0^2 - \lambda^2 > 0$, has the form

$$x(t) = e^{-\lambda t}(a\cos\omega t + b\sin\omega t) +$$

$$+ \frac{F[(\omega_0^2 - \gamma^2)\cos\gamma t + 2\lambda\gamma\sin\gamma t]}{m[(\omega_0^2 - \gamma^2)^2 + 4\lambda^2\gamma^2]},$$

where a and b are constants to be determined from the initial conditions. If we put $x(0) = \dot{x}(0) = 0$, we find finally

$$x(t) = \frac{F}{m[(\omega_0^2 - \gamma^2)^2 + 4\lambda^2\gamma^2]}\left[(\omega_0^2 - \gamma^2)(\cos\gamma t - e^{-\lambda t}\cos\omega t) + \right.$$

$$\left. + 2\lambda\gamma\left(\sin\gamma t - \frac{\omega_0^2 + \gamma^2}{2\gamma\omega}e^{-\lambda t}\sin\omega t\right)\right]. \qquad (1)$$

Let us study this solution in the region near the resonance, $\gamma = \omega_0 + \varepsilon$, $|\varepsilon| \ll \omega_0$. If there is no friction at all, that is, $\lambda = 0$, the motion of the oscillator near resonance will show beats:

$$x = \frac{F}{m\omega_0\varepsilon}\sin(\varepsilon t/2)\cdot\sin\omega_0 t, \qquad (2)$$

and the amplitude and the frequency of the beats are determined by how near resonance we are (Fig. 126a). However, when $\gamma = \omega_0$, that is, at exact resonance, we get by letting $\varepsilon \to 0$:

$$x = \frac{F}{2m\omega_0}t\sin\omega_0 t, \qquad (3)$$

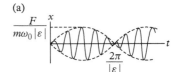

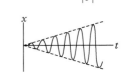

Figure 126

that is, we get oscillations with an amplitude $a(t)$ which increases indefinitely $a(t) = Ft/(2m\omega_0)$ (Fig. 126b).

When there is even a small amount of friction ($\lambda \ll \omega_0$) the character of the motion changes qualitatively. From (1) we get easily in the case when $\lambda \ll |\varepsilon|$ instead of (2)

$$x(t) = \frac{F}{2m\omega_0\varepsilon}\sqrt{1 - 2e^{-\lambda t}\cos\varepsilon t + e^{-2\lambda t}}\cos(\omega_0 t + \varphi_1(t)). \qquad (4)$$

Here $\varphi_1(t)$ is the phase of the oscillations which changes slowly with time. The amplitude of the oscillation oscillates slowly with a frequency $|\varepsilon|$ about the value $F/(2m\omega_0|\varepsilon|)$, gradually approaching that value (Fig. 127a). It is remarkable that during the transient stage the amplitude may reach twice the value of the amplitude of the steady-state oscillations.

When $|\varepsilon| \ll \lambda \ll \omega_0$,

$$x = \frac{F}{2m\omega_0\lambda}(1 - e^{-\lambda t})\sin\omega_0 t. \qquad (5)$$

(a)

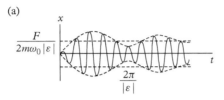

(b)

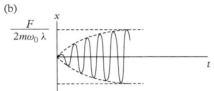

(c)

Figure 127

In that case we have the transient process with a smoothly increasing amplitude which asymptotically approaches the value $F/(2m\omega_0\lambda)$ which is determined by the friction coefficient λ (Fig. 127b). Finally, if the quantities ε and λ are of the same order of magnitude, $|\varepsilon| \sim \lambda \ll \omega_0$, we get oscillations of the amplitude around the value $F/(2\sqrt{2}m\omega_0|\varepsilon|)$ which corresponds to the steady-state oscillations which are reached very slowly (see Fig. 127c for the case $\varepsilon \approx \lambda$).

The system thus proceeds to steady-state oscillations for these three cases (Fig. 127) over a time of the order of $t \sim 1/\lambda$, as is clear from (1).

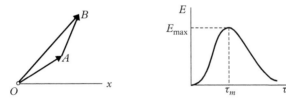

Figure 128   Figure 129

One can use a vector diagram (see Fig. 128) to study qualitative how the oscillations proceeds to a steady state (transient process) when $\lambda \ll \omega_0$. The forced oscillation is depicted by the component of the vector $\overrightarrow{OA}$, which rotates with an angular velocity γ, onto the x-axis. The vector of free oscillations, $\overrightarrow{AB}$, rotates with an angular velocity ω, and its length decreases as $e^{-\lambda t}$. In the initial moment $\overrightarrow{AB} + \overrightarrow{OA} \approx 0$.

What is the nature of the transient process, if $x(0) = 0$, $\dot{x}(0) \neq 0$?

5.12. a) The energy acquired by the oscillator,

$$E = \frac{\pi F^2}{2m} \tau^2 \exp\left[-\tfrac{1}{2}(\omega\tau)^2\right]$$

depends on how fast the force is switched on (i.e., on the parameter $\omega\tau$). For an instantaneous impact ($\omega\tau \ll 1$) or a very slow switching on, ($\omega\tau \gg 1$), the energy transfer is small, while the maximum energy transfer, $E_{\max} = \pi F^2/(m\omega^2 e)$, is reached when $\tau_m = \sqrt{2}/\omega$ (Fig. 129).

b) If $x \to a\cos(\omega t + \varphi)$, as $t \to -\infty$,[1] we have

$$\Delta E = E(+\infty) - E(-\infty) =$$

$$= \frac{\pi F^2}{2m} \tau^2 e^{-(\omega\tau)^2/2} - \sqrt{\pi}\, a\omega\tau F e^{-(\omega\tau)^2/4} \sin\varphi.$$

Depending on the value of φ, the oscillator gains or loses energy. This change in energy is similar to the absorption or stimulated emission of light by an atom.

When we average over the phase φ we get the same result as in point 5.12a.

5.13. a) $x(t) = \frac{1}{2\mu}[\xi_1(t) - \xi_2(t)]$, where

$$\xi_{1,2} = e^{\pm\mu t}\left[\int\limits_0^t \frac{1}{m} F(\tau)e^{\mp\mu\tau}\, d\tau + \dot{x}_0 \pm \mu x_0\right].$$

b) $x(t) = \frac{1}{\omega}\,\mathrm{Im}\left\{e^{i\omega t - \lambda t}\left[\int\limits_0^t \frac{1}{m} F(\tau)e^{\lambda\tau - i\omega\tau}\, d\tau + \dot{x}_0 + (i\omega + \lambda)x_0\right]\right\},$

where $\omega = \sqrt{\omega_0^2 - \lambda^2}$.

5.14. The force

$$\mathbf{F}(t) = -\frac{\partial}{\partial\mathbf{r}} U(|\mathbf{r} - \mathbf{r}_0(t)|), \tag{1}$$

---

[1] The meaning of φ is that of the "impact phase", that is, the phase which the oscillator would have had at $t = 0$, if there were no force acting upon it.

is acting upon the oscillator; here $\mathbf{r}(t)$ is the deflection of the oscillator and $\mathbf{r}_0(t)$ the radius vector of the impinging particle. Assuming that the particle is little deflected, we can put $\mathbf{r}_0(t) = \boldsymbol{\rho} + \mathbf{v}t$ with $\boldsymbol{\rho}$ being the impact parameter (vectors $\boldsymbol{\rho}$ and $\mathbf{v}$ are mutually orthogonal). Assuming also that the amplitude of the vibrations of the oscillator is small we can put $\mathbf{r} = 0$ in (1) (after differentiating) and then

$$\mathbf{F}(t) = -2\varkappa^2 V(\boldsymbol{\rho} + \mathbf{v}t) \cdot \exp(-\varkappa^2\rho^2 - \varkappa^2 v^2 t^2).$$

The oscillations along $\boldsymbol{\rho}$ and $\mathbf{v}$ are independent and the corresponding energy excitations are equal to

$$\frac{1}{2m} \left| \int_{-\infty}^{+\infty} F_\rho(t) e^{-i\omega t} \, dt \right|^2 \quad \text{and} \quad \frac{1}{2m} \left| \int_{-\infty}^{+\infty} F_v(t) e^{-i\omega t} \, dt \right|^2, \tag{2}$$

where F_v and F_ρ are the components of $\mathbf{F}$ in the directions of $\mathbf{v}$ and $\boldsymbol{\rho}$. The total energy excitation of the oscillator is

$$\varepsilon = \varepsilon_1 \left(1 + \frac{x}{a}\right) e^{-x}, \quad \varepsilon_1 = \frac{\pi V^2}{2E} a e^{-a}, \tag{3}$$

where

$$E = \tfrac{1}{2} m v^2, \quad a = \tfrac{1}{2} \left(\frac{\omega}{\varkappa v}\right)^2, \quad x = 2(\varkappa\rho)^2.$$

The cross-section for exciting the oscillator to an energy between ε and $\varepsilon + d\varepsilon$ is

$$d\sigma = \pi \sum_k |d\rho_k^2| = \frac{\pi}{2\varkappa^2} \frac{d\varepsilon}{\varepsilon} \sum_k \left| \frac{a + x_k(\varepsilon)}{1 - a - x_k(\varepsilon)} \right|, \tag{4}$$

where the x_k are the roots of (3).

For a further consideration, it is most helpful to solve (3) graphically, in the same way as was done in problem 3.11a. When $\varepsilon \ll \varepsilon_1$, we get $d\sigma = \frac{\pi}{2\varkappa^2} \frac{d\varepsilon}{\varepsilon}$ (in (4) we assume that $x_k(\varepsilon) \gg 1$, $x_k \gg a$). For large ε the result depends on the value of a. If $a > 1$, we can only have $\varepsilon < \varepsilon_1$ (see Fig. 130a; for the cross-section see Fig. 130a). However, if $a < 1$ we can have $\varepsilon < \varepsilon_2 = \frac{\pi V^2}{2Ee}$ (see Fig. 131b) and at $\varepsilon = \varepsilon_1$ the function $d\sigma/d\varepsilon$ will have a discontinuity while for $\varepsilon_2 - \varepsilon \ll \varepsilon_2$ it has an integrable singularity,

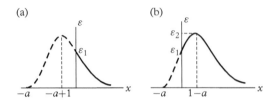

Figure 130

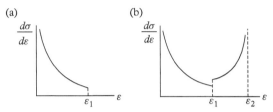

Figure 131

$$d\sigma = \frac{\pi}{\sqrt{2}\varkappa^2} \frac{d\varepsilon}{\varepsilon_2} \frac{1}{\sqrt{1 - \varepsilon/\varepsilon_2}}$$

(see Fig. 131b).

5.15. If the oscillator has an "impact phase" φ (see problem 5.12b), we get, by repeating the calculations of the preceding problem, for the energy of the oscillator the expression (ε_0 is the initial energy of the oscillator)

$$\varepsilon = \varepsilon_1 e^{-2(\varkappa\rho)^2} + 2\sqrt{\varepsilon_1\varepsilon_0}e^{-(\varkappa\rho)^2}\cos\varphi + \varepsilon_0, \tag{1}$$

where

$$\varepsilon_1 = \frac{\pi}{4E}\left(\frac{V\omega}{\varkappa v}\right)^2 \exp\left[-\frac{\omega^2}{2(\varkappa v)^2}\right].$$

When $\cos\varphi > 0$, we have $\varepsilon > \varepsilon_0$ for all ρ, while we can also have values $\rho_{1,2}$ such that $\varepsilon < \varepsilon_0$, if $\cos\varphi < 0$. Solving (1) for ρ^2, we find

$$\rho^2 = \frac{1}{\varkappa^2}\ln\frac{\sqrt{\varepsilon_1/\varepsilon_0}}{-\cos\varphi + \sqrt{(\varepsilon/\varepsilon_0) - \sin^2\varphi}}, \qquad \text{when } \varepsilon > \varepsilon_0,$$

$$\rho_{1,2}^2 = \frac{1}{\varkappa^2}\ln\frac{\sqrt{\varepsilon_1}}{\sqrt{\varepsilon_0}|\cos\varphi| \pm \sqrt{\varepsilon - \varepsilon_0\sin^2\varphi}}, \qquad \text{when } \cos\varphi < 0$$

and $\varepsilon_0 > \varepsilon > \varepsilon_{\min} = \varepsilon_0\sin^2\varphi$.

Hence we have

$$do = \pi \left| \frac{d\rho^2}{d\varepsilon} \right| d\varepsilon = \frac{\pi}{2\varkappa^2} \frac{d\varepsilon}{\varepsilon - \varepsilon_0 \sin^2 \varphi - \cos \varphi \cdot \sqrt{\varepsilon \varepsilon_0 - \varepsilon_0^2 \sin^2 \varphi}}, \tag{2}$$

when $\varepsilon > \varepsilon_0$, and

$$do = \pi d(-\rho_1^2 + \rho_2^2) = \frac{1}{\varkappa^2} \frac{\varepsilon_0 \pi |\cos \varphi| \, d\varepsilon}{(\varepsilon_0 - \varepsilon) \sqrt{\varepsilon \varepsilon_0 - \varepsilon_0^2 \sin^2 \varphi}} \tag{3}$$

when $\varepsilon_{\min} < \varepsilon < \varepsilon_0$ and $\cos \varphi < 0$.

Averaging over all possible phases φ for a given ε, we obtain

$$\left\langle \frac{do}{d\varepsilon} \right\rangle = \frac{\pi}{2\varkappa^2 |\varepsilon_0 - \varepsilon|} \tag{4}$$

(Fig. 132). The averaging is performed, using the formulae

$$\left\langle \frac{do}{d\varepsilon} \right\rangle = \frac{1}{2\pi} \int_0^{2\pi} \frac{do}{d\varepsilon} \, d\varphi,$$

when $\varepsilon > \varepsilon_0$ and

$$\left\langle \frac{do}{d\varepsilon} \right\rangle = \frac{1}{2\pi} \int_{\pi-\alpha}^{\pi+\alpha} \frac{do}{d\varepsilon} \, d\varphi,$$

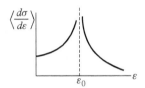

Figure 132

when $\varepsilon < \varepsilon_0$. Here $\alpha = \arcsin \sqrt{\varepsilon/\varepsilon_0}$.

The singularity of the cross-sections (2), (3), and (4), as $\varepsilon \to \varepsilon_0$, is connected with the fact that the oscillator is excited for any, however large, ρ.

What is the case of the additional singularity in (3) and why does it not appear in (4)?

5.16. We have the Lagrangian function

$$L = \tfrac{1}{2} m\dot{x}^2 - \tfrac{1}{2} m\omega^2 x^2 + xF(t).$$

We then have for the energy of the system

$$E(t) = \tfrac{1}{2} m(\mathrm{Re}\,\xi)^2 + \tfrac{1}{2} m(\mathrm{Im}\,\xi)^2 - \frac{F(t)}{\omega} \mathrm{Im}\,\xi = \tfrac{1}{2} m \left| \xi - \frac{iF(t)}{m\omega} \right|^2 - \frac{F^2(t)}{2m\omega^2}$$

where

$$\xi = \dot{x} + i\omega x = e^{i\omega t} \int_{-\infty}^{t} e^{-i\omega\tau} \frac{1}{m} F(\tau)\, d\tau \qquad (1)$$

(see [1], § 22). Although the expression for the energy has a well-definite limit as $t \to \infty$, the integral defining $\xi(t)$ has no limit as $t \to \infty$ (since $F(\tau) \to F_0$ as $\tau \to \infty$). Integrating (1) by parts, we obtain

$$\xi(t) = \frac{iF(t)}{m\omega} - \frac{ie^{i\omega t}}{m\omega} \int_{-\infty}^{t} F'(\tau) e^{-i\omega\tau}\, d\tau, \qquad (2)$$

where $F'(\tau) \to 0$, as $\tau \to \infty$ and the integral converges as $t \to \infty$. It is clear from (2) that as $t \to \infty$ the motion of the oscillator in this case is a harmonic oscillation (the second term in (2)), around a new equilibrium position $x_0 = F_0/(m\omega^2)$ (the first term in (2)). The energy transferred to the oscillator is in accordance with this given by

$$E(+\infty) = \frac{F_0^2}{2m\omega^2} + \frac{1}{2m\omega^2} \left| \int_{-\infty}^{\infty} F'(t) e^{-i\omega t}\, dt \right|^2 .$$

5.17.

$$\Delta E = E(+\infty) - E(-\infty) = \frac{F_0^2}{2m\omega^2} + \frac{\lambda^4 F_0^2}{2m\omega^2(\lambda^2 + \omega^2)^2} - \frac{a\lambda^2 F_0 \cos\varphi}{\lambda^2 + om^2},$$

$E_0 = \frac{1}{2} m\omega^2 a^2$ and φ is the "impact phase" (see footnote to problem 5.12b).

5.18. If in the formula (see [1], § 22)

$$\xi(\tau) = \xi(0)e^{i\omega\tau} + e^{i\omega\tau} \int_{0}^{\tau} \frac{F(t)}{m} e^{-i\omega t}\, dt$$

we integrate by parts n times, we get

$$\xi(\tau) = \xi(0)e^{i\omega\tau} + \frac{iF_0}{m\omega} + \frac{F^{(n)}(+0)e^{i\omega\tau} - F^{(n)}(\tau-0)}{m(i\omega)^{(n+1)}} +$$

$$+ \frac{e^{i\omega\tau}}{m(i\omega)^{(n+1)}} \int_{0}^{\tau} F^{(n+1)}(t) e^{-i\omega t}\, dt.$$

Here $|\xi(0)| = a_0\omega$, where a_0 is the amplitude of the oscillations at the moment the force is switched on. The penultimate term in this formula is of order of magnitude $\frac{F_0}{m\omega}(\omega\tau)^{-n}$, while the last tem is, generally speaking, considerable smaller – provided $F^{(n+1)}(t)$ changes smoothly. The square of the amplitude of the oscillations

$$\frac{1}{\omega^2}\left|\xi - \frac{iF_0}{m\omega}\right|^2$$

is for $t > \tau$ of the order of magnitude

$$\left(a_0 + \frac{F_0}{m\omega^2}\frac{1}{(\omega\tau)^n}\right)^2.$$

Thus, if the force $F(t)$ is switched on slowly and smoothly, the energy transferred is very small.

5.19. a) During the time interval $0 \leqslant t \leqslant \tau$ the oscillations have the form

$$x = \frac{Ft}{m\omega^2\tau} + B\sin\omega t + C\cos\omega t.$$

The oscillations will be stable if

$$x(\tau) = x(0), \quad \dot{x}(\tau) = \dot{x}(0).$$

These conditions lead to the following set of equations

$$\frac{F}{m\omega^2} + B\sin\omega\tau + C(\cos\omega\tau - 1) = 0,$$
$$B(\cos\omega\tau - 1) - C\sin\omega\tau = 0,$$
$$\tag{1}$$

which determines the constants B and C. Thus we have for $0 \leqslant t \leqslant \tau$

$$x(t) = \frac{F}{m\omega^2}\left[\frac{t}{\tau} - \frac{\sin(\omega t - \omega\tau/2)}{2\sin(\omega\tau/2)}\right]. \tag{2}$$

However, if t lies in the interval $n\tau \leqslant t \leqslant (n+1)\tau$ (where n is an integer), we must replaced t on the right-hand side of (2) by $t' = t - n\tau$ ($0 \leqslant t' \leqslant \tau$).

When $\omega\tau$ is close to an integer number times 2π the second term in (2) turns out to be very large – a case which is close to resonance. When $\omega\tau = 2\pi l$ (l an integer), there can be no stable oscillations – the set (1) is inconsistent.[1]

b) $x(t) = \frac{1}{\omega}\,\mathrm{Im}\left[\frac{iF}{m\omega} + \frac{F}{m(\lambda+i\omega)}e^{-\lambda t} + Ae^{i\omega t}\right]$, when $0 \leqslant t \leqslant \tau$; here

$$A = -\frac{F}{m(\lambda+i\omega)}\frac{1-e^{-\lambda\tau}}{1-e^{i\omega\tau}};$$

while for $n\tau \leqslant t \leqslant (n+1)\tau$ we must replace on the right-hand side t by $t' = t - n\tau$.

c) When $\omega_0 = (\mathscr{L}C)^{-1/2} > \lambda = R/(2\mathscr{L})$ the stable current is

$$I(t) = -\frac{1}{\sqrt{\omega_0^2 - \lambda^2}}\frac{V}{\mathscr{L}}\,\mathrm{Im}\left(\frac{e^{\alpha t}}{1-e^{\alpha\tau}} + \frac{1}{\alpha\tau}\right), \quad \alpha = -\lambda + i\sqrt{\omega_0^2 - \lambda^2} \quad (1)$$

for $0 \leqslant t \leqslant \tau$. When $n \leqslant \frac{t}{\tau} \leqslant n+1$, we must in the formula for the current replace t by $t' = t - n\tau$.

Can one use (1) to obtain an expression for the stable current when $\omega_0 < \lambda$ or when $\omega_0 = \lambda$?

5.20. a) $\quad A = \frac{1}{T}\int_0^T F(t)\dot{x}(t)\,dt =$

$$= \frac{\lambda\omega^2}{m}\left[\frac{f_1^2}{(\omega^2 - \omega_0^2)^2 + 4\lambda^2\omega^2} + \frac{4f_2^2}{(\omega_0^2 - 4\omega^2)^2 + 16\lambda^2\omega^2}\right],$$

that is, the two harmonics in the force both transfer energy, independently of one another (the period $T = 2\pi/\omega$).

b) $A = \frac{4\lambda\omega^2}{m}\sum_{n=1}^{\infty}\frac{|a_n|^2 n^2}{(\omega_0^2 - n^2\omega^2)^2 + 4\lambda^2\omega^2 n^2}.$

c) $\langle A \rangle = \frac{\lambda}{m}\left[\frac{f_1^2\omega_1^2}{(\omega_0^2 - \omega_1^2)^2 + 4\lambda^2\omega_1^2} + \frac{f_2^2\omega_2^2}{(\omega_0^2 - \omega_2^2)^2 + 4\lambda^2\omega_2^2}\right].$

---

[1] If we write the force as a Fourier series

$$F(t) = \frac{1}{2}F - \sum_{l=1}^{\infty}\frac{F}{\pi l}\sin(2\pi lt/\tau),$$

we see that each harmonic term in the force which is acting can cause a resonance build-up. When $\tau = 2\pi l/\omega$ we have for sufficiently large t (how large?)

$$x(t) \sim \frac{Ft}{2\pi m\omega l}\cos\omega t.$$

When we average over a long time interval, $T \gg 2\pi/\omega_{1,2}$, it turns out that each of the two forces $f_1 \cos\omega_1 t$ and $f_2 \cos\omega_2 t$ acts independently on the oscillator. This is because only the squares of trigonometric functions have non-vanishing averages,

$$\frac{1}{T}\int_0^T \sin^2\omega_1 t \, dt = \frac{1}{2} + \frac{1}{4\omega_1 T}(1 - \sin 2\omega_1 T) \to \frac{1}{2},$$

as $T \to \infty$, while the average values of cross-products such as $\sin\omega_1 t \cdot \cos\omega_1 t$, $\sin\omega_1 t \cdot \cos\omega_2 t$, and so on vanish: for instance,

$$\frac{1}{T}\int_0^T \sin\omega_1 t \cdot \cos\omega_2 t \, dt = \frac{1 - \cos(\omega_1 - \omega_2)t}{2(\omega_1 - \omega_2)T} + \frac{1 - \cos(\omega_1 + \omega_2)T}{2(\omega_1 + \omega_2)T} \to 0,$$

as $T \to \infty$.

d) The displacement of the oscillator is

$$x = \int_{-\infty}^{\infty} \frac{\psi(\omega)e^{i\omega t}d\omega}{\omega_0^2 - \omega^2 + 2i\lambda\omega},$$

so that the total work done by the force $F(t)$ is equal to

$$A = \int_{-\infty}^{\infty} \dot{x}(t)F(t)\,dt = \frac{8\pi\lambda}{m}\int_0^{\infty} \frac{\omega^2|\psi(\omega)|^2}{(\omega_0^2 - \omega^2)^2 + 4\lambda^2\omega^2}\,d\omega. \tag{1}$$

To prove this equality, one uses the inverse Fourier transform

$$\int_{-\infty}^{\infty} F(t)e^{i\omega t}\,dt = 2\pi\,\psi^*(\omega).$$

When $\lambda \ll \omega_0$ the main contribution to the integral (1) comes from the vicinity of the eigen-frequency of the oscillator $\omega = \omega_0$. We have thus

$$A \approx \frac{4\pi|\psi(\omega_0)|^2\omega_0}{m}\left[\lambda\int_0^{\infty} \frac{d\omega^2}{(\omega_0^2 - \omega^2)^2 + 4\lambda^2\omega_0^2}\right].$$

The factor inside the square brackets, which can easily be evaluated, is independent of λ, when $\lambda \to 0$:

$$A = \frac{|2\pi \psi(\omega_0)|^2}{2m}$$

(cf. [1], (22.12)).

5.21. Taking into account that the displacement of an oscillator x is a small value of the first order in F, we get

$$\Delta E = \int_{-\infty}^{\infty} F(x, t)\dot{x}\, dt \approx \int_{-\infty}^{\infty} f(t)\dot{x}\, dt,$$

$$\Delta P = \int_{-\infty}^{\infty} F(x, t)\, dt \approx \int_{-\infty}^{\infty} \left[f(t) - \dot{f}(t)\frac{x}{V}\right] dt.$$

Integrating the second term by parts, we find

$$\Delta P = \int_{-\infty}^{\infty} f(t)\, dt + \frac{\Delta E}{V}.$$

In particular, when $\int_{-\infty}^{\infty} f(t)\, dt = 0$, we have $\Delta E = V\Delta p$.

The condition of smallness of x can be demonstrated on the example of the action of wave groups $f(t) = fe^{-|t|/\tau} \cos \gamma t$ on an oscillator. A small parameter in expansion of $F(x, t)$ is x/λ where $\lambda = 2\pi V/\gamma$ is a characteristic wave length, that is,

$$\frac{|x|}{\lambda} = \frac{f\gamma}{2\pi m|\omega^2 - \gamma^2|} \ll 1.$$

§6

Small oscillations of systems with several degrees of freedom

6.1. Let x_i be the displacement of the ith particle from the equilibrium position $(i = 1, 2)$. The Lagrangian function of the system is

$$L = \tfrac{1}{2} m \left(\dot{x}_1^2 + \dot{x}_2^2 \right) - \tfrac{1}{2} k [x_1^2 + (x_1 - x_2)^2]. \tag{1}$$

The equations of motion

$$m\ddot{x}_1 + k(2x_1 - x_2) = 0, \quad m\ddot{x}_2 + k(x_2 - x_1) = 0$$

can be reduced to a set of algebraic equations through the substitution $x_i = A_i \cos(\omega t + \varphi)$:

$$(-m\omega^2 + 2k)A_1 - kax2 = 0, \quad -kA_1 + (-m\omega^2 + k)A_2 = 0. \tag{2}$$

This set has a non-trivial solution only when its determinant vanishes:

$$(-m\omega^2 + 2k)(-m\omega^2 + k) - k^2 = 0. \tag{3}$$

From here we get the eigen-frequencies

$$\omega_{1,2}^2 = \frac{3 \mp \sqrt{5}}{2} \frac{k}{m}.$$

Of the two equations (2), only one is independent, because of (3). Substituting the values ω_1 and ω_2 in (2), we obtain the ratios of the amplitudes

$$A_1 = \frac{2}{\sqrt{5}+1} A_2 \equiv A \qquad \text{for } \omega = \omega_1,$$

$$A_1 = -\frac{2}{\sqrt{5}-1} A_2 \equiv B \qquad \text{for } \omega = \omega_2.$$

Exploring Classical Mechanics: A Collection of 350+ Solved Problems for Students, Lecturers, and Researchers. First Edition.
Gleb L. Kotkin and Valeriy G. Serbo, Oxford University Press (2020). © Gleb L. Kotkin and Valeriy G. Serbo 2020.
DOI: 10.1093/oso/9780198853787.001.0001

The eigen-vibrations of the system are thus

$$x_1 = A\cos(\omega_1 t + \varphi_1) + B\cos(\omega_2 t + \varphi_2),$$

$$x_2 = \frac{\sqrt{5}+1}{2}A\cos(\omega_1 t + \varphi_1) - \frac{\sqrt{5}-1}{2}B\cos(\omega_2 t + \varphi_2). \tag{4}$$

The constant A, B, φ_1, and φ_2 are determined by the initial conditions.

The free oscillations (4) completely describe the motion of the system. However, when solving various problems, such as, for instance, problems with extra force (see problems 6.2b and 6.25), or when developing the perturbation theory (see problem 6.35), or when quantizing the system, it is more convenient to use the normal coordinates. This is because the normal coordinates q_i, defined by the equalities

$$x_1 = q_1 + q_2,$$

$$x_2 = \frac{\sqrt{5}+1}{2}q_1 - \frac{\sqrt{5}-1}{2}q_2, \tag{5}$$

reduce the Lagrangian function (1) to a sum of squares:

$$L = \frac{5+\sqrt{5}}{4}m(\dot{q}_1^2 - \omega_1^2 q_1^2) + \frac{5-\sqrt{5}}{4}m(\dot{q}_2^2 - \omega_2^2 q_2^2), \tag{6}$$

while the equations of motion become one-dimensional:

$$\ddot{q}_i + \omega_i^2 q_i = 0, \quad i = 1, 2.$$

This is similar how we consider the problem of the motion of two interacting particles. The latter can be reduced to the problems of the centre-of-mass motion and the motion of a particle with the reduce mass in the given field of force.

Finally, we note that a more general case of a system of N particles with one suspension point is considered in problem 7.2.

6.2. The Lagrangian function of the system is (cf. the preceding problem)

$$L = \frac{1}{2}m\left(\dot{x}_1^2 + \dot{x}_2^2\right) - \frac{1}{2}k\left[(x_1 - a(t))^2 + (x_1 - x_2)^2\right].$$

If we discard the term $-\frac{1}{2}ka^2(t)$, which is the total derivative with respect to the time, the function L can be rewritten as

$$L = L_0 + \Delta L, \quad \Delta L = x_1 ka(t), \tag{1}$$

where L_0 is the Lagrangian function of a system with the fixed suspension point (see (1) from the preceding problem). This way of writing the Lagrangian function is more convenient as it enables us to immediately write the external force "vector":

$$\begin{pmatrix} F_{x_1} \\ F_{x_2} \end{pmatrix} = \begin{pmatrix} ka(t) \\ 0 \end{pmatrix}.$$

a) The equations of motion

$$m\ddot{x}_1 + k(2x_1 - x_2) = ka\cos\gamma t,$$
$$m\ddot{x}_2 + k(x_2 - x_1) = 0,$$

$$(2)$$

can be reduce to a inhomogeneous linear set of two equations to determine A and B through the substitution[1]

$$x_1 = A\cos\gamma t, \quad x_2 = B\cos\gamma t.$$

Thus, we get

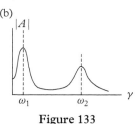

(a)

$$A = \frac{ak(-m\gamma^2 + k)}{m^2(\gamma^2 - \omega_1^2)(\gamma^2 - \omega_2^2)},$$

$$B = \frac{ak^2}{m^2(\gamma^2 - \omega_1^2)(\gamma^2 - \omega_2^2)},$$

where $\omega_{1,2}^2 = \frac{3 \mp \sqrt{5}}{2}\frac{k}{m}$ are the eigen-frequencies of the system. The frequency dependence of the amplitudes A and B is shown in Fig. 133a.

When the frequency goes through the resonance values $\gamma = \omega_{1,2}$, the amplitude A and B change sign; this corresponds to a change of π in the phase of the oscillations. At the frequency $\gamma = \sqrt{k/m}$ the oscillations of the upper particle are completely damped: $A = 0$.

In Fig. 133b, we have shown qualitatively how $|A|$ depends on the frequency of the applied force when there is friction present.

Figure 133

---

[1] The general solution of the set of equations in (2) is a superposition of free and forced oscillations. As soon as there is even a smallest amount of friction present, the free oscillations are damped, so that after a long time interval the solution of (2) is independent of the initial conditions and consist of the forced oscillations (3).

At what frequencies γ will the oscillations of the upper particle be damped, if we suspend from the lower particle another particle on the same spring?

b) Introducing the normal coordinates $q_{1,2}$ (see formula (5) from the preceding problem), we can write the Lagrangian function (1) in the form

$$L = L_1(q_1, \dot{q}_1) + L_2(q_2, \dot{q}_2),$$

$$L_{1,2} = \frac{5 \pm \sqrt{5}}{4} m(\dot{q}_{1,2}^2 - \omega_{1,2}^2 q_{1,2}^2) + q_{1,2} ka(t)$$

(cf. (6) of the preceding problem).

The problem is thus reduced to finding the stable oscillations of two independent harmonic oscillators on each of which a sawtoothed force acts (see problem 5.19a).

Certainly, in point 6.2a, it was possible to solve the problem by using normal coordinates (cf. problem 6.25).

6.3. First, we introduce a system of Cartesian coordinates with the centre in the point of suspension and the y-axis directed down along the vertical. As generalized coordinates, we choose the coordinates x_1 and x_2 of the points A and B. In the expression for the potential energy $U = -mgy_1 - mgy_2$ we substitute $y_1(x_1)$ and $y_2(x_1, x_2)$ and take into account all terms up to second order of magnitude in $x_{1,2}/l$:

$$y_1 = \sqrt{4l^2 - x_1^2} \approx 2l - \frac{x_1^2}{4l},$$

$$y_2 = y_1 + \sqrt{l^2 - (x_2 - x_1)^2} \approx 3l - \frac{x_1^2}{4l} - \frac{(x_2 - x_1)^2}{2l},$$

while in the expression for the kinetic energy

$$T = \tfrac{1}{2} m(\dot{x}_1^2 + \dot{y}_1^2 + \dot{x}_2^2 + \dot{y}_2^2)$$

we substitute $\dot{y}_1$ and $\dot{y}_2$ up to the first order:

$$\dot{y}_1 = -\frac{x_1}{2l}\dot{x}_1 \approx 0, \quad \dot{y}_2 = \dot{y}_1 - \frac{(x_2 - x_1)}{l}(\dot{x}_2 - \dot{x}_1) \approx 0.$$

After that, the Lagrangian function

$$L = \tfrac{1}{2} m(\dot{x}_1^2 + \dot{x}_2^2) - \frac{mg}{2l}x_1^2 - \frac{mg}{2l}(x_2 - x_1)^2 + 5mgl$$

coincides with the Lagrangian function of the system considered in problem 6.1, if we take $k = mg/l$ and discard the nonessential constant $5mgl$. Therefore, the functions $x_1(t)$ and $x_2(t)$ found in problem 6.1 are also valid for the double pendulum.

If the pendulum suspension point moves according to the law $x_0 = a(t) \ll l$, it is easily demonstrate that we return to the Lagrangian function considered in problem 6.2.

6.4. Let φ and ψ be the deflection angles of the upper and lower particles from the vertical. The normal oscillations are $\psi = 2\varphi$ with the frequency $\sqrt{4g/(5l)}$ and $\psi = -2\varphi$ with the frequency $\sqrt{4g/(3l)}$.

6.5. The motion is described by the equations

$$x = a\cos(\omega_1 t + \varphi), \quad y = b\cos(\omega_2 t + \psi).$$

Constant a, b, φ, and ψ are determined by the initial conditions. The orbit lies inside the rectangle (Fig. 134):

$$-a \leqslant x \leqslant a, \quad -b \leqslant y \leqslant b.$$

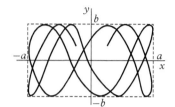

Figure 134

Generally speaking, the orbit "fills" the whole rectangle. More precisely, if the ratio ω_1/ω_2 is irrational, the orbit comes arbitrary close to any point inside this rectangle. The motion of the point is not periodic in this case, although the motion of its components along the two coordinate axis is. However, if the ratio ω_1/ω_2 is rational ($l\omega_1 = n\omega_2$ with l and n integers), the orbit is a closed curve, a so-called Lissajous figure. The motion in this case is periodic, and the period is $2\pi n/\omega_1$.

6.6. a) The transition to normal coordinates is for this system simply a rotation in the (x, y)-plane (Fig. 135):

$$x = Q_1\cos\varphi - Q_2\sin\varphi, \quad y = Q_1\sin\varphi + Q_2\cos\varphi. \tag{1}$$

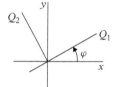

Figure 135

Indeed, the kinetic energy does not change its form under the rotation while the coefficient of $Q_1 Q_2$ in the potential energy, which is equal to

$$-\frac{1}{2}\left(\omega_1^2 - \omega_2^2\right)\sin 2\varphi - \alpha\cos 2\varphi,$$

can be made to vanish, if we determine the parameter φ from the condition

$$\cot 2\varphi = \frac{\omega_2^2 - \omega_1^2}{2\alpha}.$$

The dependence of φ on ω_1 is shown in Fig. 136; the region where φ changes from $\varphi = 0$ to $\varphi = \pi/2$ has a width of the order of α/ω_2.

When the coupling is weak, $\alpha \ll |\omega_1^2 - \omega_2^2|$, the normal oscillations are localized, that is, when $\omega_1 < \omega_2$, we have $\varphi \approx 0$ and $x \approx Q_1, y \approx Q_2$, while for $\omega_1 > \omega_2, \varphi \approx \pi/2$, and $x \approx -Q_2, y \approx Q_1$.

When $|\omega_1^2 - \omega_2^2| \ll \alpha$ the normal oscillations no longer localized:

$$\varphi \approx \frac{\pi}{4}, \quad x \approx \frac{1}{\sqrt{2}}(Q_1 - Q_2), \quad y \approx \frac{1}{\sqrt{2}}(Q_1 + Q_2)$$

(see [1], § 23, problem 1).

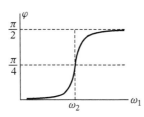

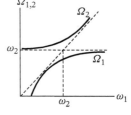

Figure 136   Figure 137

The normal frequencies,

$$\Omega_{1,2}^2 = \frac{1}{2}\left[\omega_1^2 + \omega_2^2 \mp \sqrt{(\omega_2^2 - \omega_1^2)^2 + 4\alpha^2}\right], \tag{2}$$

lie outside the range of partial frequencies[1]; that is, $\Omega_1 < \omega_1$ and $\Omega_2 > \omega_2$ (to fix the ideas we assume $\omega_1 < \omega_2$). Relations of this kind for systems with many degrees of freedom are known as "Rayleigh theorem" (see [15] and problem 6.24).

Fig. 137 shows how $\Omega_{1,2}$ depends on ω_1. It is clear from this figure that the normal frequencies $\Omega_{1,2}$ differ little from the partial frequencies $\omega_{1,2}$ (just as the normal coordinates Q_1 and Q_2 differ little from coordinates x and y) when α is small, everywhere, except the degeneracy region, where

$$|\omega_1^2 - \omega_2^2| \lesssim \alpha.$$

When ω_1 becomes sufficiently small, one of the normal frequencies becomes imaginary: the system ceases to be stable.

In terms of the coordinates Q_1 and Q_2 the law of motion and the orbits are the same as in the preceding problem.

---

[1] We follow Mandelstam [9] in calling those frequency which are obtained from the original system when $x \equiv 0$ (or $y \equiv 0$) *partial frequencies*.

b) The normal coordinates can in this case be obtained from the results of the preceding problem, simply by replaced $\omega^2_{1,2} \to m_{1,2}$ and $\alpha \to -\beta$, while the normal frequencies are the reciprocal of the normal frequencies $\Omega_{1,2}$ of the preceding problem. Why is this the case?

Can one see from the form of the Lagrangian function that the normal frequencies are independent of the sign α (or β) without finding $\Omega_{1,2}$ explicitly?

6.7. a) The Lagrangian function of the system is (see problem 4.23)

$$L = \tfrac{1}{2}\left(\mathscr{L}_1\dot{q}_1^2 + \mathscr{L}_2\dot{q}_2^2\right) - \tfrac{1}{2}\left[\frac{q_1^2}{C_1} + \frac{q_2^2}{C_2} + \frac{(q_1+q_2)^2}{C}\right],$$

where q_1 and q_2 are the charges on the upper plates of the capacitors C_1 and C_2. Introducing new variables $\sqrt{\mathscr{L}_1}q_1 = x$ and $\sqrt{\mathscr{L}_2}q_2 = y$, we obtain the Lagrangian function of the problem 6.6a with the parameters

$$\omega_1^2 = \frac{1}{\mathscr{L}_1}\left(\frac{1}{C} + \frac{1}{C_1}\right), \quad \omega_2^2 = \frac{1}{\mathscr{L}_2}\left(\frac{1}{C} + \frac{1}{C_2}\right), \quad \alpha = \frac{1}{C\sqrt{\mathscr{L}_1\mathscr{L}_2}}.$$

b) By the change of variables $q_1 = \sqrt{C_1}x$ and $q_2 = \sqrt{C_2}y$ we can change the Lagrangian function of the this system to the Lagrangian function of problem 6.6b with the parameters

$$m_{1,2} = (\mathscr{L} + \mathscr{L}_{1,2})C_{1,2}, \quad \beta = \mathscr{L}\sqrt{C_1 C_2}.$$

Can these systems become unstable?

6.8. Let x_1 and x_2 be the displacements of the particles m_1 and m_2 from their equilibrium positions. By the change of variables $\sqrt{m_1}x_1 = x$ and $\sqrt{m_2}x_2 = y$, we obtain for the system the same Lagrangian function as in problem 6.6a.

We can obtain the answer to the problem in various limiting cases without solving it. For instance, if all $k_i = k$ and $m_1 \ll m_2$, we can have a normal vibration with a very low frequency $\Omega_1^2 = 3k/(2m_2)$, $x_1 = x_2/2$ (the particle m_1 is, so to say, part of the spring, while the particle m_2 vibrates between springs of stiffness $k/2$ on the left and k on the right) and one with a very high frequency $\Omega_2^2 = 2k/m_1$ (when the particle m_2 is almost at rests). One can find the amplitude of the oscillations of the second particle considering its motion as being under the influence of an extra force kx_1 with a high frequency (see [1], formula (22.4)): $x_2 = -\frac{m_1}{2m_2}x_1$.

It is of interest to consider in a similar way the cases

a) $m_1 = m_2$, $k_1 = k_2 \ll k_3$;

b) all stiffnesses are different, but of the same order of magnitude, and $m_1 \ll m_2$;

c) $k_2 \gg k_1 = k_3$, and the masses m_1 and m_2 of the same order of magnitude.

6.9. a) $x_{1,2} = \frac{v}{2\omega_1} \sin\omega_1 t \pm \frac{v}{2\omega_2} \sin\omega_2 t$, when $k_1 \ll k$, the oscillations have the form of beats:

$$x_1 = \frac{v}{\omega} \cos\varepsilon t \cdot \sin\omega t, \quad x_2 = -\frac{v}{\omega} \sin\varepsilon t \cdot \cos\omega t.$$

b) $x_{1,2} = \frac{1}{2} a(\cos\omega_1 t \pm \cos\omega_2 t)$; when $k_1 \ll k$, we have

$$x_1 = a\cos\varepsilon t \cdot \cos\omega t, \quad x_2 = a\sin\varepsilon t \cdot \sin\omega t.$$

Everywhere in this problem $\omega_1^2 = \frac{k}{m}$ and $\omega_2^2 = \frac{2k_1 + k}{m}$, $\varepsilon = \frac{k_1}{2k}\omega_1$, $\omega = \frac{1}{2}(\omega_1 + \omega_2)$.

6.10. The energy transferred from the first to the second particle during the time dt is equal to the work done by the force $F = k_1(x_1 - x_2)$:

$$dE = k_1(x_1 - x_2)\, dx_2 = k_1(x_1 - x_2)\dot{x}_2\, dt,$$

and the energy flux is $dE/dt = k_1(x_1 - x_2)\dot{x}_2$. In the limiting case when $k_1 \ll k$ in problem 6.9a, the flux of energy averaged over the fast oscillations is equal to

$$\tfrac{1}{2}\varepsilon m v^2 \sin 2\varepsilon t.$$

6.11. The equations of motion

$$m\ddot{x}_1 + k_1(x_1 - x_2) + kx_1 + \alpha\dot{x}_1 = 0,$$
$$m\ddot{x}_2 + k_1(x_2 - x_1) + kx_2 + \alpha\dot{x}_2 = 0$$

split into two equations for the normal coordinates, when we use the substitution $x_{1,2} = (q_1 \pm q_2)/\sqrt{2}$:

$$\ddot{q}_1 + \omega_1^2 q_1 + 2\lambda\dot{q}_1 = 0,$$
$$\ddot{q}_2 + \omega_2^2 q_2 + 2\lambda\dot{q}_2 = 0,$$

where $\omega_1^2 = k/m$, $\omega_2^2 = (k + 2k_1)/m$, $2\lambda = \alpha/m$.
  We have thus, when $\lambda < \omega_{1,2}$ (see [1], § 25)

$$x_{1,2} = e^{-\lambda t}[a\cos(\gamma_1 t + \varphi_1) \pm b\cos(\gamma_2 t + \varphi_2)],$$

where $\gamma_{1,2} = \sqrt{\omega_{1,2}^2 - \lambda^2}$.
  The characteristic equation for the system of Fig. 25 is no longer biquadratic when there is friction present, but of the fourth degree, and it is therefore considerable more complicated to find the eigen-vibrations.

6.12. The Lagrangian function of the double pendulum is

$$L = \tfrac{1}{2} M \dot{x}_1^2 + \tfrac{1}{2} m \dot{x}_2^2 - \frac{Mg}{2l} x_1^2 - \frac{mg}{2l} \left[x_1^2 + (x_2 - x_1)^2 \right],$$

where x_1 and x_2 are the displacements of M and m points from the vertical passing through the suspension point (cf. problem 6.3). The replacements

$$x_1 = \frac{x}{\sqrt{M}}, \quad x_2 = \frac{y}{\sqrt{m}}$$

reduce the Lagrangian function to the form considered in problem 6.6a, with

$$\omega_1^2 - \omega_2^2 = \frac{2mg}{Ml} \ll \alpha = \frac{g}{l}\sqrt{\frac{m}{M}}.$$

In this case we have

$$x = \frac{1}{\sqrt{2}}(Q_1 - Q_2), \quad y = \frac{1}{\sqrt{2}}(Q_1 + Q_2),$$

where $Q_i = a_i \cos \Omega_i t + b_i \sin \Omega_i t$,

$$\Omega_{1,2} = \sqrt{\frac{g}{l} \mp \frac{\gamma}{2}}, \quad \gamma = \sqrt{\frac{mg}{Ml}}.$$

Taking into account the initial conditions $Q_{1,2}(0) = l\beta \sqrt{m/2}$, $\dot{Q}_{1,2}(0) = 0$, we get

$$x_1 = l\beta \sqrt{\frac{m}{M}} \sin \gamma t \sin \sqrt{\frac{g}{l}} t,$$

$$x_2 = l\beta \cos \gamma t \cos \sqrt{\frac{g}{l}} t.$$

Thus, the pendulums oscillate "in turn", and the amplitude of the upper pendulum is $\sqrt{M/m}$ times smaller than that of the lower one.

6.13.

$$x_1 = \frac{ak(k_1 + k - m\gamma^2)}{m^2(\gamma^2 - \omega_1^2)(\gamma^2 - \omega_2^2)} \cos \gamma t,$$

$$x_2 = \frac{akk_1}{m^2(\gamma^2 - \omega_1^2)(\gamma^2 - \omega_2^2)} \cos \gamma t,$$

where $\omega_1^2 = k/m$, $\omega_2^2 = (k + 2k_1)/m$.

6.14. $x_1 = x_2 = \dfrac{ak}{k - m\gamma^2} \cos \gamma t$, where the x_i is the displacement along the ring from the equilibrium positions of ith particle. Resonance is possible only at one of the normal frequencies when $\gamma^2 = k/m$ (see problem 6.25).

6.15. Let x_i be the displacement of ith particle along the ring from the equilibrium position and then

$$x_1 = x_3 = \frac{ak(\omega_2^2 - \gamma^2)}{m(\gamma^2 - \omega_1^2)(\gamma^2 - \omega_3^2)} \cos \gamma t,$$

$$x_2 = \frac{2ak^2}{m^2(\gamma^2 - \omega_1^2)(\gamma^2 - \omega_3^2)} \cos \gamma t,$$

where the eigen-frequencies ω_i are

$$\omega_{1,3}^2 = \left(2 \mp \sqrt{2}\right) \frac{k}{m}, \quad \omega_2^2 = \frac{2k}{m}.$$

Note that when $\gamma = \omega_2$ we have the displacement $x_1 = x_3 = 0$, and $x_2 = -a \cos \gamma t$.
Why is the number of the resonances in system less than the number of the normal frequencies?

6.16. The equations of motion are (cf. problem 6.13)

$$m\ddot{x}_1 + \alpha \dot{x}_1 + kx_1 + k_1(x_1 - x_2) = ka \, \mathrm{Re} \, e^{i\gamma t},$$

$$m\ddot{x}_2 + \alpha \dot{x}_2 + kx_2 + k_1(x_2 - x_1) = 0.$$

We look for solution of these equations in the form

$$x_1 = \mathrm{Re}\left(Ae^{i\gamma t}\right), \quad x_2 = \mathrm{Re}\left(Be^{I\gamma t}\right).$$

For A and B we get

$$(-m\gamma^2 + 2im\lambda\gamma + k + k_1)A - k_1 B = ka,$$

$$-k_1 A + (-m\gamma^2 + 2im\lambda\gamma + k + k_1)B = 0$$

with $2m\lambda = \alpha$, whence

$$A = \frac{ak(k + k_1 - m\gamma^2 + 2i\lambda m\gamma)}{m^2(\gamma^2 - 2i\lambda\gamma - \omega_1^2)(\gamma^2 - 2i\lambda\gamma - \omega_2^2)},$$

$$B = \frac{akk_1}{m^2(\gamma^2 - 2i\lambda\gamma - \omega_1^2)(\gamma^2 - 2i\lambda\gamma - \omega_2^2)},$$

$$x_1 = \frac{ak\sqrt{\left(\gamma^2 - \frac{1}{2}\omega_1^2 - \frac{1}{2}\omega_2^2\right)^2 + 4\lambda^2\gamma^2}\cos(\gamma t + \varphi_1 + \varphi_2 + \psi)}{m\sqrt{[(\gamma^2 - \omega_1^2)^2 + 4\lambda^2\gamma^2][(\gamma^2 - \omega_2^2)^2 + 4\lambda^2\gamma^2]}},$$

$$x_2 = \frac{akk_1\cos(\gamma t + \varphi_1 + \varphi_2)}{m^2\sqrt{[(\gamma^2 - \omega_1^2)^2 + 4\lambda^2\gamma^2][(\gamma^2 - \omega_2^2)^2 + 4\lambda^2\gamma^2]}},$$

$$\omega_1^2 = \frac{k}{m}, \quad \omega_2^2 = \frac{k + 2k_1}{m}, \quad \tan\varphi_{1,2} = \frac{2\lambda\gamma}{\gamma^2 - \omega_{1,2}^2}, \quad \tan\psi = \frac{4\lambda\gamma}{\omega_1^2 + \omega_2^2 - 2\gamma^2}.$$

There is a phase difference ψ between the oscillations of the two particles; the oscillations of the first particle are never completely damped. As function of a frequency of the applied force, γ, the amplitudes of the oscillations have either one or two maxima, depending on the ratio of the parameters ω_1, ω_2, and λ.

6.17. If x and y are the displacements from the equilibrium position of the first and the second particle, respectively, we have for the Lagrangian function of the system

$$L = \tfrac{1}{2}m\left(\dot{x}^2 + \dot{y}^2 - \frac{k_1 + k_2}{m}x^2 - \frac{k_2 + k_3}{m}y^2 + \frac{2k_2}{m}xy\right) + k_1ax\cos\gamma t;$$

it differs from the Lagrangian function considered in problem 6.6a only in the term $xk_1a\cos\gamma t$, corresponding to a force $k_1a\cos\gamma t$ acting upon the first particle. We shall use here the notations of problem 6.6a. The partial frequency $\omega_{1,2} = \sqrt{(k_{1,3} + k_2)/m}$ corresponds to the eigen-frequency of a system which we obtain by fixing the second (first) particle, that is, by putting $y = 0$ (respectively, $x = 0$). When we change to the normal coordinates Q_1 and Q_2, the Lagrangian function becomes

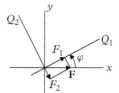

Figure 138

$$L = \tfrac{1}{2}m\left(\dot{Q}_1^2 - \Omega_1^2 Q_1^2 + \dot{Q}_2^2 - \Omega_2^2 Q_2^2\right) + (F_1 Q_1 + F_2 Q_2)\cos\gamma t,$$

where $F_1 = k_1 a \cos\varphi$ and $F_2 = -k_2 a \sin\varphi$ are the components of the amplitude of the force $F = k_1 a$ onto the normal coordinates Q_1 and Q_2 (Fig. 138). For the coordinates $Q_{1,2}$, we get the equations of motion of the harmonic oscillators with frequencies $\Omega_{1,2}$ under the action of the applied forces $F_{1,2} \cos\gamma t$. The initial conditions are $Q_i(0) = \dot{Q}_i(0) = 0$. We get

$$Q_{1,2} = \frac{F_{1,2}(\cos\gamma t - \cos\Omega_{1,2} t)}{m(\Omega_{1,2}^2 - \gamma^2)}.$$

It is interesting to consider resonance at the second eigen-frequency for the system in the weak coupling approximation,

$$\frac{k_2}{k_3 - k_1} \equiv \varepsilon \ll 1$$

(to fix the ideas, we have assume that $k_1 < k_3$). If we put $\gamma = \Omega_2(1 + \varepsilon_1)$, we have

$$Q_1 = \frac{k_1 a}{k_1 - k_3}(\cos\omega_2 t - \cos\omega_1 t),$$

$$Q_2 = -\frac{k_1 a \varepsilon}{m \omega_2^2 \varepsilon_1} \sin\left(\varepsilon_1 \frac{\omega_2}{2} t\right) \sin\omega_2 t \qquad \text{when } |\varepsilon_1| \ll 1,$$

$$Q_2 = -\frac{k_1 a}{2 m \omega_2} \varepsilon t \sin\omega_2 t \qquad\qquad \text{when } \varepsilon_1 = 0.$$

We see thus that for even the smallest coupling the amplitude Q_2 can become large or even steadily increases with time, although the rate of change will then be small. Since the angle of rotation is small ($\sin\varphi = \varepsilon$), the displacements are $x = Q_1 - \varepsilon Q_2$ and $y = Q_2$.

What is the rate of change of the amplitude of the vibrations for resonance near the first frequency $\gamma = \Omega_1$?

How will the character of the vibrations change if a small friction force acts on both particles, proportional to the velocity (cf. problem 5.11)?

6.18.

$$\text{a)} \quad x = \frac{F_0 \cos\varphi}{m(\omega_1^2 - \gamma^2)} \cos\gamma t, \quad y = \frac{F_0 \sin\varphi}{m(\omega_2^2 - \gamma^2)} \cos\gamma t,$$

where $\omega_1^2 = 2k_1/m$, $\omega_2^2 = 2k/m$, φ is the angle between the vector $\mathbf{F}_0$ and the AB-axis and x and y are the displacements from the equilibrium position along AB and CD. The particle oscillates along a strait line through the centre.

It is interesting that when $\gamma^2 = \omega_1^2 \sin^2\varphi + \omega_2^2 \cos^2\varphi$, this strait line is at right angles to the vector $\mathbf{F}_0$. In this case the work done by the acting force is equal to zero. Therefore, it seems that even the smallest amount of friction must lead to a damping of the oscillations. Explain this situation.

b) $x = \dfrac{F}{m(\omega_1^2 - \gamma^2)}\cos\gamma t, \quad y = \dfrac{F}{m(\omega_2^2 - \gamma^2)}\sin\gamma t.$

The trajectory is an ellipse with semi-axes

$$a = \frac{F}{m|\omega_1^2 - \gamma^2|}, \qquad b = \frac{F}{m|\omega_2^2 - \gamma^2|}.$$

If the quantities $(\omega_1^2 - \gamma^2)$ and $(\omega_2^2 - \gamma^2)$ have opposite signs, the particle moves clockwise along the ellipse, while the force vector rotates counterclockwise.

How does the picture given here of the motion of the particle change when the tension in the springs in the equilibrium position differs from zero?

6.19. Let x_i be the displacement of the ith particle along the ring from the equilibrium position. The three particles can rotate on the ring with a constant angular velocity:

$$x_1 = x_2 = x_3 = Ct + C_1 = q_1(t), \quad \omega_1 = 0. \tag{1}$$

The oscillations where the particles 1 and 2 move towards one another with equal amplitude,

$$x_1 = -x_2 = A\cos(\omega_2 t + \alpha) = q_2(t), \quad x_3 = 0, \quad \omega_2 = \sqrt{3k/m}, \tag{2}$$

have the same frequency as the oscillations where the particles 2 and 3 move towards one another:

$$x_1 = 0, \quad x_2 = -x_3 = B\cos(\omega_3 t + \beta) = q_3(t), \quad \omega_3 = \omega_2. \tag{3}$$

We introduce the "displacement vector"

$$\mathbf{r} = \begin{pmatrix} x_1 \\ x_2 \\ x_3 \end{pmatrix},$$

and we can then write the oscillations (1)–(3) as three vectors (see Fig. 139),

$$\mathbf{r}_1 = \begin{pmatrix} 1 \\ 1 \\ 1 \end{pmatrix} q_1, \quad \mathbf{r}_2 = \begin{pmatrix} 1 \\ -1 \\ 0 \end{pmatrix} q_2, \quad \mathbf{r}_3 = \begin{pmatrix} 0 \\ 1 \\ -1 \end{pmatrix} q_3.$$

Any linear superposition of the $\mathbf{r}_2$ and $\mathbf{r}_3$ vectors is again an oscillation with the frequency ω_2. Thus, in a space with Cartesian coordinates $x_1, x_2,$ and x_3, a set of solutions that represent the oscillations with twice degenerate frequency $\omega_2 = \omega_3$ defines a plane

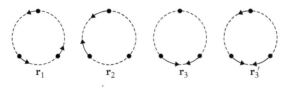

Figure 139

which passes through the vectors $\mathbf{r}_2$ and $\mathbf{r}_3$.[1] As easy to see from (4), both of these vectors—and, hence, all vectors lying in this plane—are orthogonal to the vector $\mathbf{r}_1$ (the general orthogonality condition seen in problem 6.23).

The Lagrangian function of the system is

$$L = \tfrac{1}{2} m \left(\dot{x}_1^2 + \dot{x}_2^2 + \dot{x}_3^2 \right) - \tfrac{1}{2} k \left[(x_1 - x_2)^2 + (x_2 - x_3)^2 + (x_3 - x_1)^2 \right]. \tag{5}$$

The normal coordinates must simultaneously diagonalize two quadratic forms—the kinetic energy and the potential energy. The kinetic energy in (5) is already proportional to the sum of the squares of the velocities. Therefore, the transformation from x_i to the normal coordinates which does not changing its form must be an orthogonal transformation. Hence, the vectors of the corresponding normal oscillations must be mutually orthogonal. The vectors $\mathbf{r}_i$ are independent, but not mutually orthogonal: $\mathbf{r}_1\mathbf{r}_2 = \mathbf{r}_1\mathbf{r}_3 = 0$, but $\mathbf{r}_2\mathbf{r}_3 \neq 0$. To obtain the normal coordinates, we must choose two mutually orthogonal vectors in the plane of vectors $\mathbf{r}_2$ and $\mathbf{r}_3$. These can be, for example, the vector $\mathbf{r}_2$ and the orthogonal vector $\mathbf{e}q_3$ (the unit vector $\mathbf{e}$ has been found from the condition $\mathbf{e}\mathbf{r}_1 = \mathbf{e}\mathbf{r}_2 = 0$). As a result, a set of normalized vectors[2]

$$\mathbf{r}_1' = \frac{\mathbf{r}_1}{\sqrt{3}}, \quad \mathbf{r}_2' = \frac{\mathbf{r}_2}{\sqrt{2}}, \quad \mathbf{r}_3' = \mathbf{e}q_3 = \frac{1}{\sqrt{6}} \begin{pmatrix} 1 \\ 1 \\ -2 \end{pmatrix} q_3 \tag{6}$$

allow us to define the normal coordinates:

$$x_1 = \frac{1}{\sqrt{3}} q_1 + \frac{1}{\sqrt{2}} q_2 + \frac{1}{\sqrt{6}} q_3,$$

$$x_2 = \frac{1}{\sqrt{3}} q_1 - \frac{1}{\sqrt{2}} q_2 + \frac{1}{\sqrt{6}} q_3, \tag{7}$$

$$x_3 = \frac{1}{\sqrt{3}} q_1 - \frac{2}{\sqrt{6}} q_3,$$

---

[1] Note that the linear combination in this plane of the form $\alpha\mathbf{r}_1(t) + \beta\mathbf{r}_2(t)$ is an oscillation either along a strait line (when $\alpha = \beta$, $\beta + \pi$) or along an ellipse (when $\alpha \neq \beta$).

[2] The factors $1/\sqrt{3}$ and $1/\sqrt{2}$ are introduced so that the vectors $\mathbf{r}_i$ are normalized by the condition $\mathbf{r}_i\mathbf{r}_k = \delta_{ik}q_i^2$. Under this condition, the transformation (7) is orthogonal.

which reduce the Lagrangian function (5) to the form

$$L = \tfrac{1}{2} m(\dot{q}_1^2 + \dot{q}_2^2 - \omega_2^2 q_2^2 + \dot{q}_3^2 - \omega_3^2 q_3^2). \tag{8}$$

Of course, any coordinates obtained from q_2, q_3 by an orthogonal transformation (i.e., by a simply rotation around $\mathbf{r}_1$) are also the normal coordinates.

6.20. The initial conditions for the displacements x_i along the ring are

$$x_1(0) = a, \quad x_2(0) = x_3(0) = \dot{x}_i(0) = 0.$$

We have thus for the normal coordinates q_i (see formula (7) of the preceding problem) the initial conditions:

$$q_1(0) = \frac{a}{\sqrt{3}}, \quad q_2(0) = \frac{a}{\sqrt{2}}, \quad q_3(0) = \frac{a}{\sqrt{6}}, \quad \dot{q}_i(0) = 0.$$

Therefore, we find

$$q_1 = \frac{a}{\sqrt{3}}, \quad q_2 = \frac{a}{\sqrt{2}} \cos \omega_2 t, \quad q_3 = \frac{a}{\sqrt{6}} \cos \omega_3 t,$$

and taking into account that $\omega_2 = \omega_3$, we get finally

$$x_1 = \frac{a}{3} + \frac{2a}{3} \cos \omega_2 t, \quad x_2 = x_3 = \frac{a}{3} - \frac{a}{3} \cos \omega_2 t.$$

6.21. We use the notations of problem 6.19. The Lagrangian function of the system is

$$L = \tfrac{1}{2} m \left(\dot{x}_1^2 + 2\dot{x}_2^2 + 3\dot{x}_3^2 \right) - \tfrac{1}{2} k \left[2(x_1 - x_2)^2 + 6(x_2 - x_3)^2 + 3(x_3 - x_1)^2 \right]. \tag{1}$$

The equations of motion are reduced to the system of three algebraic equations by substitution $x_i = A_i \cos(\omega t + \varphi)$:

$$\begin{aligned}
(-m\omega^2 + 5k)A_1 - 2kA_2 - 3kA_3 &= 0, \\
-2kA_1 + (-2m\omega^2 + 8k)A_2 - 6kA_3 &= 0, \\
-3kA_1 - 6kA_2 + (-3m\omega^2 + 9k)A_3 &= 0.
\end{aligned} \tag{2}$$

This system has a non-trivial solution when its determinant is equal to zero:

$$\omega^2 (m\omega^2 - 6k)^2 = 0.$$

From here we find the eigen-frequencies of the system:

$$\omega_1 = 0, \quad \omega_2 = \omega_3 = \sqrt{6k/m}.$$

The value of $\omega_1 = 0$ corresponds to the evident solution—the rotation along the ring with the constant angular velocity

$$\mathbf{r}_1 = \begin{pmatrix} 1 \\ 1 \\ 1 \end{pmatrix} q_1, \quad q_1(t) = Ct + C_1. \tag{3}$$

For matching frequencies $\omega_2 = \omega_3$, only one equation in set (2) is independent:

$$A_1 + 2A_2 + 3A_3 = 0. \tag{4}$$

Any set of values of A_i that satisfies the condition (4) gives us the oscillations with the frequency ω_2. In particular, it is possible to choose such oscillations where either the first, second, or the third particle is at rest:

$$\mathbf{r}_2 = \begin{pmatrix} 0 \\ 3 \\ -2 \end{pmatrix} q_2, \quad \mathbf{r}_3 = \begin{pmatrix} 3 \\ 0 \\ -1 \end{pmatrix} q_3, \quad \mathbf{r}_4 = \begin{pmatrix} 2 \\ -1 \\ 0 \end{pmatrix} q_4. \tag{5}$$

$$q_i = C_i \cos(\omega_2 t + \varphi_i), \quad i = 2, 3, 4.$$

According to (4), any linear combination of vectors (5) is orthogonal to the vector

$$\begin{pmatrix} 1 \\ 2 \\ 3 \end{pmatrix}.$$

It is easy to verify that the set of vectors

$$\mathbf{r}_1, \quad \mathbf{r}_2, \quad \mathbf{r}_3' = \begin{pmatrix} 5 \\ -1 \\ -1 \end{pmatrix} q_3 \tag{6}$$

allows us, as in problem 6.19, to determine the normal coordinates which lead the Lagrangian function (1) to the diagonal form. The vectors in (6) satisfy not the ordinary condition of orthogonality (as in problem 6.19), but the condition of the "weight orthogonality" (see problem 6.23).

6.22. The vectors of the normal oscillations are

$$
\mathbf{r}_1 = \frac{1}{\sqrt{2}} \begin{pmatrix} 1 \\ 0 \\ -1 \\ 0 \end{pmatrix} q_1, \qquad \mathbf{r}_2 = \frac{1}{\sqrt{2}} \begin{pmatrix} 0 \\ 1 \\ 0 \\ -1 \end{pmatrix} q_2,
$$

$$
\mathbf{r}_3 = \frac{1}{2} \begin{pmatrix} 1 \\ -1 \\ 1 \\ -1 \end{pmatrix} q_3, \qquad \mathbf{r}_4 = \frac{1}{2} \begin{pmatrix} 1 \\ 1 \\ 1 \\ 1 \end{pmatrix} q_4,
$$

$$
q_l = A_l \cos(\omega_l t + \varphi_l), \quad l = 1, 2, 3; \quad q_4 = A_4 t + A_5,
$$

$$
\omega_1 = \omega_2 = \sqrt{2k/m}, \quad \omega_3 = 2\sqrt{k/m}.
$$

The Lagrangian function of the system reads

$$
L = \tfrac{1}{2} m \left(\dot{q}_1^2 + \dot{q}_2^2 + \dot{q}_3^2 + \dot{q}_4^2 - \omega_1^2 q_1^2 - \omega_2^2 q_2^2 - \omega_3^2 q_3^2 \right).
$$

This is, of course, not the only possible choice. Any vectors obtained from the ones given here through a rotation in the plane determined by the vectors $\mathbf{r}_1$ and $\mathbf{r}_2$ will also be vectors of the normal oscillations. For instance,

$$
\mathbf{r}_1' = \frac{1}{2} \begin{pmatrix} 1 \\ 1 \\ -1 \\ -1 \end{pmatrix} q_1', \quad \mathbf{r}_2' = \frac{1}{2} \begin{pmatrix} 1 \\ -1 \\ -1 \\ 1 \end{pmatrix} q_2', \quad \mathbf{r}_3' = \mathbf{r}_3, \quad \mathbf{r}_4' = \mathbf{r}_4 \tag{2}
$$

(rotation over $\pi/4$). However, the vectors $\mathbf{r}_1$, $\mathbf{r}_2'$, $\mathbf{r}_3$, $\mathbf{r}_4$ will not reduced the Lagrangian function to a sum of squares, although they also are independent.

6.23. The amplitude of the normal oscillations satisfy the equations

$$
-\omega_l^2 \sum_j m_{ij} A_j^{(l)} + \sum_j k_{ij} A_j^{(l)} = 0, \tag{1}
$$

$$
-\omega_s^2 \sum_j m_{ij} A_j^{(s)} + \sum_j k_{ij} A_j^{(s)} = 0. \tag{2}
$$

Multiply (1) by $A_i^{(s)}$ and (2) by $A_i^{(l)}$ and summing both equations over i, we get instead of (1) and (2)

$$-\omega_l^2 \sum_{ij} m_{ij} A_j^{(l)} A_i^{(s)} + \sum_{ij} k_{ij} A_j^{(l)} A_i^{(s)} = 0, \qquad (3)$$

$$-\omega_s^2 \sum_{ij} m_{ij} A_j^{(s)} A_i^{(l)} + \sum_{ij} k_{ij} A_j^{(s)} A_i^{(l)} = 0. \qquad (4)$$

Subtracting (4) from (3) and using the fact that $m_{ij} = m_{ji}$ and $k_{ij} = k_{ji}$, we obtain

$$(\omega_s^2 - \omega_l^2) \sum_{ij} m_{ij} A_i^{(s)} A_j^{(l)} = 0;$$

that is, when $\omega_s \neq \omega_l$,

$$\sum_{ij} m_{ij} A_i^{(s)} A_j^{(l)} = 0, \qquad (5)$$

and at the same time from (3) we obtain

$$\sum_{ij} k_{ij} A_i^{(s)} A_j^{(l)} = 0. \qquad (6)$$

It is helpful to use the terminology from linear algebra. We shall call the set of amplitudes of a given oscillation the amplitude vector $\mathbf{A}^{(l)} = (A_1^{(l)}, A_2^{(l)}, \ldots, A_n^{(l)})$. The relations (5) and (6) which we have just prove mean that the amplitudes $\mathbf{A}^{(s)}$ and $\mathbf{A}^{(l)}$ are mutually orthogonal, provided that the scalar product is defined by means of the metric tensors m_{ij} or k_{ij}.

In the case of degeneracy (if $\omega_s = \omega_l$), the amplitude $\mathbf{A}^{(s)}$ and $\mathbf{A}^{(l)}$ do not necessary satisfy (5) and (6). However, one can in that case always choose—indeed, in several was—such amplitudes that they satisfy (5) and (6) and also reduce the Lagrangian function to a sum of squares.

6.24. Introducing the normal coordinates

$$x_i = \sum_l A_i^{(l)} q_l,$$

we transform the constraint equations to the form

$$\sum_l b_l q_l = 0, \quad b_l = \sum_i a_i A_i^{(l)}.$$

Equations of motion with an indefinite Lagrangian multiplier λ

$$M_l(\ddot{q}_l + \Omega_l^2 q_l) = b_l \lambda$$

can be solved by putting

$$q_l = C_l \cos(\omega t + \varphi), \quad \lambda = \Lambda \cos(\omega t + \varphi).$$

Expressing C_l from

$$M_l(\Omega_l^2 - \omega^2)C_l = b_l \Lambda,$$

and substituting it into the constraint equations, we obtain for new frequencies

$$\Lambda \sum_l \frac{b_l^2}{M_l(\Omega_l^2 - \omega^2)} = 0.$$

To study this equation, it is helpful to use the graph (Fig. 140)

$$y(\omega^2) = \sum_l \frac{b_l^2}{M_l(\Omega_l^2 - \omega^2)} = 0.$$

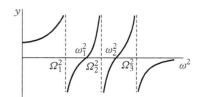

Figure 140

Note that the function $y(\omega^2)$ changes the sign by passing through the infinite value when $\omega^2 = \Omega_l^2$.
After that, the location of the roots ω_l becomes quite clear. If any of the coefficients b_l is zero, then the corresponding normal oscillation (and its frequency) do not change when one applies a constraint.

The fact considered in this problem allows for a simple geometric interpretation (see [8], § 24).

6.25. Substituting $x_j = \sum_l \lambda^{(l)} a_j^{(l)} \cos \gamma t$ into the equations of motion,

$$\sum_j m_{ij} \ddot{x}_j + \sum_j k_{ij} x_j = f_i \cos \gamma t,$$

we obtain the following equations to determine the coefficients $\lambda^{(l)}$:

$$-\gamma^2 \sum_{l,j} m_{ij} \lambda^{(l)} A_j^{(l)} + \sum_{j,l} k_{ij} \lambda^{(l)} A_j^{(l)} = f_i. \tag{2}$$

The simplest way to solve these equations is by using the orthogonality relations (5) and (6) of problem 6.23.

To do this, we multiply the equations in (2) by $A_i^{(s)}$ and sum over i; then we get

$$\lambda^{(s)} = \frac{F_s}{M_s(\omega_s^2 - \gamma^2)},$$

where

$$F_s = \sum_i A_i^{(s)} f_i, \quad M_s = \sum_{i,j} m_{ij} A_i^{(s)} A_j^{(s)}, \quad K_s = \sum_{i,j} k_{ij} A_i^{(s)} A_j^{(s)}.$$

The quantity $\omega_s = \sqrt{K_s/M_s}$ is the sth eigen-frequency of the system, in accordance with (4) of problem 6.23.[1] The γ-dependence of the $\lambda^{(s)}$ shows resonances.

For the normal coordinates q_s, introduced by the the equations

$$x_i = \sum_s A_i^{(s)} q_s(t), \tag{3}$$

instead of the equations (1), we have the following equations of motion:

$$M_s \ddot{q}_s + K_s q_s = F_s \cos \gamma t. \tag{4}$$

Hence, if the force vector f_i is orthogonal to the amplitude vector of the sth normal oscillation,

$$\sum_i A_i^{(s)} f_i = 0,$$

the corresponding normal coordinate satisfies the equation for free oscillations, and there is no resonance for $\omega = \omega_s$. Note that the work done by the applied force is in this case equal to zero:

$$\sum_i f_i \, dx_i = \sum_i f_i A_i^{(s)} \, dq_s = 0.$$

---

[1] If some of the eigen-frequencies are degenerate, we shall assume that the amplitudes of the eigen-oscillations corresponding to them are chosen in such a way that they satisfy the orthogonality relations (5) and (6) of problem 6.23.

Consider the case where the force vector is parallel to the amplitude vector of one of the normal oscillation,

$$\frac{f_i}{A_i^{(s)}} = \text{const} \quad (i = 1, 2, \ldots, N).$$

Can such a force excite other normal vibrations?

6.26. The stable oscillations can be presented in the form (see preceding problem)

$$x_i = \sum_j \beta_{ij} f_j,$$

where

$$\beta_{ij} = \sum_l \frac{A_i^{(l)} A_j^{(l)}}{M_l(\omega_l^2 - \gamma^2)}.$$

The reciprocity theorem reflects that $\beta_{ij} = \beta_{ji}$.

How will the formulation of this theorem change if the coordinates x_i and x_j have different dimensions (e.g. for the case of an electromechanical system)?

6.27. The normal oscillations are

$$\begin{pmatrix} 1 \\ 1 \\ 1 \\ 1 \end{pmatrix} q_1, \quad \begin{pmatrix} 1 \\ 0 \\ -1 \\ 0 \end{pmatrix} q_2, \quad \begin{pmatrix} 0 \\ 1 \\ 0 \\ -1 \end{pmatrix} q_3, \quad \begin{pmatrix} 1 \\ -m/M \\ 1 \\ -m/M \end{pmatrix} q_4,$$

where

$$q_1 = At + B, \quad q_i = A_i \cos(\omega_i t + \alpha_i), \quad i = 2, 3, 4;$$

$$\omega_2^2 = \frac{2k}{m}, \quad \omega_3^2 = \frac{2k}{M}, \quad \omega_4^2 = \frac{2k(M + m)}{mM}.$$

The first three oscillations are easily guessed, and the last one can be obtained using the orthogonality conditions to the first three. Since the particle masses are different, the orthogonality condition of the two normal oscillations with the amplitudes **A** and **B** have the form

$$mA_1 B_1 + MA_2 B_2 + mA_3 B_3 + MA_4 B_4 = 0$$

(see problem 6.23).

6.28. Let x_i be the displacement of the ith particle along the ring. We can easily guess two normal vibrations:

$$\mathbf{r}_1 = \begin{pmatrix} 1 \\ 1 \\ 1 \\ 1 \end{pmatrix} q_1(t), \quad \mathbf{r}_2 = \begin{pmatrix} 1 \\ 0 \\ -1 \\ 0 \end{pmatrix} q_2(t), \tag{1}$$

$$q_1(t) = C_1 t + C_2, \quad q_2(t) = A_2 \cos(\omega_2 t + \varphi_2), \quad \omega_2 = \sqrt{2k/m}.$$

The two other vectors must be orthogonal to the vectors (1) in the metric determined by the coefficients of the quadratic form of the kinetic energy (see problem 6.23); that is, they must have the form

$$\mathbf{r} = \begin{pmatrix} a \\ b \\ a \\ -a - \frac{1}{2}b \end{pmatrix} q(t). \tag{2}$$

Substituting (2) into the equations of motion for the first and the second particles,

$$m\ddot{x}_1 + k(2x_1 - x_4 - x_2) = 0, \quad m\ddot{x}_2 + k(2x_2 - x_1 - x_3) = 0,$$

we get two equations to determine the values of a, b and frequencies:

$$(-m\omega^2 + 3k)a - \tfrac{1}{2}kb = 0,$$
$$-2ka + (-m\omega^2 + 2k)b = 0. \tag{3}$$

Solving (3), we find

$$\omega_{3,4}^2 = \frac{5 \mp \sqrt{5}}{2} \frac{k}{m}, \quad b_{3,4} = (1 \pm \sqrt{5})a_{3,4}$$

or

$$\mathbf{r}_{3,4} = \begin{pmatrix} 1 \\ 1 \pm \sqrt{5} \\ 1 \\ -\frac{1}{2}(3 \pm \sqrt{5}) \end{pmatrix} q_{3,4}(t),$$

$$q_{3,4} = A_{3,4} \cos(\omega_{3,4} t + \varphi_{3,4}).$$

6.29. a) Let x_i, y_i, and z_i be the displacements of the ith particle from its equilibrium position. The Lagrangian function of the system has the form (see problem 5.7)

$$L = L_1(x, \dot{x}) + L_1(y, \dot{y}) + L_1(z, \dot{z}),$$

$$L_1(x, \dot{x}) = \tfrac{1}{2} m \left(\dot{x}_1^2 + \dot{x}_2^2 + \dot{x}_3^2 + \dot{x}_4^2 + \dot{x}_5^2 \right) - \tfrac{1}{2} k \Big[x_1^2 + (x_1 - x_5)^2 +$$

$$+ (x_5 - x_3)^2 + x_3^2 + x_2^2 + (x_2 - x_5)^2 + (x_5 - x_4)^2 + x_4^2 \Big];$$

therefore, the oscillations in the x-, y-, and z-directions proceed independently. We can easily guess three of the normal oscillations in the x-direction:

$$\mathbf{r}_1 = \begin{pmatrix} 1 \\ 0 \\ -1 \\ 0 \\ 0 \end{pmatrix} q_1, \quad \mathbf{r}_2 = \begin{pmatrix} 0 \\ 1 \\ 0 \\ -1 \\ 0 \end{pmatrix} q_2, \quad \mathbf{r}_3 = \begin{pmatrix} 1 \\ -1 \\ 1 \\ -1 \\ 0 \end{pmatrix} q_3, \tag{1}$$

$$q_i = A_i \cos(\omega_i t + \varphi_i), \quad \omega_1 = \omega_2 = \omega_3 = \sqrt{2k/m}.$$

The other two normal oscillations must be orthogonal to the vectors (1) and therefore have the form[1]

$$\mathbf{r}_{4,5} = \begin{pmatrix} a \\ a \\ a \\ a \\ d \end{pmatrix} q_{4,5}.$$

Substituting this vector into the equations of motion for the first and fifth particle,

$$m\ddot{x}_1 + k(2x_1 - x_5) = 0,$$
$$m\ddot{x}_5 + k(4x_5 - x_1 - x_2 - x_3 - x_4) = 0,$$

we obtain two equations to determine the unknown parameters a, d, and the frequencies $\omega_{4,5}$:

$$(-\omega^2 m + 2k)a - kd = 0,$$
$$-4ka + (-m\omega^2 + 4k)d = 0. \tag{3}$$

---

[1] Let $\mathbf{r}_{4,5} = (a, b, c, e, d) \equiv \mathbf{r}$; then the orthogonality conditions $(\mathbf{r}, \mathbf{r}_1) = (\mathbf{r}, \mathbf{r}_2) = (\mathbf{r}, \mathbf{r}_3) = 0$ lead to the equations $a = b = c = e$.

Solving (3), we find $\omega_{4,5} = \sqrt{(3 \mp \sqrt{5})k/m}$ and $d_{4,5} = (-1 \pm \sqrt{5})a_{4,5}$, and finally

$$\mathbf{r}_{4,5} = \begin{pmatrix} 1 \\ 1 \\ 1 \\ 1 \\ -1 \pm \sqrt{5} \end{pmatrix} q_{4,5}. \tag{4}$$

The results for oscillations along the y- and z-axes are the same as for those along the x-axis. Thus, there are altogether three different frequencies in the system: $\omega_1 = \sqrt{2k/m}$ which is ninefold degenerate and two frequencies $\omega_{4,5} = \sqrt{(3 \mp \sqrt{5})k/m}$ which are threefold degenerate (see problem 6.42 about the lifting of the degeneracy).

b) The oscillations along the z-axis can easily be guessed:

$$\mathbf{r}_1 = \begin{pmatrix} 1 \\ 0 \\ -1 \\ 0 \end{pmatrix} q_1, \quad \mathbf{r}_2 = \begin{pmatrix} 0 \\ 1 \\ 0 \\ -1 \end{pmatrix} q_2, \quad \mathbf{r}_{3,4} = \begin{pmatrix} 1 \\ \mp 1 \\ 1 \\ \mp 1 \end{pmatrix} q_{3,4},$$

$$q_i = A_i \cos(\omega_i t + \varphi_i), \quad \omega_1 = \omega_2 = \omega_3 = \sqrt{\frac{2f}{ml}}, \quad \omega_4 = \sqrt{\frac{f}{ml}},$$

where f is the tension in the springs, and l the length of one of the spring at equilibrium.

If $f = kl$, then the oscillations in the x- or y-directions will have the same form as in the z-direction if we put $\mathbf{r} = (x_1, x_2, x_3, x_4)$ (or $\mathbf{r} = (y_2, y_1, y_4, y_3)$). If, however, $f \neq kl$, the degeneration is lifted. Two of the normal oscillations with frequencies $\omega_1 = \sqrt{2k/m}$ and $\omega_2 = \sqrt{2f/(ml)}$ are the same as $\mathbf{r}_1$ and $\mathbf{r}_2$. The two other must have the form

$$\begin{pmatrix} a \\ b \\ a \\ b \end{pmatrix} \cos(\omega t + \varphi)$$

because of the orthogonality condition. To find them, it is sufficient to consider the equations of motion of two particles:

$$m\ddot{x}_1 + k(2x_1 - x_5) = 0, \quad m\ddot{x}_2 + \frac{f}{l}(2x_2 - x_5) = 0.$$

Here

$$x_5 = \frac{k(x_1 + x_3) + (f/l)(x_2 + x_4)}{2k + 2f/l}$$

is the coordinate of the point where the springs are joined together which is determined from the condition that the potential energy be a minimum for given values of $x_{1,2,3,4}$.
    Solving these equations, we obtain

$$\omega_3^2 = \frac{f + kl}{ml}, \qquad b_3 = -\frac{f}{kl} a_3;$$

$$\omega_4^2 = \frac{2kf}{m(f + kl)}, \qquad b_4 = \frac{kl}{f} a_4.$$

6.30. In this case, the result can be obtained by a simple generalization of the results of the problem 6.21 without an explicit calculation of the eigen-frequencies ω_i.
    Let $\omega_1 = 0$ and the degeneracy $\omega_2 = \omega_3$ be in the system. The frequency $\omega_1 = 0$ corresponds to the rotation of the particles along the ring

$$\mathbf{r}_1 = \begin{pmatrix} 1 \\ 1 \\ 1 \end{pmatrix} (Ct + C_1).$$

Due to the degeneracy of the frequency ω_2 any vector

$$\mathbf{r}_1 = \begin{pmatrix} A_1 \\ A_2 \\ A_3 \end{pmatrix} \cos(\omega t + \varphi), \tag{1}$$

which satisfies the condition

$$m_1 A_1 + m_2 A_2 + m_3 A_3 = 0, \tag{2}$$

is a normal oscillation with the frequency $\omega = \omega_2$. (The equality (2) is the orthogonality condition for the vector $\mathbf{r}_1$ in the metric defined by the coefficients of the quadratic form of the kinetic energy—see problem 6.23). In particular, one can chose such a normal oscillation (1) that the first particle will be at rest:

$$A_1 = 0, \quad m_2 A_2 + m_3 A_3 = 0. \tag{3}$$

If we substitute (3) in the equations of motion

$$(m_1 \omega^2 - k_2 - k_3) A_1 + k_3 A_2 + k_2 A_3 = 0,$$
$$k_3 A_1 + (m_2 \omega^2 - k_1 - k_3) A_2 + k_1 A_3 = 0,$$
$$k_2 A_1 + k_1 A_2 + (m_3 \omega^2 - k_1 - k_2) A_3 = 0,$$

we see that they can have a non-zero solution only when

$$k_3 A_2 + k_2 A_3 = 0. \tag{4}$$

Comparing (3) and (4), we find that $m_2 k_2 = m_3 k_3$. Repeating the similar reasoning for the cases where the second or the third particle is at rest, we find that the coefficients k_i must satisfy the condition

$$m_1 k_1 = m_2 k_2 = m_3 k_3 \tag{5}$$

for the case of the frequency degeneracy. On the other hand, it is clear from the previous consideration that (5) is a sufficient condition for the degeneracy of the frequencies. Indeed, if the condition (5) is satisfied, then there are three different normal oscillations with the frequencies other than zero. Of these three oscillations only two are linearly independent due to (2). From here it follows that these three oscillations have the same frequency.

Thus, (5) is the necessary and sufficient condition for the degeneracy of the frequencies.

6.31. It is helpful to use for the solution the method explained in problem 6.28.

   a) The normal oscillations are

$$\mathbf{r}_1 = \begin{pmatrix} 1 \\ 1 \\ 1 \end{pmatrix} (C_1 t + C_2), \quad \mathbf{r}_2 = \begin{pmatrix} 1 \\ -1 \\ 0 \end{pmatrix} A_2 \cos(\omega_2 t + \varphi_2),$$

$$\mathbf{r}_3 = \begin{pmatrix} 1 \\ 1 \\ -2 \end{pmatrix} A_3 \cos(\omega_3 t + \varphi_3), \tag{1}$$

$$\omega_2^2 = \frac{(3 + 2\varepsilon)k}{m}, \quad \omega_3^2 = \frac{3k}{m}, \quad \varepsilon = \frac{\delta k}{k};$$

for small ε, they are close to the oscillations (6) of problem 6.19: the amplitudes of the oscillation are the same, but all frequencies are different. Therefore, although in problem 6.19 any superposition of the vectors $\mathbf{r}_2'$ and $\mathbf{r}_3'$ again gave normal oscillations, now the choice of the vectors $\mathbf{r}_2$ and $\mathbf{r}_3$ is completely unambiguous.

   b) The normal oscillations

$$\mathbf{r}_1 = \begin{pmatrix} 1 \\ 1 \\ 1 \end{pmatrix} (C_1 t + C_2), \quad \mathbf{r}_2 = \begin{pmatrix} 1 \\ -1 \\ 0 \end{pmatrix} A_2 \cos(\omega_2 t + \varphi_2), \tag{2}$$

$$\mathbf{r}_3 = \begin{pmatrix} 1 \\ 1 \\ -\dfrac{2}{1+\varepsilon} \end{pmatrix} A_3 \cos(\omega_3 t + \varphi_3),$$

(2)

$$\omega_2^2 = \frac{3k}{m}, \quad \omega_3^2 = \frac{3+\varepsilon}{1+\varepsilon} \frac{k}{m}, \quad \varepsilon = \frac{\delta m}{m}$$

are for small ε close to the oscillations (6) of problem 6.19. If the extra mass had been added to particle 2, the normal oscillations

$$\mathbf{r}_1 = \begin{pmatrix} 1 \\ 1 \\ 1 \end{pmatrix} (C_1 t + C_2), \quad \mathbf{r}_2 = \begin{pmatrix} 1 \\ 0 \\ -1 \end{pmatrix} a_2 \cos(\omega_2 t + \varphi_2),$$

$$\mathbf{r}_3 = \begin{pmatrix} 1 \\ -\dfrac{2}{1+\varepsilon} \\ 1 \end{pmatrix} A_3 \cos(\omega_3 t + \varphi_3),$$

would be close to the superposition of the normal oscillations (6) of problem 6.19.

c) $\mathbf{r}_1 = \begin{pmatrix} 1 \\ 1 \\ 1 \end{pmatrix} (C_1 t + C_2),$

$$\mathbf{r}_{2,3} = \left\{ a_{2,3} \begin{pmatrix} 1 \\ 0 \\ -1 - \varepsilon_1 \end{pmatrix} + b_{2,3} \begin{pmatrix} 0 \\ 1 \\ -1 - \varepsilon_2 \end{pmatrix} \right\} \cos(\omega_{2,3} t + \varphi_{2,3}),$$

where

$$\frac{b_{2,3}}{a_{2,3}} = \frac{1}{\varepsilon_2} \left(\varepsilon_2 - \varepsilon_1 \pm \sqrt{\varepsilon_1^2 + \varepsilon_2^2 - \varepsilon_1 \varepsilon_2} \right),$$

$$\omega_{2,3}^2 \approx \frac{k}{m} \left(3 - \varepsilon_1 - \varepsilon_2 \mp \sqrt{\varepsilon_1^2 + \varepsilon_2^2 - \varepsilon_1 \varepsilon_2} \right), \quad \varepsilon_i = \frac{\delta m_i}{m}.$$

6.32. a) We expand the initial displacement

$$\mathbf{r}(0) = \begin{pmatrix} a \\ 0 \\ -a \end{pmatrix}$$

in terms of the vectors $\mathbf{r}_i$ (see (1) of the preceding problem) taken at $t = 0$:

$$\mathbf{r}(0) = \mathbf{r}_1(0) + \mathbf{r}_2(0) + \mathbf{r}_3(0).$$

(1)

We do the same for the initial velocities:

$$\dot{\mathbf{r}}(0) = \dot{\mathbf{r}}_1(0) + \dot{\mathbf{r}}_2(0) + \dot{\mathbf{r}}_3(0). \tag{2}$$

From the set of equations in (1) and (2), we find for the constants the following values: $A_2 = A_3 = a/2$, $C_1 = C_2 = \varphi_2 = \varphi_3 = 0$, or

$$\mathbf{r} = \frac{a}{2} \begin{pmatrix} \cos\omega_2 t + \cos\omega_3 t \\ -\cos\omega_2 t + \cos\omega_3 t \\ -2\cos\omega_3 t \end{pmatrix} \approx a \begin{pmatrix} \cos(\varepsilon\omega_3 t/6)\cos\omega_3 t \\ \sin(\varepsilon\omega_3 t/6)\sin\omega_3 t \\ -\cos\omega_3 t \end{pmatrix}.$$

The motion of the particles 1 and 2 thus shows beats with a frequency which is determined by the perturbation δk, while the particle 3 performs a simple oscillation with frequency ω_3. We emphasize that even a very small perturbation δk leads to secular changes which become appreciable after sufficiently long times (cf. problem 2.40).

6.33. a), b)

$$\mathbf{r}_1 = \begin{pmatrix} 1 \\ -1 \\ -1 \\ 1 \end{pmatrix} q_1(t), \quad \mathbf{r}_2 = \begin{pmatrix} 1 \\ 1 \\ -1 \\ -1 \end{pmatrix} q_2(t), \quad \mathbf{r}_{3,4} = \begin{pmatrix} 1 \\ \mp 1 \\ 1 \\ \mp 1 \end{pmatrix} q_{3,4}(t);$$

c) the same as in problem 6.22, (1).

6.34. $x_{1,2} = -x_{3,4} = \pm\frac{1}{2}a\cos\sqrt{(2k+2\delta k)/m}\,t + \frac{1}{2}a\cos\sqrt{2k/m}\,t$;

the oscillations of the particles show beats (cf. problem 6.32).

6.35. We can expect that the changes of the frequencies and of the vectors of the normal oscillations will be small, so we use the method of successive approximations. It is helpful to use the normal coordinates of the original system (see problem 6.25)

$$x_i = \sum_l A_i^{(l)} q_l.$$

In this case, δL takes the form

$$\delta L = \frac{1}{2}\sum_{l,s}(\delta M_{ls}\dot{q}_l\dot{q}_s - \delta K_{ls}q_l q_s), \tag{1}$$

where

$$\delta M_{ls} = \sum_{i,j}\delta m_{ij}A_i^{(l)}A_j^{(s)}, \quad \delta K_{ls} = \sum_{i,j}\delta k_{ij}A_i^{(l)}A_j^{(s)}; \tag{2}$$

the equations of motion are

$$M_l(\ddot{q}_l + \omega_l^2 q_l) = -\sum_s (\delta M_{ls}\ddot{q}_s + \delta K_{ls}q_s).$$ (3)

Assuming that only the oscillation q_n is excited in zero approximation, we can leave in the right-hand side of (3) only the term $s = n$.

To determine the correction to the frequency ω_n, it is enough to write the single equation (with $l = n$):

$$(M_n + \delta M_{nn})\ddot{q}_n + (M_{nn}\omega_n^2 + \delta K_{nn})q_n = 0,$$

where

$$(\omega_n + \delta\omega_n)^2 = \frac{M_{nn}\omega_n^2 + \delta K_{nn}}{M_n + \delta M_{nn}},$$

as a result,

$$\delta\omega_n = \frac{\delta K_{nn}}{2\omega_n M_n} - \frac{\omega_n \delta M_{nn}}{2M_n}.$$ (4)

The equations with $l \neq n$ allow us to find corrections to the vector of the normal oscillation. In this case, the right-hand sides of the equations can be considered as the given forces with the frequency ω_n. The excitation of the oscillations q_l, as we have expected, is weak, as these "forces" are small.

We can also get the following approximation for the ω_n and the vectors of the normal oscillations (e.g. see [14], Ch. 1, § 5).

It is worth noting that the value δM_{nn} in (2) represents a supplement to the doubled kinetic energy of the system provided that the velocities $\dot{x}_i = A_i^{(n)}$. From this, it follows, in particular, that with increasing of particles' masses, the quantity $\delta M_{nn} \geqslant 0$ and, according to (4), $\delta\omega_n \leqslant 0$. In the similar way, it is easy to see that if one increases the stiffness of the springs, the eigen-frequencies can only increase (see [8], § 24 and [15]).

It is important to understand what will change when we look for a correction to the degenerate frequency (let $\omega_p = \omega_n$). In this case, the "force" in the right-hand sides of the equations (3) becomes resonant. Therefore, the coordinate $q_p(t)$ increases with time, and it must be taken into account in the right-hand sides of the equations (3). Therefore, in this case, one must use the equations (3) together with $l = n, p$, leaving in the right-hand sides only the terms with $s = n, p$

$$M_n(\ddot{q}_n + \omega_n^2 q_n) = -\delta M_{nn}\ddot{q}_n - \delta M_{np}\ddot{q}_p - \delta K_{nn}q_n - \delta K_{np}q_p,$$
$$M_p(\ddot{q}_p + \omega_p^2 q_p) = -\delta M_{pn}\ddot{q}_n - \delta M_{pp}\ddot{q}_p - \delta K_{pn}q_n - \delta K_{pp}q_p.$$ (5)

It is clear that this can be applied to the case $\omega_p \approx \omega_n$ as well.

Thus, to determine the corrections to all the eigen-frequencies, including the degenerate ones, one can drop all terms in the ΔL (see (1)) which contain products of the normal coordinates related to the different frequencies of the original system.

6.36. We use the notions and the results of problems 6.28 and 6.35. It is obvious that $\delta\omega_1 = 0$. For other frequencies, we have the correction

$$\delta\omega_n = -\frac{\omega_n}{2} \frac{\displaystyle\sum_{i,j} (\mathbf{r}_n)_i \delta m_{ij} (\mathbf{r}_n)_j}{\displaystyle\sum_{i,j} (\mathbf{r}_n)_i m_{ij} (\mathbf{r}_n)_j}, \quad n = 2, 3, 4. \tag{1}$$

The matrix m_{ij} of the kinetic energy is diagonal; moreover,

$$m_{11} = m_{22} = m_{33} = m, \qquad m_{44} = 2m. \tag{2}$$

The matrix δm_{ij} has a single non-zero element;

$$\delta m_{11} = \Delta m. \tag{3}$$

Substituting expressions (2) and (3) into (1), as well as the components of the vectors of the normal oscillations $\mathbf{r}_n$, which have been found in problem 6.28, we obtain

$$\delta\omega_2 = -\tfrac{1}{4}\varepsilon\omega_2, \quad \delta\omega_{3,4} = -\frac{3 \mp \sqrt{5}}{40} \varepsilon\omega_{3,4}.$$

6.37. We choose the vector potential to be

$$\mathbf{A}(\mathbf{r}) = \tfrac{1}{2} B(-y, x, 0);$$

the Lagrangian function is then

$$L = \tfrac{1}{2} m \left(\dot{x}^2 + \dot{y}^2 + \dot{z}^2 \right) - \tfrac{1}{2} m \left(\omega_1^2 x^2 + \omega_2^2 y^2 + \omega_3^2 z^2 \right) + \tfrac{1}{2} m\omega_B (x\dot{y} - y\dot{x}),$$

where $\omega_B = \frac{eB}{mc}$. For x and y, we get

$$\ddot{x} + \omega_1^2 x - \omega_B \dot{y} = 0,$$
$$\ddot{y} + \omega_2^2 y + \omega_B \dot{x} = 0.$$

It is helpful to look for oscillations in the form

$$x = \mathrm{Re}(A e^{i\Omega t}), \quad y = \mathrm{Re}(B e^{i\Omega t}).$$

The set of equations

$$(\omega_1^2 - \Omega^2)A - i\omega_B\Omega B = 0,$$
$$i\omega_B\Omega A + (\omega_2^2 - \Omega^2)B = 0$$

leads to the oscillations

$$x = \mathrm{Re}\left(A_k e^{i\Omega_k t}\right) = a_k \cos(\Omega_k t + \varphi_k),$$

$$y = \mathrm{Re}\left(A_k \frac{-i\omega_B\Omega_k}{\omega_2^2 - \Omega_k^2} e^{i\Omega_k t}\right) = a_k \frac{\omega_B\Omega_k}{\omega_2^2 - \Omega_k^2} \sin(\Omega_k t + \varphi_k),$$

$$A_k = a_k e^{i\varphi_k}, \quad k = 1, 2$$

with frequency

$$\Omega_{1,2}^2 = \tfrac{1}{2}\left[\omega_1^2 + \omega_2^2 + \omega_B^2 \pm \sqrt{\left(\omega_1^2 + \omega_2^2 + \omega_B^2\right)^2 - 4\omega_1^2\omega_2^2}\right],$$

for which

$$\Omega_1\Omega_2 = \omega_1\omega_2$$

is true.

To fix our ideas, let $\omega_1 > \omega_2, \omega_B > 0$. The first of the oscillations which we have found is then a clockwise the motion along an ellipse with its major axis along the x-axis, while the second oscillation is a counterclockwise motion along an ellipse with its major axis along the y-axis.

The motion along the z-axis turns out to be a harmonic motion which is independent of the magnetic field,

$$z = a_3 \cos(\omega_3 t + \varphi_3).$$

The free motion of the oscillator is a superposition of the oscillations obtained. We call these oscillations normal oscilliations, just generalizing the concept of normal oscillations: the motion in the x- and y-directions occur with the same frequency, but with a shift in phase. It is impossible to reduce the Lagrangian function to diagonal form using only a linear transformation of the coordinates as the transition to normal coordinates in the present case is connected with a canonical transformation (see problem 11.8–11.10).

a) If the magnetic field is weak, $\omega_B \ll \omega_1 - \omega_2$, the ellipses of the normal oscillations are strongly elongated, and the frequencies

$$\Omega_{1,2} \approx \omega_{1,2} \pm \frac{\omega_B^2\omega_{1,2}}{2(\omega_1^2 - \omega_2^2)}$$

are close to $\omega_{1,2}$. The orbit of the oscillator without a magnetic field fills a rectangle with sides parallel to the coordinate axes (see problem 6.5); the influence of a weak magnetic field is merely to slightly deform the region filled by the orbit. The Larmor theorem (see [7], § 17.3) is not applicable as the field $U(\mathbf{r})$ is not symmetric with respect to the z-axis.

b) In a strong magnetic field, $\omega_B \gg \omega_{1,2}$, the normal oscillation with frequency $\Omega_1 \approx \omega_B$ takes place along circle, and the normal oscillation with frequency $\Omega_2 \approx \omega_1 \omega_2 / \omega_B$ along an ellipse with axes which are in the x- and y-directions and which stand in the ratio ω_2/ω_1. The motion is thus along the circle with a centre which moves slowly along the ellipse.

It is well-known that if a charged particle moves in a strong uniform magnetic field in a plane at right angles to the field, the occurrence of a weak, quasi-uniform field $U(\mathbf{r})$ (i.e. such that the force $\mathbf{F} = -\dfrac{\partial U}{\partial \mathbf{r}}$ changes little within the circular orbit) leads to a slow displacement (drift) of the centre of the orbits in a direction at right angles to $\mathbf{F}$ (i.e. along the equipotential lines of $U(\mathbf{r})$) (see [2], § 22). Note that in our case a similar drift occurs also in a strongly inhomogeneous oscillator field.

c) If $\omega_1 = \omega_2$, the normal oscillations in the (x, y)-plane correspond to motions along a circles in opposite senses with frequencies $\Omega_{1,2} = \widetilde{\omega} \pm \frac{1}{2}\omega_B$, where $\widetilde{\omega} = \sqrt{\omega_1^2 + \frac{1}{4}\omega_B^2}$. In a system rotating with the angular velocity $(-\frac{1}{2}\omega_B)$ the frequencies of both motions thus turns out to be equal to $\widetilde{\omega}$. Such motions are the normal oscillations of an isotropic oscillator with frequency $\widetilde{\omega}$. Indeed, the sum and difference of such oscillations with equal amplitudes,

$$\begin{pmatrix} \cos\widetilde{\omega}t \\ -\sin\widetilde{\omega}t \end{pmatrix} \pm \begin{pmatrix} \cos\widetilde{\omega}t \\ \sin\widetilde{\omega}t \end{pmatrix}$$

are linear oscillations in the x- or y-directions (if we neglect the motion in the direction of the magnetic field).

If the magnetic field is weak, $\omega_B \ll \omega_1$, we have $\widetilde{\omega} \approx \omega_1$, and the whole effect of the field on the motion of the oscillator reduces to a rotation ("precession") around the z-axis with a frequency $(-\frac{1}{2}\omega_B)$ (Larmor theorem, compare [7], § 17.3 and [2], § 45). If, however, $\omega_B \gtrsim \omega_1$, there is no longer any obvious use for the rotating system which we have employed.

6.38. We can solve the equations of motion,

$$\ddot{x} + \omega_1^2 x = \omega_z \dot{y},$$
$$\ddot{y} + \omega_2^2 y = -\omega_z \dot{x} + \omega_x \dot{z},$$
$$\ddot{z} + \omega_3^2 z = -\omega_x \dot{y},$$

with

$$\omega_x = \frac{eB_x}{mc}, \qquad \omega_z = \frac{eB_z}{mc}$$

by the method of successive approximations. We look for the coordinates in the form $x = x^{(1)} + x^{(2)}$, $y = y^{(1)} + y^{(2)}$, $z = z^{(1)} + z^{(2)}$, where $x^{(2)}, y^{(2)}$, and $z^{(2)}$ are small compared to $x^{(1)}, y^{(1)}$, and $z^{(1)}$. In first approximation we neglect small terms in the right-hand sides of the equations:

$$x^{(1)} = A\cos(\omega_1 t + \alpha),$$
$$y^{(1)} = B\cos(\omega_2 t + \beta),$$
$$z^{(1)} = C\cos(\omega_3 t + \gamma).$$

We get

$$x^{(2)} = \frac{-\omega_z \omega_2 B \sin(\omega_2 t + \beta)}{\omega_1^2 - \omega_2^2},$$
$$y^{(2)} = \frac{\omega_1 \omega_z A \sin(\omega_1 t + \alpha)}{\omega_2^2 - \omega_1^2} - \frac{\omega_x \omega_3 C \sin(\omega_3 t + \gamma)}{\omega_2^2 - \omega_3^2}, \qquad (2)$$
$$z^{(2)} = \frac{\omega_x \omega_2 B \sin(\omega_2 t + \beta)}{\omega_3^2 - \omega_2^2}.$$

The corrections turns out to be small, provided $|\omega_z| \ll |\omega_1 - \omega_2|$ and $|\omega_x| \ll |\omega_2 - \omega_3|$. The normal oscillations are oscillations along ellipses which are strongly alongated along the coordinate axes.

If, however, for example, $|\omega_z| \gtrsim |\omega_1 - \omega_2|$, $|\omega_x| \ll |\omega_2 - \omega_3|$, according to (2) $x^{(2)}$ and $y^{(2)}$ are no longer small. This is connected with the fact that the frequencies of the "forces" $\omega_z \dot{y}^{(1)}$ and $-\omega_z \dot{x}^{(1)}$ in (1) turn out to lie close to the eigen-frequencies of the oscillator. In that case we must retain the resonant terms in the first approximation equations:

$$\ddot{x}^{(1)} + \omega_1^2 x^{(1)} - \omega_z \dot{y}^{(1)} = 0,$$
$$\ddot{y}^{(1)} + \omega_2^2 x^{(1)} + \omega_z \dot{x}^{(1)} = 0, \qquad (3)$$
$$\ddot{z}^{(1)} + \omega_3^2 z^{(1)} = 0;$$

that is, it is necessary to take the effect of B_z on the motion exactly into account. We consider the set (3) in problem 6.37. For the second-order corrections, we have the equations

$$\ddot{x}^{(2)} + \omega_1^2 x^{(2)} = 0,$$
$$\ddot{y}^{(2)} + \omega_2^2 y^{(2)} = \omega_x \dot{z}^{(1)},$$
$$\ddot{z}^{(2)} + \omega_3^2 z^{(2)} = -\omega_x \dot{y}^{(1)}.$$

For the sake of simplicity, we restrict ourselves to the case $\omega_1 = \omega_2 \equiv \omega \gg \omega_z$ and we have then the following normal oscillations:

$$
\begin{pmatrix} x \\ y \\ z \end{pmatrix} = \mathrm{Re} \left\{ A_1 \begin{pmatrix} 1 \\ i \\ \dfrac{\omega\omega_x}{\omega_3^2 - \omega^2} \end{pmatrix} e^{i(\omega+\omega_z/2)t} + \right.
$$

$$
\left. + A_2 \begin{pmatrix} 1 \\ -i \\ \dfrac{-\omega\omega_x}{\omega_3^2 - \omega^2} \end{pmatrix} e^{i(\omega-\omega_z/2)t} + A_3 \begin{pmatrix} 0 \\ \dfrac{i\omega_x\omega_3}{\omega_3^2 - \omega^2} \\ 1 \end{pmatrix} e^{i\omega_3 t} \right\}. \tag{4}
$$

In the approximation used, the normal oscillations (4) with frequencies $\omega \pm \frac{1}{2}\omega_z$, take place along a circle in planes which make angles $\mp \omega_x\omega/(\omega_3^2 - \omega^2)$ with the (x, y)-plane, while the oscillation with frequency ω_3 is along an ellipse in the (y, z)-plane which is strongly elongated in the z-direction.

6.39. We assume that the oscillations of the pendulum are small oscillations: we reckon the angle φ counterclockwise from the vertical, and as the second coordinate we choose the charge q on the right-hand plate. When the pendulum is deflected over an angle φ, the magnetic flux through the circuit is equal to $\Phi = \mathrm{const} - \frac{1}{2}Bl^2\varphi$, so that the Lagrangian function of the system is (see problem 4.24)

$$
L = \frac{1}{2} \left(ml^2\dot{\varphi}^2 + \mathscr{L}\dot{q}^2 - mgl\varphi^2 - \frac{q^2}{C} - Bl^2\varphi\dot{q} \right).
$$

If we introduce the coordinates $x = l\varphi$ and $y = \sqrt{\mathscr{L}/m}\, q$, the Lagrangian function of our system differs from the one considered in problem 6.37 (with parameters $\omega_1^2 = \frac{g}{l}$, $\omega_2^2 = \frac{1}{\mathscr{L}C}, \omega_B = -\dfrac{Bl}{2\sqrt{m\mathscr{L}}}$ and $z = 0$) only by a total derivative with respect to the time: $\frac{1}{2}m\omega_B\dfrac{d(xy)}{dt}$. The equations of motion of problem 6.37 and their solutions are thus also valid in the present case.

6.40. Let

$$
\mathbf{r} = \mathbf{A}\cos(\omega t + \varphi), \quad \mathbf{A} = (A_1, A_2, \ldots, A_N) \tag{1}
$$

be any normal oscillation. Since the replacement

$$
x_i \to \sum_j S_{ij}x_j
$$

does not change the form of the Lagrangian function, a normal oscillation of the form

$$\hat{S}\mathbf{r} = \hat{S}\mathbf{A}\cos(\omega t + \varphi), \quad (\hat{S}\mathbf{A})_i = \sum_j S_{ij}A_j \tag{2}$$

must be valid together with (1). Here $\hat{S}$ is a matrix with elements S_{ij}, which have the properties:

$$\hat{S}^T = \hat{S}, \quad \hat{S}\hat{S} = \hat{E}, \tag{3}$$

where $\hat{E}$ is a unit matrix and $\hat{S}^T$—the transposed matrix $\hat{S}$.

a) If the given frequency ω is not degenerate, the solution (2) can differ from (1) except for a common factor: $\hat{S}\mathbf{r} = c\mathbf{r}$, moreover,

$$\hat{S}\hat{S}\mathbf{r} = c\hat{S}\mathbf{r} = c^2\mathbf{r}. \tag{4}$$

Since $\hat{S}\hat{S} = \hat{E}$, we obtain from (4) that $\mathbf{r} = c^2\mathbf{r}$, or $c^2 = 1$ and $c = \pm 1$. Therefore, for a non-degenerate frequency, we get

$$\text{or} \quad \hat{S}\mathbf{r} = +\mathbf{r}, \quad \text{or} \quad \hat{S}\mathbf{r} = -\mathbf{r}.$$

b) If the frequency ω is degenerate, then oscillations (1) and (2) may not coincide. But their sum and difference

$$\mathbf{r} \pm \hat{S}\mathbf{r} = (\mathbf{A} \pm \hat{S}\mathbf{A})\cos(\omega t + \varphi)$$

are the normal oscillations with the same frequency and have all necessary properties of symmetry.

c) The addition to the Lagrangian function has the form $\Delta L = \sum_i f_i x_i$, where

$$\mathbf{f} = (f_1, f_2, \ldots, f_N)$$

is the external force acting upon the system.

Let the force $\mathbf{f}$ be symmetric, while the normal oscillation $\mathbf{r}_a$ (1) be anti-symmetric with respect to transformation S; that is,

$$\hat{S}\mathbf{f} = +\mathbf{f}, \quad \hat{S}\mathbf{r}_a = -\mathbf{r}_a. \tag{5}$$

This force does not affect the oscillation $\mathbf{r}_a$ if the vectors $\mathbf{f}$ and $\mathbf{r}_a$ are mutually orthogonal (see problem 6.25):

$$(\mathbf{f}, \mathbf{r}_a) = 0. \tag{6}$$

It follows from (5) that

$$(\hat{S}\mathbf{f}, \hat{S}\mathbf{r}_a) = -(\mathbf{f}, \mathbf{r}_a). \tag{7}$$

On the other hand, the left-hand side of (7) can be rewritten as

$$(\hat{S}\mathbf{f}, \hat{S}\mathbf{r}_a) = (\mathbf{f}, \hat{S}^T \hat{S}\mathbf{r}_a) = +(\mathbf{f}, \mathbf{r}_a) \tag{8}$$

due to (3). Then, comparing (7) and (8), we obtain (6).

Do parts a)–c) of the problem remain unchanged if we do not require in advance the condition $\hat{S}^T = \hat{S}$?

6.41. Let x_i be the displacement of the ith particle along the ring from the equilibrium position; for definiteness, we consider a counterclockwise displacement as positive. The system is clearly symmetric with respect to rotation over an angle $180°$ around the *AB*-axis which passes through the equilibrium position of the second particle and the centre of the ring. Therefore, the Lagrangian function of the system

$$L = \tfrac{1}{2}m\left(\dot{x}_1^2 + \dot{x}_2^2 + \dot{x}_3^2\right) + \tfrac{1}{2}M\left(\dot{x}_4^2 + \dot{x}_5^2\right) - \tfrac{1}{2}k\left[\sum_{i=1}^{4}(x_i - x_{i+1})^2 + (x_5 - x_1)^2\right]$$

does not change its form under the replacement corresponding to such a rotation

$$x_2 \to -x_2 \quad x_1 \to x_3, \quad x_3 \to -x_1, \quad x_4 \to -x_5, \quad x_5 \to x_4. \tag{1}$$

Using the symmetry considerations (see the preceding problem) and the orthogonality conditions, it is easy to reduce this problem with five degrees of freedom to the two independent problems with two degrees of freedom each.

Indeed, the vectors of the normal oscillation, which are symmetric and anti-symmetric with respect to the transformation (1), have the form

$$\mathbf{r}_s = \begin{pmatrix} a \\ 0 \\ -a \\ b \\ -b \end{pmatrix} \cos(\omega_s t + \varphi_s), \qquad \mathbf{r}_a = \begin{pmatrix} c \\ d \\ c \\ f \\ f \end{pmatrix} \cos(\omega_a t + \varphi_a).$$

In addition, one anti-symmetric "oscillation" can be easily guessed—it is a rotation of all the particles around the ring

$$
\mathbf{r}_{a1} = \begin{pmatrix} 1 \\ 1 \\ 1 \\ 1 \\ 1 \end{pmatrix} (Ct + C_1), \quad \omega_{a1} = 0.
$$

Two other (besides $\mathbf{r}_{a1}$) anti-symmetric oscillations must be orthogonal to $\mathbf{r}_{a1}$ with the metric tensor determined by the coefficients of the kinetic energy, that is,

$$
m(2c + d) + 2Mf = 0. \tag{2}
$$

As a result, the oscillations $\mathbf{r}_a$ and $\mathbf{r}_s$ have only two undefined coefficients. To determine them, it is enough to use only two equations of motion out of five, for example, those for the first and fifth particles:

$$
\begin{aligned}
m\ddot{x}_1 + k(2x_1 - x_2 - x_5) &= 0, \\
M\ddot{x}_5 + k(2x_5 - x_4 - x_1) &= 0.
\end{aligned} \tag{3}
$$

Substituting the explicit form of $\mathbf{r}_s$, we find for the two symmetrical oscillations

$$
b_{1,2} = \left(\frac{m}{k} \omega_{s1,2}^2 - 2 \right) a_{1,2},
$$

$$
\omega_{s1,2}^2 = \frac{k}{2mM} \left(2M + 3m \mp \sqrt{4(M - m)^2 + 5m^2} \right).
$$

Similarly, substituting in (3) the vector $\mathbf{r}_a$ and taking into account (2), we find

$$
c_{2,3} = \left(1 - \frac{M}{k} \omega_{a2,3}^2 \right) f_{2,3}, \quad d_{2,3} = -2c_{2,3} - \frac{2M}{m} f_{2,3},
$$

$$
\omega_{a2,3}^2 = \frac{k}{2mM} \left(4M + m \mp \sqrt{\frac{1}{2}(4M - m)^2 + \frac{1}{2}m^2} \right).
$$

6.42. The considered system is close to the one studied in problem 6.29a; the Lagrangian function in our problem differs by a small quantity

$$
\delta L = \delta L_1(x, \dot{x}) + \delta L_2(y, \dot{y}) + \delta L_3(z, \dot{z}),
$$

$$
\delta L_1(x, \dot{x}) = \frac{1}{2} \varepsilon k \left[x_2^2 + (x_2 - x_5)^2 + (x_4 - x_5)^2 + x_4^2 \right],
$$

where $\varepsilon = (l - l_1)/l \ll 1$. The oscillations in $x, y,$ and z occur independently; in the following we are only interested in the x-oscillations.

To determine the corrections to the frequencies, it is helpful to use the method of successive approximations (see problem 6.35). The frequencies $\omega_{3,4}$ are non-degenerate; therefore, we can directly use (4) of problem 6.35 to these oscillations.

The frequency ω_1 of the original problem (6.29a) is threefold degenerate; therefore, it may seem that to determine the corrections to the frequencies and vectors of the normal oscillation we will have to consider the set of equations of (5) from problem 6.35. However, the symmetry properties of the system enable us to indicate the vectors of the normal oscillations of the original system, which change a little when we add δL. These are the vectors (1) from problem 6.29, because only they have the certain symmetry properties. Namely, the oscillation $\mathbf{r}_3$ is symmetric with respect to the AB-axis and anti-symmetric with respect to the CD-axis, and the oscillation $\mathbf{r}_1$ is symmetric and $\mathbf{r}_2$, anti-symmetric with respect to both axes. Corrections to the frequencies of these oscillations can also be calculated using (4) from problem 6.35.

Substituting $x_1 = -x_3 = 1, x_2 = x_4 = x_5 = 0$, we find

$$\delta K_{11} = -2\delta L_1 = 0;$$

therefore, $\delta\omega_1 = 0$. Similarly,

$$\delta K_{22} = -2\delta L_1(x_1 = x_3 = x_5 = 0, x_2 = -x_4 = 1) = -4\varepsilon k,$$
$$M_2 = 2L_1(\dot{x}_1 = \dot{x}_3 = \dot{x}_5 = 0, \dot{x}_2 = -\dot{x}_4 = 1, x_i = 0) = 2m,$$

and $\delta\omega_2 = -\varepsilon\sqrt{k/(2m)}$;

$$\delta K_{33} = -4\varepsilon k, \quad M_3 = 4m, \quad \delta\omega_3 = -\tfrac{1}{2}\varepsilon\sqrt{k/(2m)}.$$

Representing the vector of the initial displacements

$$\mathbf{r}(0) = \begin{pmatrix} a \\ 0 \\ 0 \\ -a \\ 0 \end{pmatrix}$$

and the vector of the initial velocities $\dot{\mathbf{r}}(0) = 0$ as $\mathbf{r}(0) = \sum_i \mathbf{r}_i(0)$ and $\dot{\mathbf{r}}(0) = \sum_i \dot{\mathbf{r}}_i(0)$, respectively, we find that

$$A_1 = A_2 = A_3 = \tfrac{1}{2}a, \quad A_4 = A_5 = \varphi_i = 0.$$

Thus, in the given approximation, the fourth and fifth normal oscillations are not excited, and the oscillations of the particles

$$x_{1,3} = \tfrac{1}{2}a(\pm\cos\omega_1 t + \cos\omega_3 t),$$
$$x_{2,4} = \tfrac{1}{2}a(\pm\cos\omega_2 t - \cos\omega_3 t), \quad x_5 = 0$$

are beats (see problem 6.32).

6.43. In this problem, it is helpful to use the method of successive approximations (see problem 6.35). The change in mass leads to the correction to the Lagrangian function of the form

$$\delta L = \tfrac{1}{2}(\delta m_1\,\dot{x}_1^2 + \delta m_2\,\dot{x}_2^2).$$

This correction must be expressed in terms of the normal coordinates of the original system (see problem 6.22). In this case, the coefficient in front of the product of the generalized velocities $\dot{q}_1\dot{q}_2$ corresponding to the degenerate frequencies is zero. Other products $\dot{q}_l\dot{q}_s$ (for $\omega_l \neq \omega_s$) can be omitted as this issue has been noted in problem 6.35. We get

$$\delta L_1 = \tfrac{1}{4}\delta m_1\,\dot{q}_1^2 + \tfrac{1}{4}\delta m_2\,\dot{q}_2^2 + \tfrac{1}{8}(\delta m_1 + \delta m_2)(\dot{q}_3^2 + \dot{q}_4^2).$$

The Lagrangian function $L + \delta L_1$, as well as the Lagrangian function of the original system, is reduced to the terms, each of which contains only one of the coordinates q_l. As a result, the coordinates q_l remain the normal ones, and to calculate the corrections to the frequencies, one can use (4) of problem 6.35:

$$\delta\omega_1 = -\tfrac{1}{4}\varepsilon_1\omega_1, \quad \delta\omega_2 = -\tfrac{1}{4}\varepsilon_2\omega_3, \quad \delta\omega_3 = -\tfrac{1}{8}(\varepsilon_1 + \varepsilon_2)\omega_3,$$
$$\delta\omega_4 = 0, \quad \varepsilon_i = \delta m_i/m.$$

All eigen-frequencies of the system become different; the ambiguity in choice of the vectors of the normal oscillations disappears since with the accuracy up to ε_i these vectors are the vectors (1) from problem 6.22.

It is noteworthy that when $\delta m_1 = \delta m_2$, the frequencies $\omega_1 + \delta\omega_1$ and $\omega_2 + \delta\omega_2$ coincide with each other again (up to the second-order corrections $|\delta\omega_1 - \delta\omega_2| \sim \varepsilon_1^2\omega_1$). In this case, the Lagrangian function $L + \delta L_1$ once more leads to the ambiguous choice of the vectors of the normal oscillations. However, in the exact solution of the problem at $\delta m_1 = \delta m_2$, the vectors of the normal oscillations have the form

$$\begin{pmatrix} 1 \\ 1 \\ -1-\varepsilon \\ -1-\varepsilon \end{pmatrix}, \quad \begin{pmatrix} 1 \\ -1 \\ \tfrac{1}{2}(3\mp\sqrt{4+4\varepsilon+9\varepsilon^2}) \\ -\tfrac{1}{2}(3\mp\sqrt{4+4\varepsilon+9\varepsilon^2}) \end{pmatrix}, \quad \begin{pmatrix} 1 \\ 1 \\ 1 \\ 1 \end{pmatrix}$$

and at small ε they are close to the vectors (2) of problem 6.22 (up to the normalization factors, omitted here). The sharp change in the form of the normal oscillations occurs in a very narrow range of masses $|\delta m_1 - \delta m_2| \lesssim \varepsilon^2 m$ (cf. problem 6.6a). To determine the vectors of the normal oscillations in this range of values of δm_1 and δm_2, it would be possible to use the next approximation in the method of successive approximations.

6.44. Obviously, the motion of particles in the direction of the AA- and BB-axes is independent. We will consider only the motion along the AA-axis.

For the first and fourth particles, we assume the displacements to be positive to the left, while for the second and third, positive to the right. According to the results of problem 6.40, the normal oscillations $\mathbf{r} = (x_1, x_2, x_3, x_4)$ can be chosen either symmetric or anti-symmetric with respect to the AA- and BB-axes. The oscillations which are symmetric with respect to the AA-axis have $x_1 = x_4, x_2 = x_3$. If they are also symmetric with respect to the BB-axis, then $x_1 = x_2$, $x_3 = x_4$, so for this twofold symmetric oscillation we have $\mathbf{r}_{ss} = (1, 1, 1, 1)q_{ss}$.

For the oscillations which are symmetric with respect to the AA-axis and anti-symmetric with respect to the BB-axis, we have $x_1 = x_4, x_2 = x_3$ and $x_1 = -x_2, x_4 = -x_3$, so that

$$\mathbf{r}_{sa} = (1, -1, -1, 1)q_{sa}.$$

Similarly, we find

$$\mathbf{r}_{as} = (1, 1, -1, -1)q_{as}, \quad \mathbf{r}_{aa} = (1, -1, 1, -1)q_{aa}.$$

The vectors of the normal oscillations in the vertical direction can be found in the same way.

The oscillation frequencies can be found by substituting the vectors obtained in the equations of motion.

In the presence of degeneration, there are a lot of normal oscillations (except for the ones as previously described) which do not have the previously described symmetry properties. It is easy to figure out, for example, that the frequencies of the ss and aa oscillations coincide, $\omega_{ss} = \omega_{aa} = 2\sqrt{k/m}$, if the tension of the springs is not arbitrarily, but equals kl (where l is the length of each of the springs in the equilibrium position). In that case, however, any superposition of vectors $\mathbf{r}_{aa}$ and $\mathbf{r}_{ss}$ will be the normal oscillations, for example, the vector $(1, 0, 1, 0)q_{ss}$.

Similarly, we can find the normal oscillations along the BB-axis.

6.45. Taking into account the symmetry properties allows us to reduce this system with seven degrees of freedom to a few simple properties with not more than two degrees of freedom each. Indeed, due to the symmetry of the system with respect to a plane perpendicular to the plane of the scales, all the normal oscillations can be chosen either symmetric or anti-symmetric with respect to this plane. In addition, the normal oscillations are divided into those which displace the particles from the plane of the scales and those which do not. We will consider only the latter ones.

Let α, β, and γ be the deviation angles from the vertical of the centre of the frame, threads BA, and DE, respectively. Apart from the obvious symmetric oscillations $\alpha = 0$, $\beta = -\gamma$ with frequency $\sqrt{g/(3l)}$, there are two anti-symmetric oscillations for which $\beta = \gamma$. Since the contributions of different normal oscillations to the Lagrangian function are additive, then to find the anti-symmetric oscillations, it will suffice to know only term

$$L_a = ml^2(2\dot{\alpha}^2 + 9\dot{\beta}^2 + 3\dot{\alpha}\dot{\beta}) - mgl(\alpha^2 + 3\beta^2),$$

corresponding to this type of oscillations. As a result, we obtain two anti-symmetric oscillations: $\alpha = \beta = \gamma$ with frequency $\sqrt{2g/(7l)}$ and $\alpha = -3\beta = -3\gamma$ with frequency $\sqrt{2g/(3l)}$.

To describe the oscillations that displace the particles from the plane of the scales, we use the Cartesian coordinates of the particle displacements from the equilibrium position along of the x-axis, which in turn is directed perpendicular to the equilibrium plane of the scales. It is obvious that the symmetric oscillations

$$x_A = x_E, \quad x_B = x_D$$

coincide with the oscillations of the coplanar double pendulum[1]; therefore,

$$\omega_s^2 = \left(7 \mp \sqrt{37}\right)\frac{g}{3l}, \quad x_A = (6 \pm \sqrt{37})x_B.$$

Of the two anti-symmetric oscillations

$$x_A = -x_E, \quad x_B = -x_D,$$

one is obvious: it is a rotation of the scales around the vertical axis: $x_A = x_B$. The other anti-symmetric oscillation is orthogonal to the one we found, and therefore, we have $x_A = -x_B = x_D = -x_E$. But this oscillation does not shift the centre of each thread in the first approximation, so the frequency of such oscillations coincides with the frequency of the pendulum of the length $3l/2$; that is, it equals $\sqrt{2g/(3l)}$.

6.46. Obviously, the motions of particles along the AA- and BB-axes and in the direction perpendicular to the plane of the frame are independent. As an example, we will consider the oscillations along the AA-axis.

---

[1] See [1], § 23, problem 2 with parameters

$$m_1 = m_2 = 2m, \ l_1 = \tfrac{1}{2}l, \ l_2 = 3l, \ \varphi_1 = \frac{2x_B}{l}, \ \varphi_2 = \frac{x_A - x_B}{3l}.$$

It is helpful to place the components of the oscillation vector in the table similar in shape to the frame. For particles on the left from the axis-BB, we take the displacements to the left as positive, while for the particles located on the right we take the displacements to the right as positive. The oscillations ss symmetric with respect to both AA- and BB-axes have the form

$$\begin{pmatrix} x & X & X & x \\ x & X & X & x \end{pmatrix}.$$

The oscillations x, X are reduced to the oscillations of the system considered in problem 6.8 with $m_1 = m$, $m_2 = M$, where we must also put $k_1 = k + k'$, $k_2 = k$, $k_3 = 2k + k'$, $k' = f/l$. Thus, there are two ss oscillations.

The sa-oscillations, which are symmetric with respect to the AA-axis and anti-symmetric with respect to the BB-axis, have the form

$$\begin{pmatrix} x & X & -X & -x \\ x & X & -X & -x \end{pmatrix},$$

where, in this case, $k_3 = k'$, $k_1 = k + k'$.

Similarly, we have for as-and aa-oscillations:

$$\begin{pmatrix} x & X & X & x \\ -x & -X & -X & -x \end{pmatrix}, \quad k_1 = k + 3k', \quad k_3 = 2k + 3k',$$

$$\begin{pmatrix} x & X & -X & -x \\ -x & -X & X & x \end{pmatrix}, \quad k_1 = k + 3k', \quad k_3 = 3k'.$$

The remaining sixteen normal oscillations can be found in the similar manner.

6.47. In the notation of the preceding problem, the force vector is

$$\begin{pmatrix} -k - k', & -k', & k', & k + k' \\ -k - k', & -k', & k', & k + k' \end{pmatrix} a \cos \gamma t.$$

It is orthogonal to all the normal oscillations, except for symmetric–anti-symmetric (sa). Therefore, the resonance only occurs for two frequencies $\gamma = \omega_{1,2}^{sa}$, where

$$(\omega_{1,2}^{sa})^2 = \frac{1}{2mM} \left\{ k(2M + m) + \frac{f}{l}(M + m) \pm \right.$$

$$\left. \pm \sqrt{\left[k(2M - m) + \frac{f}{l}(M - m) \right]^2 + 4k^2 Mm} \right\}.$$

6.48. The linear molecule C_2H_2 has only seven normal oscillations: two flexural oscillations in two mutually perpendicular planes and three longitudinal oscillations (e.g. see [1], §24).

The problem of determining the two symmetric longitudinal oscillations $x_1 = -x_4$, $x_2 = -x_3$ is reduced to problem 6.8 with parameters $k_2 = k_{HC}$, $k_3 = 2k_{CC}$, $k_1 = 0$, where k_{HC} and k_{CC} are stiffness of the HC and CC bonds. The anti-symmetric longitudinal oscillation $x_1 = x_4$, $x_2 = x_3$, $m_H x_1 + m_C x_2 = 0$ has the frequency

$$\omega_{a1} = \sqrt{\frac{k_{HC}(m_H + m_C)}{m_H m_C}}$$

(the same as the "molecule" CH). Note that the longitudinal displacement of the molecule as a whole, $x_1 = x_2 = x_3 = x_4$, can be considered as the second anti-symmetric oscillation, and the condition of its orthogonality to the first anti-symmetric oscillation coincides with the requirement that the total linear momentum of the molecule is zero.

The transverse symmetric oscillation reads $y_1 = y_4$, $y_2 = y_3$,

$$m_H y_1 + m_C y_2 = 0, \quad \omega_{s3}^2 = \frac{k_{HCC}(m_H + m_C)}{l_{HC}^2 m_H m_C}.$$

Here k_{HCC} is the stiffness of the molecule at the flexure; that is, the potential energy increases by the term $k_{HCC}\delta^2/2$ when the bond HCC bends over an angle δ.

The transverse anti-symmetric oscillation reads $y_1 = -y_4$, $y_2 = -y_3$,

$$m_C l_{CC} y_2 + m_H (l_{CC} + 2l_{HC}) y_1 = 0,$$

$$\omega_{a2}^2 = \frac{k_{HCC}[m_C l_{CC}^2 + m_H (2l_{HC} + l_{CC})^2]}{m_H m_C l_{HC}^2 l_{CC}^2},$$

where l_{HC} and l_{CC} are the equilibrium distances between the H–C and C–C atoms, respectively. Note that the relation between the displacements of the atoms of this oscillation can be found from the requirement that the total angular momentum of the molecule is zero (or due to orthogonality to the vector of the rotation of a molecule as a whole $y_1 = -y_4$, $y_2 = -y_3$, $y_2(2l_{HC} + l_{CC}) = y_1 l_{CC}$, which can be considered as the second anti-symmetric oscillation).

6.49. We denote the displacements of the particle from the equilibrium position in the direction BD through x_1 and x_2, and in the direction CF—through y_1 and y_2. The rotation around the axis of symmetry CF over the angle $180°$ leads to the replacements:

$$x_1 \rightleftarrows -x_2, \quad y_1 \rightleftarrows y_2;$$

that is, the transformation $\hat{S}$, which is the rotation of the oscillation vector $\mathbf{r} = (x_1, y_1, x_2, y_2)$, has the form

$$\hat{S}\mathbf{r} = (-x_2, y_2, -x_1, y_1).$$

Due to the symmetry of the system, the normal oscillations of two types are possible in it: $\mathbf{r}_s$ – symmetric with respect to the CF-axis (for which $\hat{S}\mathbf{r}_s = \mathbf{r}_s$) and anti-symmetric $\mathbf{r}_a$ (for which $\hat{S}\mathbf{r}_a = -\mathbf{r}_a$). For the symmetric oscillations, we have $(-x_2, y_2, -x_1, y_1) = (x_1, y_1, x_2, y_2)$, from where

$$\mathbf{r}_s = (x, y, -x, y).$$

Analogously,

$$\mathbf{r}_a = (x, y, x, -y).$$

For the anti-symmetric oscillations, the rotation around the axis of symmetry is equivalent to the change of the oscillation phase by π.

Two symmetric and two anti-symmetric oscillations are possible; to determine the frequency and vectors of each pair of oscillations, it is sufficient to use only two equations of motion.

We can obtain some more information about the type of normal oscillations without knowing stiffness of the springs. The condition of orthogonality of the oscillation vectors $\mathbf{r}_{s1}$ and $\mathbf{r}_{s2}$ can be reduced to equality

$$2m(x_{s1}x_{s2} + y_{s1}y_{s2}) = 0,$$

which shows that the vectors of the displacements of, for example, particle 1 at each of the two symmetrical oscillations are mutually orthogonal (Fig. 141a). The same applies to the antisymmetric oscillations (Fig. 141b).

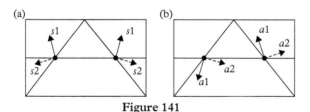

Figure 141

The direction in which the particle displaces at each of the normal oscillations cannot be determined without knowing stiffness of the springs. Indeed, if stiffness and tension springs AC and CE are small, then the particle displacements under the normal oscillations are directed almost along or across the springs BD. On the contrary, if the stiffness of springs BD is low, the normal oscillations occur almost along or across the springs AC and CE.

Of course, if the normal oscillation is degenerate, one can choose other vectors of the normal oscillations which do not possess the symmetry properties. For example, if we "switch off" spring BD, each of the particles will be able to oscillate along or across springs AC and CE.

6.50. We denote the displacement of each atom from its equilibrium position in the OD-directions through x_i, in the OA-directions through y_i, and perpendicular to the plane of the molecule through z_i.

The oscillations of the molecule C_2H_4 are divided into those which leave the molecule planar and those which output atoms of the plane of the molecule. We will start with the latter.

It is helpful to present the vector of displacement of the atoms in the form

$$\mathbf{r} = \begin{pmatrix} z_1 & & z_4 \\ & z_2\ z_5 & \\ z_3 & & z_6 \end{pmatrix}.$$

For the oscillation symmetric with respect to the AB-axis, we have

$$z_4 = z_1, \quad z_5 = z_2, \quad z_6 = z_3.$$

If, in addition, this oscillation is anti-symmetric with respect to the CD-axis, we get

$$z_3 = -z_1, \quad z_2 = -z_2, \quad z_5 = -z_5, \quad z_6 = -z_4.$$

As a result,

$$\mathbf{r}_{sa} = \begin{pmatrix} 1 & & 1 \\ & 0\ 0 & \\ -1 & & -1 \end{pmatrix} q_{sa}.$$

This "oscillation" turns out to be, in fact, the rotation around the CD-axis.

Similarly, the symmetric oscillations with respect to both axes are

$$\mathbf{r}_{ss} = \begin{pmatrix} z_1 & & z_1 \\ & z_2\ z_2 & \\ z_1 & & z_1 \end{pmatrix}.$$

One of them, in particular, is

$$\mathbf{r}_{ss,1} = \begin{pmatrix} 1 & & 1 \\ & 1\ 1 & \\ 1 & & 1 \end{pmatrix} q_{ss,1}.$$

This is a translational motion. The amplitudes of the other oscillation can be determined by taking into account its orthogonality to the vector $\mathbf{r}_{ss,1}$:

$$4mz_1 + 2Mz_2 = 0,$$

where m and M are the masses of the hydrogen and carbon atoms, respectively:

$$\mathbf{r}_{ss,2} = \begin{pmatrix} 1 & & 1 \\ & -\dfrac{2m}{M}\ -\dfrac{2m}{M} & \\ 1 & & 1 \end{pmatrix} q_{ss,2}.$$

The antisymmetric oscillation with respect to both axes,

$$\mathbf{r}_{aa} = \begin{pmatrix} 1 & & -1 \\ & 0\ 0 & \\ -1 & & 1 \end{pmatrix} q_{aa},$$

is a torsional oscillation around the CD-axis.

Finally,

$$\mathbf{r}_{as,i} = \begin{pmatrix} 1 & & -1 \\ & a_i\ -a_i & \\ 1 & & -1 \end{pmatrix} q_{as,i}.$$

Here for the rotation around the AB-axis (the oscillation $\mathbf{r}_{sa,1}$), the factor $a_1 = l_{OC}/l_{OH}$ is the ratio of the distances of the carbon atoms and hydrogen atoms from the AB-axis in the equilibrium position; for the flexural oscillation ($\mathbf{r}_{as,2}$), it is

$$a_2 = -\frac{2m}{M}\frac{l_{OH}}{l_{OC}}.$$

The similar approach can also be applied to the oscillations which do not output atoms from the plane of the molecule.

We can present the corresponding displacement vectors in the form

$$\mathbf{r} = \begin{pmatrix} x_1\,y_1 & & x_4\,y_4 \\ & x_2\,y_2\,x_5\,y_5 & \\ x_3\,y_3 & & x_6\,y_6 \end{pmatrix}.$$

The general form of the *as*-oscillations is

$$\mathbf{r}_{as} = \begin{pmatrix} x_1\ \ y_1 & & x_1\ -y_1 \\ & x_2\,0\,x_2\,0 & \\ x_1\,-y_1 & & x_1\ \ y_1 \end{pmatrix}.$$

One of the *as*-"oscillations" is the translational motion of the molecule as a whole along the *x*-axis ($x_1 = x_2, y_1 = 0$). The other two *as*-oscillations are shown in Fig. 142a and b.

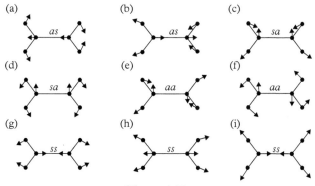

(a) (b) (c)
  as   *as*   *sa*

(d) (e) (f)
  sa   *aa*   *aa*

(g) (h) (i)
  ss   *ss*   *ss*

Figure 142

To visualize the kind of displacements of atoms under these vibrations, note that the distance between the carbon atoms does not change: the C–C link "does not work". If we neglect the interaction between the relatively distant atoms (e.g., 1–4, 1–5 in Fig. 43) and also stiffness of the angles of the form 1–2–5, then the considered oscillations will coincide with those which are symmetric with respect to the *CD*-axis oscillations of two "molecules" H_2C occurring in anti-phase (cf. problem 6.48; oscillations of molecules of the form A_2B are considered in [1], §24, problem 2).

The general form of *sa*-oscillations is

$$\mathbf{r}_{sa} = \begin{pmatrix} x_1\ \ y_1 & & -x_1\,y_1 \\ & 0\,y_2\,0\,y_2 & \\ -x_1\,y_1 & & x_1\ \ y_1 \end{pmatrix}.$$

In addition to the translational motion in the direction of the *y*-axis, there are two other oscillations (Fig. 142c and d). One of them (Fig. 142c) can be presented as the oscillations of the two "molecules" H_2C which are antisymmetric with respect to the CD-axis and occur in phase. The other (Fig. 142d) is the rotations of the " molecules" H_2C in the opposite directions. If we completely neglect the constraints of the distant atoms and the stiffness of the angles 1–2–5, then the frequency of this oscillation will be zero.

Among aa-oscillations, there is a rotation of the molecule as a whole in the xy-plane, the anti-symmetric oscillations of the "molecules" H_2C in anti-phase (Fig. 142e) and their rotations in one direction (Fig. 142f).

Three ss-totally symmetric oscillations are also possible as shown in Fig. 142g–i.

6.51. The Lagrangian function is

$$L = \tfrac{1}{2} m(\dot{\mathbf{u}}_1^2 + \dot{\mathbf{u}}_2^2 + \dot{\mathbf{u}}_3^2) - \tfrac{1}{2} k \Big\{ (|\mathbf{r}_{10} - \mathbf{r}_{20} + \mathbf{u}_1 - \mathbf{u}_2| - l)^2 +$$
$$+ (|\mathbf{r}_{20} - \mathbf{r}_{30} + \mathbf{u}_2 - \mathbf{u}_3| - l)^2 - (|\mathbf{r}_{30} - \mathbf{r}_{10} + \mathbf{u}_3 - \mathbf{u}_1| - l)^2 \Big\},$$

where $l = |\mathbf{r}_{10} - \mathbf{r}_{20}| = |\mathbf{r}_{20} - \mathbf{r}_{30}| = |\mathbf{r}_{30} - \mathbf{r}_{10}|$; $\mathbf{u}_a$ is the displacement of the ath atom from its equilibrium position, which is determined by the radius vector $\mathbf{r}_{a0}$. Since $|\mathbf{u}_a| \ll l$, we have

$$L = \tfrac{1}{2} m(\dot{\mathbf{u}}_1^2 + \dot{\mathbf{u}}_2^2 + \dot{\mathbf{u}}_3^2) -$$
$$- \tfrac{1}{2} k \Big\{ (\mathbf{e}_{12}(\mathbf{u}_1 - \mathbf{u}_2))^2 + (\mathbf{e}_{23}(\mathbf{u}_2 - \mathbf{u}_3))^2 + (\mathbf{e}_{31}(\mathbf{u}_3 - \mathbf{u}_1))^2 \Big\}, \tag{1}$$

where

$$\mathbf{e}_{12} = \frac{\mathbf{r}_{10} - \mathbf{r}_{20}}{l}, \qquad \mathbf{e}_{23} = \frac{\mathbf{r}_{20} - \mathbf{r}_{30}}{l}, \qquad \mathbf{e}_{31} = \frac{\mathbf{r}_{30} - \mathbf{r}_{10}}{l}.$$

In the system of reference where the total momentum equals zero, $m(\dot{\mathbf{u}}_1 + \dot{\mathbf{u}}_2 + \dot{\mathbf{u}}_3) = 0$, the condition

$$\mathbf{u}_1 + \mathbf{u}_2 + \mathbf{u}_3 = 0 \tag{2}$$

is satisfied. Moreover, we impose upon $\mathbf{u}_a$ the condition

$$[\mathbf{r}_{10}, \mathbf{u}_1] + [\mathbf{r}_{20}, \mathbf{u}_2] + [\mathbf{r}_{30}, \mathbf{u}_3] = 0, \tag{3}$$

which is equivalent to the requirment that the angular moment of the molecule,

$$\mathbf{M} = m \sum_a [\mathbf{r}_{a0} + \mathbf{u}_a, \dot{\mathbf{u}}_a], \tag{4}$$

vanishes up to and including terms of first order in $\mathbf{u}_a$.

It is helpful for the description of the motion to introduce for each atom its own Cartesian system of coordinates (Fig. 143), thus retaining symmetry in the description of the system. Equations (2) and (3) in terms of these coordinates read[1]

Figure 143

$$y_1 + \left(-\frac{1}{2}y_2 + \frac{\sqrt{3}}{2}x_2\right) + \left(-\frac{1}{2}y_3 + \frac{\sqrt{3}}{2}x_3\right) = 0,$$

(5)

$$y_2 + \left(-\frac{1}{2}y_3 + \frac{\sqrt{3}}{2}x_3\right) + \left(-\frac{1}{2}y_1 - \frac{\sqrt{3}}{2}x_1\right) = 0,$$

(6)

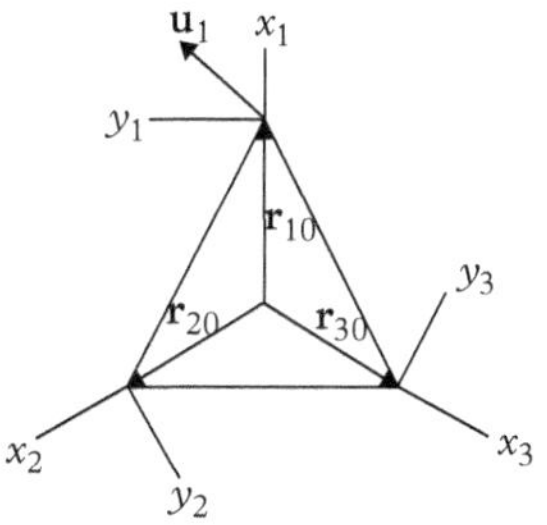

and hence

$$y_1 = \frac{x_3 - x_2}{\sqrt{3}}, \quad y_2 = \frac{x_1 - x_3}{\sqrt{3}}, \quad y_3 = \frac{x_2 - x_1}{\sqrt{3}},$$

and

$$L = \frac{5m}{6}\left(\dot{x}_1^2 + \dot{x}_2^2 + \dot{x}_3^2\right) - \frac{m}{3}(\dot{x}_1\dot{x}_2 + \dot{x}_2\dot{x}_3 + \dot{x}_3\dot{x}_1) - \frac{3}{2}k\left(x_1^2 + x_2^2 + x_3^2\right).$$

One normal oscillation (totally symmetric) is obvious:

$$x_1^{(1)} = x_2^{(1)} = x_3^{(1)} = \frac{1}{\sqrt{3}}q_1.$$

(7)

The two other oscillations are orthogonal to the first oscillation which leads to the condition[2]

$$x_1^{(s)} + x_2^{(s)} + x_3^{(s)} = 0, \quad s = 2, 3.$$

(8)

One of these oscillations is symmetric with respect to the x_1-axis: $x_2^{(2)} = x_3^{(2)}$; the other is antisymmetric: $x_1^{(3)} = 0$, $x_2^{(3)} = -x_3^{(3)}$.

---

[1] For instance, multiplying both sides of the equality (2) by $\mathbf{e}_{23}$, we get (5). Note that the vector $\mathbf{e}_{23}$ in the different coordinate systems has the coordinates

$$\mathbf{e}_{23} = (0, 1)_1 = \left(\frac{\sqrt{3}}{2}, -\frac{1}{2}\right)_2 = \left(-\frac{\sqrt{3}}{2}, -\frac{1}{2}\right)_3,$$

while $\mathbf{u}_a = (x_a, y_a)_a$. Equality (6) follows from (5) through a cyclic permutation of the indices.

[2] The metric k_{ij} is used. In (7) and (9) the factors before q are chosen so that $x_1^{(l)2} + x_2^{(l)2} + x_3^{(l)2} = q_l^2$.

Taking into account (8), we have

$$x_1^{(2)} = -2x_2^{(2)} = -2x_3^{(2)} = \sqrt{\frac{2}{3}} q_2,$$

$$x_2^{(3)} = -x_3^{(3)} = \frac{1}{\sqrt{2}} q_3. \tag{9}$$

The replacement,

$$x_1 = \frac{1}{\sqrt{3}} q_1 + \sqrt{\frac{2}{3}} q_2,$$

$$x_2 = \frac{1}{\sqrt{3}} q_1 - \frac{1}{\sqrt{6}} q_2 + \frac{1}{\sqrt{2}} q_3,$$

$$x_3 = \frac{1}{\sqrt{3}} q_1 - \frac{1}{\sqrt{6}} q_2 - \frac{1}{\sqrt{2}} q_3,$$

leads to the Lagrangian function of the form

$$L = \tfrac{1}{2} m(\dot{q}_1^2 + 2\dot{q}_2^2 + 2\dot{q}_3^2) - \tfrac{3}{2} k(q_1^2 + q_2^2 + q_3^2). \tag{10}$$

The normal oscillations corresponding to these coordinates are shown in Fig. 144. Their frequencies are

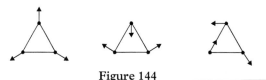

Figure 144

$$\omega_1 = \sqrt{\frac{3k}{m}}, \quad \omega_2 = \omega_3 = \sqrt{\frac{3k}{2m}}.$$

The form of the Lagrangian function (10) is retained under the rotation in the q_2, q_3-plane.

The angular momentum with the account of quadratic in $\mathbf{u}_a$ terms

$$|\mathbf{M}| = m \left| \sum_a [\mathbf{u}_a, \dot{\mathbf{u}}_a] \right| = m|q_2 \dot{q}_3 - q_3 \dot{q}_2|$$

can be non-zero if there is a phase difference between the oscillations q_2 and q_3.

It is interesting to analyse what changes are introduced in this picture if we consider now the possibility that the potential energy may depend on the angles between the bonds.

It is clear that such a dependence will not affect the frequency of the q_1 oscillations. The frequencies of the q_2 and q_3 oscillations will be changed, but a twofold degeneracy will remain. Indeed, together with a q-oscillation, we can also have another oscillation obtained from the original q-oscillation by a rotation over $2\pi/3$. Its frequency must be the same as that of the original oscillation. On the other hand, the q-oscillation will differ (only the q_1-oscillation remains the same under a rotation over $2\pi/3$). Thus, we find two independent oscillations with the same frequency. The normal coordinates must in this case satisfy only one condition: they must be orthogonal to q_1; in particular, q_2 and q_3 remain normal coordinates.

6.52. a) We introduce the coordinates of atoms B in the same way as in the preceding problem; for atom A—the coordinates will be x_4, y_4, z_4 with the axes parallel to x_1, y_1, z_1 and with the beginning in the centre of the triangle.

There are four degrees of freedom of motions which output atoms from the xy-plane. Three of them correspond to the translational motion along the z-axis and rotations around the x_4- and y_4-axes, and one to the oscillation (for which, obviously, $z_1 = z_2 = z_3$, $m_A z_4 + m(z_1 + z_2 + z_3) = 0$). The frequency of this oscillation ω_1 is not degenerate; it is only by chance that it could coincide with the frequency of any other oscillations.

Let us consider the oscillation of atoms in the xy-plane symmetric with respect to the x_4-axis. The general form of such oscillation is

$$y_1 = 0, \quad x_2 = x_3, \quad y_2 = -y_3, \quad y_4 = 0.$$

The displacement vector contains four independent parameters: x_1, x_2, y_2, x_4; that is, such motions have four degrees of freedom. One of them corresponds to the translational motion of the molecule along the x_4-axis, another one to totally symmetric oscillation

$$x_1 = x_2 = x_3, \quad y_1 = y_2 = y_3 = x_4 = y_4 = 0$$

with the frequency ω_2 (Fig. 145a), and the other two—to the oscillations with the frequencies ω_3 and ω_4 which break symmetry of the molecule (Fig. 145b and c).

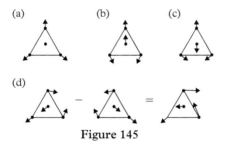

Figure 145

Of the four remaining degrees of freedom, one is the translational motion along the y_4-axis, and another is the rotation around the z_4-axis. The last two are the oscillations that can be obtained from the previously discussed oscillations with the frequencies ω_3 and ω_4 by a rotation on an angle $2\pi/3$ around the z_4-axis (cf. preceding problem).

As a result, the frequencies ω_1 and ω_2 are non-degenerate, while the frequencies ω_3 and ω_4 are twofold degenerate.

Note that the vector of the oscillation which is antisymmetric with respect to the x_4-axes can be obtained from the vector of the symmetric oscillation. For this, it is sufficient to take a certain superposition of the oscillations obtained from the latter by the rotation around the z_4-axis over angles $\pm 2\pi/3$; namely, it must be their difference (see Fig. 145d).

b) The changes in the eigen-frequencies can be determined using the perturbation theory (see problem 6.35):

$$\delta\omega = -\frac{\omega}{2}\frac{\delta M}{M}.$$

For a totally symmetric oscillation, we choose x_1 as a normal coordinate. Then quantities M and δM are defined as the coefficients in the expressions for the kinetic energy and the corrections to it, respectively:

$$\tfrac{3}{2}m\dot{x}_1^2 = \tfrac{1}{2}M\dot{x}_1^2, \quad \tfrac{1}{2}\delta m\dot{x}_1^2 = \tfrac{1}{2}\delta M\dot{x}_1^2,$$

so that

$$\delta\omega_2 = -\omega_2\frac{\delta m}{6m}.$$

For the oscillations along the z-axis, the kinetic energy and the correction to it are equal to, respectively,

$$\tfrac{3}{2}m(1+3m/m_A)\dot{z}_1^2 \quad \text{and} \quad \tfrac{1}{2}\delta m\dot{z}_1^2,$$

so that

$$\delta\omega_1 = -\tfrac{1}{2}\omega_1\frac{m_A\delta m}{3m(3m+m_A)}.$$

For example, when we replace one atom of chloride with the atomic weight of 35 by the isotope with the atomic weight of 37 in the molecule of boron chloride, it reduces the frequency ω_1 and ω_2 by 0.1% and 1%, respectively.

6.53. Let the oscillation for which the molecule retains its shape (Fig. 146a) have a frequency ω_1.

The frequency ω_2 of the oscillation which retains its form under rotations around OD over $2\pi/3$ (Fig. 146b) is, in general, different from ω_1. One can obtain another displacement of the atoms by taken a reflection in the plane BCO; we obtain oscillations which differ from the second oscillation only in that atoms A and D change roles. The frequency of that oscillation $\omega_3 = \omega_2$. Similarly, in the reflection in plane AOC, the roles of

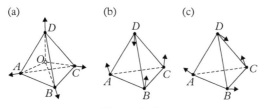

(a)   (b)   (c)

Figure 146

atoms B and D are changed, while the frequency remains the same, $\omega_4 = \omega_2$. This fourth oscillation cannot be reduced to a superposition of previous oscillations as, in contrast to those, it is not symmetric with respect to plane *AOD*.

The oscillation which is symmetric with respect to planes *AOB* and *DOC* (Fig. 147c) has a frequency ω_5 which is different from ω_1 and ω_2. A rotation over an angle $2\pi/3$ around *OD* which results in a cyclic permutation of *A, B*, and *C*, leads to an oscillation which is symmetric with respect to planes *COA* and *DOB* and its frequency $\omega_6 = \omega_5$.

The molecule has thus three eigen-frequencies which are, respectively, non-degerenate, twofold degenerate, and threefold degenerate.

In conclusion, we note that the molecules considered in problems 6.51 and 6.53 are, clearly, not to be found in nature. However, a similar approach can also be used for real molecules.

6.54. a) When a totally symmetric and twofold-degenerate oscillations, considered in the preceding problem (Fig. 146a and b) occur, the carbon atom remains at rest. There are extra two threefold degenerate frequencies. The corresponding oscillations are similar to the ones shown in Fig. 147b. The only difference is that the carbon atom oscillates either in the same direction as the atom D or in the opposite direction.

b) The addition to the Lagrangian function describing the action of the electric fields $\mathbf{E}(t)$ is

$$\delta L = \mathbf{E}(t) \sum_a e_a \mathbf{u}_a,$$

where e_a is the charge and $\mathbf{u}_a$, displacement of ath atom.

In any oscillation $m \sum_{a=1}^{4} \mathbf{u}_a + m_C \mathbf{u}_5 = 0$, so that

$$\sum_{a=1}^{5} e_a \mathbf{u}_a = -e_1 \left(\frac{m_C}{m} + 4 \right) \mathbf{u}_5.$$

For the totally symmetric and twofold-degenerate oscillations, are quantity $\sum e_a \mathbf{u}_a = 0$ and these oscillations are not excited. For the oscillation of Fig. 147b, in contrast, $\sum e_a \mathbf{u}_a \neq 0$ and the similar oscillations are excited.

As a result, resonance is possible on two frequencies.

The vector $\mathbf{E}$ can be decomposed into three terms $\mathbf{E}_j$ which are parallel to axes of symmetry of each of the three oscillations of the molecule with degenerate frequency. Each of the terms $\mathbf{E}_j$ will lead to the oscillation of the carbon atom with an amplitude proportional to $\mathbf{E}_j$, and with the same proportionality coefficient $\varkappa$. Therefore, $\mathbf{u}_5 = \varkappa\mathbf{E}$ and $|\mathbf{u}_5|$ does not depend on the orientation of the molecule. It may be shown that the oscillation amplitudes of the hydrogen atoms $|\mathbf{u}_{1,2,3,4}|$ do also not depend on the orientation of the molecule and that they are parallel to the vector $\mathbf{E}$.

§7

Oscillations of linear chains

7.1. The Lagrangian function of the system is

$$L(x, \dot{x}) = \tfrac{1}{2} m \sum_{n=1}^{N} \dot{x}_n^2 - \tfrac{1}{2} k \left[x_1^2 + \sum_{n=2}^{N} (x_n - x_{n-1})^2 + x_N^2 \right], \tag{1}$$

where the x_n is the displacement of the nth particle from its equilibrium position. We also introduce the coordinate of the equilibrium position of the nth particle, $X_n = na$, where a is the equilibrium length of one spring. The Lagrangian equations of motion

$$\begin{aligned}
&m\ddot{x}_1 + k(2x_1 - x_2) = 0, \\
&m\ddot{x}_n + k(2x_n - x_{n-1} - x_{n+1}) = 0, \quad n = 2, 3, \ldots, N-1, \\
&m\ddot{x}_N + k(2x_N - x_{N-1}) = 0
\end{aligned} \tag{2}$$

are equivalent to the set

$$m\ddot{x}_n + k(2x_n - x_{n-1} - x_{n+1}) = 0, \quad n = 1, 2, \ldots, N \tag{3}$$

with the additional condition

$$x_0 = x_{N+1} \equiv 0. \tag{4}$$

We expect from physical considerations that the normal oscillations of the system will be standing waves. It is, however, helpful to consider

$$x_n = A e^{i(\omega t \pm n\varphi)}. \tag{5}$$

The set of N the equations in (3) then reduces to the equation,

$$\omega^2 = 4\frac{k}{m} \sin^2 \frac{\varphi}{2}, \tag{6}$$

Exploring Classical Mechanics: A Collection of 350+ Solved Problems for Students, Lecturers, and Researchers. First Edition.
Gleb L. Kotkin and Valeriy G. Serbo, Oxford University Press (2020). © Gleb L. Kotkin and Valeriy G. Serbo 2020.
DOI: 10.1093/oso/9780198853787.001.0001

which determines the relationship between the frequency ω and the difference in the phase of neighbouring particles φ. The meaning of the substitution (5) consist of the choice for x_n of a solution in the form of a travelling wave with a wave vector $K = \varphi/a$ such that $n\varphi = naK = KX_n$. Equation (6) thus establishes the relation between the frequency and the wave vector.

The conditions of (4) can be satisfied by taking a superposition of the waves travelling in two directions,

$$x_n = Ae^{i(\omega t - n\varphi)} + Be^{i(\omega t + n\varphi)}.$$

The condition $x_0 = 0$ gives $A = -B$, or $x_n = 2iB\sin(n\varphi)e^{i\omega t}$, that is, a standing wave. From the condition at the other end, $x_{N+1} = 0$, we determine the spectrum of possible frequencies.

The equation $\sin(N+1)\varphi = 0$ has N independent solutions

$$\varphi_s = \frac{\pi s}{N+1}, \quad s = 1, 2, \ldots, N. \tag{7}$$

In fact, $s = 0$ and $s = N+1$ give vanishing solutions; for $s = N+l$, the phase $\varphi_{N+l} = -\varphi_{N-l+2} + 2\pi$; that is, the solution corresponding to $s = N+l$ can be expressed in terms of the solutions corresponding for $s = N-l+2$. From (6) and (7), we find N different frequencies:

$$\omega_s = 2\sqrt{\frac{k}{m}}\sin\frac{\varphi_s}{2} = 2\sqrt{\frac{k}{m}}\sin\frac{\pi s}{2(N+1)}, \quad s = 1, 2, \ldots, N. \tag{8}$$

The different frequencies are shown in Fig. 147 by the discrete points on the sine curve. The vector of the normal oscillations corresponding to the sth frequency is

$$r_s = \begin{pmatrix} x_1 \\ x_2 \\ \ldots \\ x_N \end{pmatrix} = \sqrt{\frac{2}{N+1}} \begin{pmatrix} \sin\varphi_s \\ \sin 2\varphi_s \\ \ldots \\ \sin N\varphi_x \end{pmatrix} q_s(t), \tag{9}$$

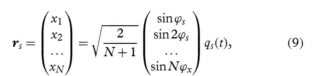

Figure 147

where

$$q_s(t) = \mathrm{Re}(2iB_s e^{I\omega_s t}) = C_s\cos(\omega_s t + \alpha_s)$$

is the sth normal coordinate, while the factor

$$\sqrt{\frac{2}{N+1}} = \left[\sum_{n=1}^{N}\sin^2 n\varphi_s\right]^{-1/2}$$

is introduced to get the normalization: $(\mathbf{r}_s, \mathbf{r}'_s) = \delta_{ss'}q_s^2$. The general solution is a superposition of all normal oscillations:

$$x_n = \sum_{s=1}^{N} \sqrt{\frac{2}{N+1}}\, q_s(t) \sin n\varphi_s.$$

The matrix, given the transition from the x_n to the q_s,

$$U_{ns} = \sqrt{\frac{2}{N+1}} \sin \frac{\pi ns}{N+1},$$

is an orthogonal matrix which reduces the Lagrangian function to a sum of squares corresponding a set of N different oscillators:

$$L = \sum_{s=1}^{N} L_s(q_s, \dot{q}_s), \quad L(q_s, \dot{q}_s) = \tfrac{1}{2} m(\dot{q}_s^2 - \omega_s^2 q_s^2).$$

7.2. The equations of motion for the given system are the same as the equations (3) of the previous problem but now with the additional conditions $x_0 = 0$, $x_N = x_{N+1}$. Thus, we get

$$\varphi_s = \frac{(2s-1)\pi}{2N+1}, \quad s = 1, 2, \ldots, N,$$

$$\omega_s = 2\sqrt{\frac{k}{m}} \sin \frac{(2s-1)\pi}{2(2N+1)},$$

$$x_n = \sum_s \sin n\varphi_s \cdot A_s \cos(\omega_s t + \alpha_s).$$

For the particular case when $N = 2$, see problem 6.1.

7.3. We use displacements of each of the particles from the vertical as the generalized coordinates (cf. problem 6.3). In such variables, the given problem is completely reduceble to problem 7.2 with $k/m = g/l$.

7.4. The equations of motion are the same as the equations (3) of problem 7.1 but now with the additional conditions $x_0 = x_N$ and $x_{N+1} = x_1$. Therefore,

$$\varphi_s = \frac{2\pi s}{N}, \quad s = 0, 1, \ldots, N-1,$$

$$\omega_s = 2\sqrt{\frac{k}{m}} \sin \frac{s\pi}{N};$$

the frequencies ω_s and ω_{N-s} are the same, and the corresponding wave vectors

$$K_s = \frac{\varphi_s}{a} = \frac{2\pi - \varphi_{N-s}}{a} = -K_{N-s} + \frac{2\pi}{a}$$

determine the travelling waves moving in opposite directions. The frequency $\omega_0 = 0$ corresponds to the motion of all particles along the ring with a constant velocity. In this system, oscillations of the form

$$x_n^{(s)} = \mathrm{Re}\, A_s e^{i(\omega_s t - n\varphi_s)}$$

are possible, that is, waves which travel along the ring. The previously mentioned twofold degeneracy of the frequencies corresponds to waves moving in opposite directions. The presence of two such waves with equal amplitudes gives a standing wave:

$$x_n^{(s)} \pm x_n^{(N-s)} = 2|A_s| \begin{cases} \cos n\varphi_s \cos(\omega_s t + \alpha_s), \\ \sin n\varphi_s \sin(\omega_s t + al_s). \end{cases} \tag{2}$$

This is also a normal oscillations (all particles move either in phase or in anti-phase).

In terms of the appropriate normal coordinates,

$$x_n = \sum_{s=1}^{R} (q_{s1} \cos n\varphi_s + q_{s2} \sin n\varphi_s) + q_0, \quad R = \tfrac{1}{2}(N-1), \quad (N: \text{ odd}), \tag{3}$$

the Lagrangian function is reduced to diagonal form:

$$L = \tfrac{1}{2} Nm \left\{ \dot{q}_0^2 + \tfrac{1}{2} \sum_{s=1}^{R} \left[\dot{q}_{s1}^2 + \dot{q}_{s2}^2 - \omega_s^2 (q_{s1}^2 + q_{s2}^2) \right] \right\}. \tag{4}$$

(If the number of particles is even, (3) and (4) must be somewhat changed as the frequency $\omega_{N/2}$ is non-degenerate; (1) defines a standing wave for $s = N/2$.)

It is interesting to note that rotations in the q_{s1}, q_{s2}-planes,

$$q_{s1} = q_{s1}' \cos \beta_s - q_{s2}' \sin \beta_s,$$
$$q_{s2} = q_{s1}' \sin \beta_s + q_{s2}' \cos \beta_s,$$

which leave the Lagrangian function (4) invariant, correspond to a displacement of a nodes of the standing waves:

$$x_n = q_0 + \sum_{s=1}^{R} [q_{s1}' \cos(n\varphi_s - \beta_s) + q_{s2}' \sin(n\varphi_s - \beta_s)].$$

The average energy flux along the ring is for the travelling waves (1) given by (cf. problem 6.10)[1]

$$S_{\mathrm{av}} = \frac{\omega}{2\pi} \int\limits_0^{2\pi/\omega} k(x_{n-1} - x_n)\dot{x}_n \, dt = \tfrac{1}{2} k|A|^2 \omega \sin\varphi,$$

while the group velocity is

$$v_{\mathrm{gr}} = \frac{d\omega}{dK} = \sqrt{\frac{k}{m}}\, a \cos\frac{\varphi}{2},$$

where a is the equilibrium length of one spring, while $K = \varphi/a$ is the wave vector. The energy is

$$E = \tfrac{1}{2} m \sum_{n=1}^N \dot{x}_n^2 + \tfrac{1}{2} k \sum_{n=1}^N (x_n - x_{n-1})^2 = 2N|A|^2 k \sin^2\frac{\varphi}{2} = \tfrac{1}{2} Nm\omega^2 |a|^2,$$

and we have thus

$$\frac{E}{Na} v_{\mathrm{gr}} = S_{\mathrm{av}}.$$

7.5. a) The equations of motion are

$$m\ddot{x}_{2n-1} + k(2x_{2n-1} - x_{2n-2} - x_{2n}) = 0,$$
$$M\ddot{x}_{2n} + k(2x_{2n} - x_{2n-1} - x_{2n+1}) = 0, \tag{1}$$

here $x_0 = x_{2N+1} = 0$, $n = 1, 2, \ldots, N$.

We shall look for a solution in the form of travelling waves with different amplitudes

$$x_{2n-1} = A e^{i[\omega t \pm (2n-1)\varphi]},$$
$$x_{2n} = B e^{i[\omega t \pm 2n\varphi]}. \tag{2}$$

To determine A and B, we get a set of homogeneous equations,

$$(-m\omega^2 + 2k)A - k(e^{-i\varphi} + e^{i\varphi})B = 0,$$
$$-k(e^{-i\varphi} + e^{i\varphi})A + (-M\omega^2 + 2k)B = 0, \tag{3}$$

---

[1] In the following, we drop the index s for the sake of simplicity. The calculation of flux S_{av} and of the energy E is very helpful carried out using complex variables (cf. [2], § 48).

which have a non-trivial solution only if its determinant vanishes. This condition determines the relations between the frequency and the difference in phase of neighbouring particles:

$$\omega^2_{(\mp)} = \frac{k}{\mu}\left(1 \mp \sqrt{1 - \frac{4\mu^2}{mM}\sin^2\varphi}\right), \quad \mu = \frac{mM}{m+M}. \tag{4}$$

The boundary conditions are satisfied only by well-defined linear combinations of the travelling waves (2), namely,

$$x_{2n-1} = A_s \sin(2n - 1)\varphi_s \cos(\omega_s t + \alpha_s),$$
$$x_{2n} = B_s \sin 2n\varphi_s \cos(\omega_s t + \alpha_s),$$

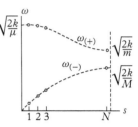

Figure 148

for which $\varphi_s = s\pi/(2N + 1)$. Since $\varphi_{2N+1-s} = \pi - \varphi_s$, we get different frequencies, having chosen one of the two signs in (4) only for $s = 1, 2, \ldots, N$. These are indicated (for this case, when $M > m$) in Fig. 148 by discrete dots on the two different branches; the lower one, $\omega_{(-)}$, is usually called the *acoustic*, and the upper one, $\omega_{(+)}$, is called the *optical* branch.

The general solution has the form

$$x_{2n-1} = \sum_{s=1}^{N} \sin(2n - 1)\varphi_s [A_{(+)s}\cos(\omega_{(+)s}t + \alpha_s) + A_{(-)s}\cos(\omega_{(-)s}t + \beta_s)],$$

$$x_{2n} = \sum_{s=1}^{N} \sin 2n\varphi_s [B_{(+)s}\cos(\omega_{(+)s}t + \alpha_s) + B_{(-)s}\cos(\omega_{(-)s}t + \beta_s)],$$

where the $A_{(\pm)s}$ and $B_{(\pm)s}$ are according to (3) connected by the relation

$$B_{(\pm)s} = \frac{2k - m\omega^2_{(\pm)s}}{2k\cos\varphi_s}A_{(\pm)s}.$$

It is remarkable that $B_{(-)s}$ and $A_{(-)s}$ corresponding to the acoustic frequencies have the same signs, while $B_{(+)s}$ and $A_{(+)s}$ for the optical frequencies have the opposite signs (i.e., neighbouring particles with masses m and M move in antiphase). The distribution of the amplitudes of the oscillations for the case $N = 8, s = 2$ is shown in Fig. 149 where the ordinate gives the numbers of the particles, and the abscissa gives the amplitudes (149a gives an optical and 149b, an acoustic mode).

How can one get from the results obtained here to the limiting case $m = M$ (see problem 7.1)?

(a)

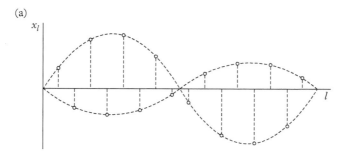

(b)

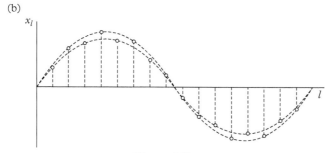

Figure 149

b) The normal oscillations are

$$x_{2n}^{(s)} = A_s \sin 2n\varphi_s \cos(\omega_s t + \alpha_s),$$

$$x_{2n-1}^{(s)} = A_s \frac{K \sin 2n\varphi_s + k \sin(2n-2)\varphi_s}{k + K - m\omega_s^2} \cos(\omega_s t + \alpha_s),$$

where

$$\omega_s^2 = \frac{1}{m}\left[K + k \mp \sqrt{(K-k)^2 + 4Kk\cos^2\varphi_s}\,\right]$$

and φ_s is determined from the equation

$$\tan(2N+1)\varphi_s = -\frac{K-k}{K+k}\tan\varphi_s, \quad s = 1, 2, \ldots, N,$$

$$0 < \varphi_s < \pi/2.$$

Fig. 150a shows the optical and acoustic branches of the frequencies for the case $K > k$. What happens in the limiting case $K = k$?

c) We have $\varphi_s = \frac{\pi}{2(N+1)}$; for $s = 1, 2 \ldots N$, we get $2N$ normal vibrations and normal frequencies which have the same form as under point 7.5b (Fig. 150b).

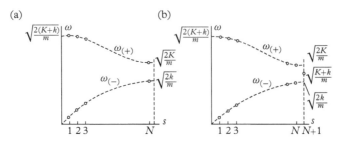

Figure 150

There is one more, $(2N+1)$, normal oscillation $x_{2n} = 0$ and $Kx_{2n-1} = -kx_{2n+1}$, the frequency of which, $\omega_0^2 = (K+k)/m$, lies in the forbidden band between the optical and acoustic branches. The distribution of the amplitudes of this oscillation is shown in Fig. 151 where the ordinate and abscissa show, respectively, the numbers of the particles and their amplitudes of the oscillations. The particles with an even numbers are not moving while neighbouring particles with odd numbers are moving in antiphase with the exponentially damped amplitudes, when we move away from the left-hand end of the chain.

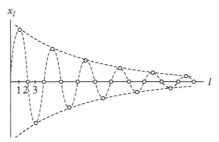

Figure 151

7.6. a) We look for a solution of the equations of motion,

$$m\ddot{x}_n + k(2x_n - x_{n-1} - x_{n+1}) = 0, \quad n = 1, 2, \ldots, N, \tag{1}$$

with the boundary conditions $x_0 = 0$ and $x_{N+1} = a\cos\gamma t$ in the form of standing waves $x_n = A\sin n\varphi \cos\gamma t$, so that the first boundary condition is immediately satisfied. From the second boundary condition, we find constant $A = a/\sin(N+1)\varphi$, while from (1) we get for the "wave vector" φ of the standing wave

$$\sin^2\frac{\varphi}{2} = \frac{m\gamma^2}{4k}.$$

When $\gamma^2 < 4k/m$, the stable oscillations

$$x_n = a \frac{\sin n\varphi}{\sin(N+1)\varphi} \cos \gamma t \qquad (2)$$

have a larger amplitude when the denominator $\sin(N+1)\varphi$ is close to zero. But this is just the condition which determined the spectrum of the eigen-frequencies ω_s (see problem 7.1); that is, we have then a near-resonance situation, $\gamma \approx \omega_s$. When

$$\gamma \ll \omega_1 = 2\sqrt{\frac{k}{m}} \sin \frac{\pi}{2(N+1)},$$

the oscillations in (2) correspond to a slow extension and compression of all springs as a whole:

$$x_n = a \frac{n}{N+1} \cos \gamma t.$$

If $\gamma^2 > 4k/m$, we change φ to $\pi - i\psi$ in (2) and get

$$x_n = (-1)^{N+1+n} a \frac{\sinh n\psi}{\sinh(N+1)\psi} \cos \gamma t,$$

where

$$\cosh^2 \frac{\psi}{2} = \frac{m\gamma^2}{4k}.$$

The oscillations are (exponentially when $n\psi \gg 1$) damped towards the left-hand end of the chain. The reasonableness of this result is particularly clear when $\gamma^2 \gg 4k/m$ when the frequency of the applied force lies appreciable above the limit of the spectrum of the normal frequencies. In that case, the particle on the extreme right oscillates with a small amplitude in anti-phase with the applied force while the $(N-1)$st particle is in first approximation at rest. Then we can consider the motion of the $(N-1)$st particle as a forced oscillation caused by an applied force of high frequency arising from the Nst particle, and so on.

We note that a similar damping of the wave occurs in the phenomenon of complete internal reflection (e.g., when short wavelength radio-waves are reflected by the iono-sphere).

What is the form of the stable oscillation when $\gamma^2 = 4k/m$?

b) To find the normal oscillations we have consider, first, the solution in the form of the travelling wave (see problem 7.1). If the wave travels to the right, to the point A, then the displacements are

$$x_n = \mathrm{Re}\left(Ae^{i(\gamma t - n\varphi)}\right),$$

where φ is determined by the equation

$$\gamma^2 = 4\frac{k}{m}\sin^2\frac{\varphi}{2}.$$

In the given problem $x_0 = a\cos\gamma t$, so we find $A = a$. Therefore,

$$x_{N+1} = \mathrm{Re}\left(Ae^{i(\gamma t - (N+1)\varphi)}\right) = a\cos(\gamma t - (N+1)\varphi).$$

The energy flux from left to right is equal to the work (averaged over period) of the spring, which connects the $(n-1)$th and nth particles, over the nth particle (cf. problem 6.10):

$$\langle k(x_{n-1} - x_n)\dot{x}_n\rangle = \tfrac{1}{2}\gamma ka^2\sin\varphi = \tfrac{1}{4}m\gamma^2 a^2\sqrt{(4k/m) - \gamma^2}.$$

When $\gamma > 2\sqrt{k/m}$, there will be no travel wave.

c) $x_n = a\dfrac{\cos\left(N - n + \tfrac{1}{2}\right)\varphi}{\cos\left(N + \tfrac{1}{2}\right)\varphi}\cos\gamma t,$

$$\sin^2\frac{\varphi}{2} = \frac{m\gamma^2}{4k}, \quad \text{when} \quad \gamma^2 < \frac{4k}{m},$$

$$x_n = (-1)^n a\frac{\sinh\left(N - n + \tfrac{1}{2}\right)\psi}{\sinh\left(N + \tfrac{1}{2}\right)\psi}\cos\gamma t,$$

$$\mathrm{ch}^2\frac{\psi}{2} = \frac{m\gamma^2}{4k}, \quad \text{when} \quad \gamma^2 > \frac{4k}{m}.$$

7.7. If the frequency of the applied force lies in the range of the acoustic eigen-frequencies, $0 < \gamma^2 < 2k/M$, or in the region of the optical eigen-frequencies, $2k/m < \gamma^2 < 2k/\mu$ (see problem 7.5a), the stable oscillations are

$$x_{2n-1} = a\frac{\sin(2n-1)\varphi}{\sin(2N+1)\varphi}\cos\gamma t,$$

$$x_{2n} = \pm\sqrt{\frac{2k - m\gamma^2}{2k - M\gamma^2}}\,a\frac{\sin 2n\varphi}{\sin(2N+1)\varphi}\cos\gamma t,$$

where

$$\cos^2 \varphi = \frac{(2k - M\gamma^2)(2k - m\gamma^2)}{4k^2},$$

while the upper (lower) sign corresponds to the frequency γ lying in the acoustic (optical) frequencies.

For frequencies lying in the "forbidden band" $2k/M < \gamma^2 < 2k/m$, we have

$$x_{2n-1} = (-1)^{N+n+1} a \frac{\cosh(2n-1)\psi}{\cosh(2N+1)\psi} \cos \gamma t,$$

$$x_{2n} = (-1)^{N+n+1} a \sqrt{\frac{2k - m\gamma^2}{M\gamma^2 - 2k}} \frac{\sinh 2n\psi}{\cosh(2N+1)\psi} \cos \gamma t,$$

$$\sinh^2 \psi = \frac{(2k - m\gamma^2)(M\gamma^2 - 2k)}{4k^2},$$

while for frequencies $\gamma^2 > 2k/\mu$, which lie above the limit of the optical branch (this is also the "forbidden band"),

$$x_{2n-1} = a \frac{\sinh(2n-1)\chi}{\sinh(2N+1)\chi} \cos \gamma t,$$

$$x_{2n} = -a \sqrt{\frac{2k - m\gamma^2}{2k - M\gamma^2}} \frac{\sinh 2n\chi}{\sinh(2N+1)\chi} \cos \gamma t,$$

$$\cosh^2 \chi = \frac{(M\gamma^2 - 2k)(m\gamma^2 - 2k)}{4k^2},$$

the oscillations are damped exponentially towards the left-hand side end of the chain.

7.8. a) We look for the solution for the equations of motion

$$m\ddot{x}_n + k(2x_n - x_{n-1} - x_{n+1}) = 0, \quad n = 1, 2, \ldots, N-1, \tag{1}$$

$$m_N \ddot{x}_N + k(2x_N - x_{N-1}) = 0 \tag{2}$$

(with the boundary condition $x_0 = 0$), in the form of standing waves:

$$x_n = A \sin n\varphi \cos(\omega t + \alpha), \quad n = 1, 2, \ldots, N-1,$$
$$x_N = B \cos(\omega t + \alpha). \tag{3}$$

From (1), we get the relation

$$\omega^2 = \frac{4k}{m} \sin^2 \frac{\varphi}{2}. \tag{4}$$

Using (3) and (4), we get from (1) and (2) the set of equations

$$A \sin N\varphi - B = 0.$$

$$-A \sin(N-1)\varphi + \left(-\frac{2m_N}{m}\sin^2\frac{\varphi}{2} + 2\right)B = 0.$$

Hence, $B = A \sin N\varphi$, while the parameter φ is determined as the solution of the transcendental equation

$$\sin N\varphi\left(\frac{4m_N}{m}\sin^2\frac{\varphi}{2} - 2 + \cos\varphi\right) = \cos N\varphi \sin\varphi. \tag{5}$$

When $m_N \gg m$, we have, apart from the obvious normal oscillations,

$$x_n^{(s)} = A_s \sin n\varphi_s \cos(\omega_s t + \alpha_s), \quad n = 1, 2, \ldots, N,$$

$$\tan N\varphi_s \approx \frac{m}{2m_N}\cot\frac{\varphi_s}{2}, \quad s = 1, 2, \ldots, N-1,$$

when the particle m_N is practically stationary ($\sin N\varphi_s \ll 1$), as well as a normal oscillation with the amplitudes of the particles which are linearly decreasing to the left-hand side of the chain:

$$x_n^{(N)} = B\frac{n}{N}\cos(\omega_N t + \alpha_N), \quad \omega_N^2 = \frac{k}{m_N}\left(1 + \frac{1}{N}\right).$$

The particle m_N then oscillates between springs of stiffness k (to the right) and k/N (to the left). That (5) has such a solution can be seen as follows. Assuming φ to be small and retaining only the main terms, we get from (5): $\varphi^2 = \frac{m}{m_N}\left(1 + \frac{1}{N}\right)$ completely in accordance with the assumption made.

When $m_N \ll m$, we have the usual oscillations characteristic for a system of $(N-1)$ particles with a spring of stiffness $k/2$ at the right-hand end (the parameter φ_s and the frequency ω_s are determined from the equation $\tan N\varphi = -\frac{\sin\varphi}{2 - \cos\varphi}$). Apart from those oscillations, there is also a normal oscillation with the amplitudes of the particles which decrease to the left-hand side of the chain:

$$x_n^{(N)} = (-1)^{N+n}B\frac{\sinh n\psi}{\sinh N\psi}\cos(\omega_N t + \alpha_N),$$

$$\cosh^2\frac{\psi}{2} = \frac{m}{2m_N} \gg 1, \quad \omega_N^2 = \frac{2k}{m_N}.$$

Formally, the value of the parameter ψ can be obtained from (5) using the substitution $\varphi = \pi - i\psi$ and assuming ψ to be large. This normal oscillation can in the first approximation be considered to be the simple oscillation of the small mass m_N

particle while the other particles are at rest, while we afterwards consider the motion of the other particles as forced oscillations under the action of high-frequency force $kx_N = kB\cos(\omega_N t + \alpha_N)$, acting upon the right-hand side of the chain of $N - 1$ identical particles (cf. problem 7.6a).

Note that in this problem the oscillations of an impurity atom in a crystal are modelled. If the mass of the impurity atom is significantly different from the mass of atoms forming the crystal, the frequency of oscillations, in which the greatest amplitude has an impurity atom, can occur in the forbidden band, and such oscillations are localized near this atom.

b) When $k_{N+1} \ll k$, the solution is the same as the solution of problem 7.2. When $k_{N+1} \gg k$, there are normal oscillations for which Nth particle is practically at rest:

$$x_n^{(s)} = A_s \sin n\varphi_s \cos(\omega_s t + \alpha_s), \quad \omega_s^2 = \frac{4k}{m}\sin^2\frac{\varphi_s}{2},$$

$$\varphi_s \approx \frac{s\pi}{N}, \quad s = 1, 2, \ldots, N-1.$$

The parameter φ_s is defined from

$$\left(2\sin^2\frac{\varphi}{2} - \frac{k_{N+1}}{k}\right)\sin N\varphi = \cos N\varphi \sin\varphi, \tag{6}$$

which in the considered approximation has the form

$$\tan N\varphi = -\frac{k}{k_{N+1}}\sin\varphi.$$

Equation (6) has yet one more solution which we can obtain by putting $\varphi = \pi - i\psi$ and assuming ψ to be large. In this case, we have

$$x_n^{(N)} = (-1)^{N+n}B_N\frac{\sinh n\psi}{\sinh N\psi}\cos(\omega_N t + \alpha_N),$$

$$\cosh^2\frac{\psi}{2} = \frac{k_{N+1}}{4k}, \quad \omega_N^2 = \frac{k_{N+1}}{m};$$

that is, the amplitudes of the particles of this oscillation decrease towards the left-hand side of the chain.

How can we obtain this last oscillation using the results of the problem 7.6?

7.9. Let φ_n be the displacement of nth pendulum from the vertical.

a) The sth normal oscillation is

$$\varphi_n = A_s\cos(n - \tfrac{1}{2})\psi_s \cdot \cos(\omega_s t + \alpha_s),$$

where the frequency spectrum (Fig. 152) begins with the value $\omega_0 = \sqrt{g/l}$:

$$\omega_s^2 = \frac{g}{l} + \frac{4k}{m}\sin^2\frac{\psi_s}{2},$$

$$\psi_s = \frac{\pi s}{N}, \qquad s = 0, 1, 2, \ldots, N-1.$$

b) In the region of the eigen-frequency of the system

$$\omega_0 < \gamma < \sqrt{\omega_0^2 + 4k/m},$$

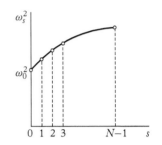

Figure 152

the forced oscillations are

$$\varphi_n = \frac{F\cos[(n-\frac{1}{2})\psi]}{2kl\sin N\psi\sin(\psi/2)}\sin\gamma t;$$

$$\gamma^2 = \frac{g}{l} + \frac{4k}{m}\sin^2\frac{\psi}{2}, \qquad 0 < \psi < \pi.$$

When $\gamma \to \omega_s$, the resonances arise because $\sin N\psi \to \sin N\psi_s \to 0$.

In the region of low frequencies $\gamma < \omega_0$, all pendulums oscillate in the same phase:

$$\varphi_n = \frac{F\cosh\left(n-\frac{1}{2}\right)\chi}{2kl\sinh N\chi\sinh(\chi/2)}\sin\gamma t; \qquad \gamma^2 = \frac{g}{l} - \frac{4k}{m}\sinh^2\frac{\chi}{2} > 0.$$

If, at the same time, the stiffness of the springs is small

$$\frac{k}{m\left(\omega_0^2 - \gamma^2\right)} = \varepsilon \ll 1,$$

then the oscillation amplitudes rapidly decrease to the left-hand side

$$\varphi_n = \varphi_N \varepsilon^{N-n}.$$

In the high-frequency region $\gamma > \sqrt{\omega_0^2 + 4k/m}$, the neighbouring pendulums oscillate in anti-phase

$$\varphi_n = \frac{(-1)^{N-n+1}F\sinh\left(n-\frac{1}{2}\right)\chi}{2kl\sinh N\chi\cosh(\chi/2)}\sin\gamma t, \qquad \gamma^2 = \frac{g}{l} + \frac{4k}{m}\operatorname{ch}^2\frac{\chi}{2}.$$

At a very high frequency $m\gamma^2/k \gg 1$, the oscillation amplitudes are also decrease rapidly to the left-hand side:

$$\varphi_n = \left(-\frac{k}{m\gamma^2}\right)^{N-n} \varphi_N.$$

c) It is clear that in the linear approximation at $b - a = 0$ all pendulums oscillate independently with the frequency $\omega_0 = \sqrt{g/l}$.

With the growth of the parameter $b - a$, the springs first decrease the restoring force of gravity and then begins to "push apart" the neighbouring pendulums, causing the instability of the small oscillations of the pendulums near the vertical.

The Lagrangian function of the system is

$$L = \tfrac{1}{2}ml^2 \sum_{n=1}^{2N} \dot{\varphi}_n^2 - U, \qquad U = \sum_{n=1}^{2N}\left(-mgl\cos\varphi_n + \tfrac{1}{2}kr_n^2\right),$$

where elongation of the nth spring is (we assume $l|\Delta_n| \ll a$)

$$r_n = \sqrt{a^2 + 4l^2\sin^2(\Delta_n/2)} - b \approx a - b + \frac{l^2}{2a}\Delta_n^2 - \frac{l^2(a^2+3l^2)}{24a^3}\Delta_n^4,$$

$$\Delta_n = \varphi_n - \varphi_{n+1}.$$

With the accuracy up to terms of the φ_n^4 order of magnitude inclusively, we have

$$U = \tfrac{1}{2}mgl\sum_n\left(\varphi_n^2 - \alpha\Delta_n^2 - \tfrac{1}{12}\varphi_n^4 + \beta\Delta_n^4\right) + \text{const},$$

$$\alpha = \frac{(b-a)kl}{amg}, \qquad \beta = \frac{1}{12}\alpha + \frac{kbl^3}{4mga^3}.$$

(1)

The equations of motion in the linear in φ_n approximation

$$\ddot{\varphi}_n + \frac{g}{l}[\varphi_n - \alpha(2\varphi_n - \varphi_{n+1} - \varphi_{n-1})] = 0$$

have solutions in the form of the travelling waves

$$\varphi_n = Ae^{i(\omega t \pm n\psi)} \qquad (2)$$

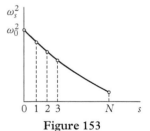

Figure 153

with the frequencies (see Fig. 153)

$$\omega_s^2 = \frac{g}{l}\left(1 - 4\alpha\sin^2\frac{\psi_s}{2}\right), \qquad \psi_s = \frac{\pi s}{N}, \qquad s = 0, 1, \ldots, N,$$

where the frequencies ω_0 and ω_N are non-generate, while the rest frequencies are twofold degenerate.

It can be seen that when

$$4\alpha - 1 > 0, \quad \text{or} \quad b - a > \frac{mga}{4kl}, \tag{3}$$

the oscillations are unstable; that is, some ω_s^2 become negative. The frequency ω_N is the first to vanish, which corresponds to $\psi_N = \pi$. The normal oscillation is of the type of "accordion", in which the neighbouring particles oscillate in anti-phase: $\varphi_n = -\varphi_{n-1}$. It is natural, therefore, to look for a new equilibrium position φ_{n0} in the form of "accordion":

$$\varphi_{10} = -\varphi_{20} = \varphi_{30} = -\varphi_{40} = \ldots = -\varphi_{2N0} = \varphi. \tag{4}$$

The value φ can be found from the equilibrium condition $\dfrac{\partial U}{\partial \varphi_n} = 0$ or

$$\varphi_n - \alpha(2\varphi_n - \varphi_{n+1} - \varphi_{n-1}) - \tfrac{1}{6}\varphi_n^3 + 2\beta(\varphi_n - \varphi_{n+1})^3 + 2\beta(\varphi_n - \varphi_{n-1})^3 = 0. \tag{5}$$

It gives

$$\varphi = \pm\sqrt{\frac{6(4\alpha - 1)}{192\beta - 1}}.$$

Let us now consider the small oscillations near the new equilibrium position (4). We will introduce the small displacements

$$x_n = \varphi_n - \varphi_{n0},$$

and then the potential energy (1) will be equal to the expression

$$U = \tfrac{1}{2}mgl\sum_n\left[\left(1 - \tfrac{1}{2}\varphi^2\right)x_n^2 - (\alpha - 24\beta\varphi^2)(x_n - x_{n+1})^2\right] + \text{const} \tag{6}$$

up to terms x_n^2 inclusively. By comparing (1) and (6), it is easy to see that x_n has a solution in the same form as φ_n (2) with frequencies

$$\omega_s^2 = \frac{g}{l}\left[1 - \tfrac{1}{2}\varphi^2 - 4(\alpha - 24\beta\varphi^2)\sin^2(\psi_s/2)\right].$$

However, now for the small $\varphi^2 < \dfrac{\alpha}{24\beta} < \tfrac{1}{2}$, all frequencies ω_s^2 are positive (see (3)):

$$\omega_s^2 > \frac{g}{l}\left[1 - \tfrac{1}{2}\varphi^2 - 4(\alpha - 24\beta\varphi^2)\right] = \frac{2g}{l}(4\alpha - 1) > 0;$$

that is, the small oscillations near the new equilibrium position (4) are stable.

Thus, with the increase of the parameter α, the original configuration of the vertical pendulum is replaced by the "accordion". This change in symmetry of the system is similar to the change of symmetry of the thermodynamic systems due to the phase transitions of the second kind. In this case, the external parameters such as temperature, magnetic field, and so on will be analogous to the α (e.g., [20]).

Of course, (5) can have other non-zero solutions, except the one found in (4). For example, this equation is satisfied by the value $\varphi_n = \sqrt{6}$, which, however, is not physical, as it corresponds to the large deflection angles, while the decomposition (1) itself is valid only for small φ.

7.10. a) Let we denote the current in nth coil as $\dot{q}_n$. The Lagrangian function is

$$L = \tfrac{1}{2} \sum_{n=1}^{N} \left[\mathscr{L} \dot{q}_n^2 - \frac{1}{C}(q_n - q_{n+1})^2 \right] + \tfrac{1}{2} \mathscr{L}_0 \dot{q}_{N+1}^2 + U q_1 \cos \gamma t$$

(the current through the Z chain is $\dot{q}_{N+1}$). The resistance R can be introduced into equations of motion using the dissipative function

$$F = \tfrac{1}{2} R \dot{q}_{N+1}^2.$$

The equations of motion are

$$\mathscr{L} \ddot{q}_1 + \frac{1}{C}(q_1 - q_2) = U \cos \gamma t, \tag{1}$$

$$\mathscr{L} \ddot{q}_n + \frac{1}{C}(2q_n - q_{n-1} - q_{n+1}) = 0, \quad n = 2, 3, \ldots, N, \tag{2}$$

$$\mathscr{L}_0 \ddot{q}_{N+1} + \frac{1}{C}(q_{N+1} - q_N) = -R \dot{q}_{N+1}. \tag{3}$$

We are looking for a solution in the form

$$q_n = \operatorname{Re}\{A e^{i\gamma t - in\varphi}\},$$

and we can consider, without losing generality, that $-\pi \leqslant \varphi \leqslant \pi$. From (2) and (3), we obtain

$$\gamma^2 = \frac{4}{\mathscr{L}C} \sin^2(\varphi/2), \tag{4}$$

$$-\gamma^2 \mathscr{L}_0 + \frac{1}{C}(1 - e^{i\varphi}) = -i\gamma R. \tag{5}$$

Hence,

$$R = \frac{\sin \varphi}{\gamma C}, \quad \mathscr{L}_0 = \frac{1 - \cos \varphi}{C \gamma^2} = \tfrac{1}{2} \mathscr{L}.$$

Since $R > 0$, there must be $\varphi > 0$; that is, the wave travels in the direction of the $\mathscr{L}R$ chains. The amplitude can then be determined from the equations (1).

For $\gamma^2 > \mathscr{L}C/4$, the propagation of travelling waves in the artificial line is impossible (cf. problem 7.6a).

b) The equations of motion

$$\mathscr{L}_1\ddot{q}_{2n-1} + \frac{1}{C}(2q_{2n-1} - q_{2n-2} - q_{2n}) = 0, \quad n = 2, 3, \ldots, N,$$

$$\mathscr{L}_2\ddot{q}_{2n} + \frac{1}{C}(2q_{2n} - q_{2n-1} - q_{2n+1}) = 0, \quad n = 1, 2, \ldots, N, \tag{6}$$

coincide, up to notations, with (1) in problem 7.5; besides,

$$\mathscr{L}_1\ddot{q}_n + \frac{1}{C}(q_1 - q_2) = U\cos\gamma t, \tag{7}$$

$$\mathscr{L}_0\ddot{q}_{2N+1} + \frac{1}{C}(q_{2N+1} - q_{2N}) = -R\dot{q}_{2N+1}. \tag{8}$$

We are looking for a solution in the form

$$q_{2n-1} = Ae^{i\gamma t - i(2n-1)\varphi},$$

$$q_{2n} = Be^{i\gamma t - i2n\varphi}. \tag{9}$$

Without losing generality, we can take $-\pi \leqslant \varphi \leqslant \pi$. From (6), we get

$$(1 - \gamma^2/\gamma_1^2)A - \cos\varphi \cdot B = 0,$$

$$\cos\varphi \cdot A - (1 - \gamma^2/\gamma_2^2)B = 0,$$

$$\gamma_{1,2}^2 = \frac{2}{\mathscr{L}_{1,2}C}, \tag{10}$$

and

$$\cos^2\varphi = (1 - \gamma^2/\gamma_1^2)(1 - \gamma^2/\gamma_2^2).$$

Let, for example, $\gamma_1 < \gamma_2$. The condition $0 \leqslant \cos^2\varphi \leqslant 1$ is satisfied when $0 \leqslant \gamma \leqslant \gamma_1$, which is the region of "acoustic" waves (cf. problem 7.5) and when $\gamma_2 \leqslant \gamma \leqslant \sqrt{\gamma_1^2 + \gamma_2^2}$, which is the region of "optic" waves. Outside these regions, the propagation of travelling waves is impossible (cf. problem 7.7).

From (8), we obtain

$$R + i\gamma\mathscr{L}_0 = \frac{i}{\gamma C} - \frac{i}{\gamma C}\frac{B}{A}e^{i\varphi} = \frac{i}{\gamma C}\left(1 - \frac{B}{A}\cos\varphi\right) + \frac{B}{A}\frac{\sin\varphi}{\gamma C} \tag{11}$$

with the condition $(B/A)\sin\varphi > 0$. In the region $\gamma \leqslant \gamma_1$, the amplitudes A and B have the same signs, so that $\varphi > 0$. In the region $\gamma_2 \leqslant \gamma \leqslant \sqrt{\gamma_1^2 + \gamma_2^2}$, on the contrary, $B/A < 0$ and $\varphi < 0$. Substituting the values of B/A, $\cos\varphi$ and $\sin\varphi$ in (11), we obtain as a result

$$R = \sqrt{\frac{\mathscr{L}_1 + \mathscr{L}_2}{2C}\left(1 - \frac{\mathscr{L}_1\mathscr{L}_2 C}{\mathscr{L}_1 + \mathscr{L}_2}\cdot\frac{\gamma^2}{2}\right)\frac{2 - \mathscr{L}_1 C\gamma^2}{2 - \mathscr{L}_2 C\gamma^2}}, \quad \mathscr{L}_0 = \tfrac{1}{2}\mathscr{L}_1.$$

The negative value of φ in the region of "optical" oscillations means that the phase velocity of the travelling wave is directed from the Z chain to the voltage source. The group velocity, on the contrary, has the opposite direction (e.g. see, Fig. 148 where $v_{\mathrm{gr}} \propto \frac{d\omega_{(+)}}{ds} < 0$; cf. problem 7.5). The group velocity is the velocity of the wave packet. If the wave packet moves along the chain, then the oscillation energy is localized in the region where the wave packet is at the moment, so naturally that the group velocity determines the flow of energy.

7.11. The equation for the oscillations of the discrete system (see (3) of problem 7.1) can be written in the form

$$\ddot{x}_n - ka\frac{a}{m}\left(\frac{x_{n+1} - x_n}{a} - \frac{x_n - x_{n-1}}{a}\right)\frac{1}{a} = 0. \tag{1}$$

The quantity $\frac{m}{a} = \frac{Nm}{Na}$ becomes in the limit the linear density of the rod ρ. The relative extension of the section a; that is, the quantity $(x_n - x_{n-1})/a$ is proportional to the force acting upon it,

$$F = ka\frac{x_n - x_{n-1}}{a},$$

and ka thus becomes in the limit the elastic modulus $\varkappa$ of the rod. Equations (1) in the limit thus becomes the wave equation

$$\frac{\partial^2 x(\xi, t)}{\partial t^2} - v^2\frac{\partial^2 x(\xi, t)}{\partial \xi^2} = 0, \tag{2}$$

where $v = \sqrt{\varkappa/\rho}$ is the phase velocity of the wave.

Instead of a set of N coupled ordinary differential equations, we have obtained one partial differential equation (cf. problem 4.32).

Note that in our derivation we had to make the important assumption that the function $x_n(t)$ tends to a well-defined limit $x(\xi, t)$ which is a sufficiently smooth function.

7.12. If a is small, we can approximate the displacement as

$$x_n = x(\xi, t),$$

$$x_{n\pm1} = x(\xi \pm a, t) = x(\xi, t) \pm a\frac{\partial x(\xi, t)}{\partial \xi} + \frac{a^2}{2}\frac{\partial^2 x(\xi, t)}{\partial \xi^2} \pm$$

$$\pm \frac{a^3}{6}\frac{\partial^3 x(\xi, t)}{\partial \xi^3} + \dots$$

Therefore, (2) of problem 7.11 changes to

$$\frac{\partial^2 x}{\partial t^2} - \frac{\varkappa}{\rho}\frac{\partial^2 x}{\partial \xi^2} - \frac{\varkappa a^2}{12\rho}\frac{\partial^4 x}{\partial \xi^4} = 0. \qquad (1)$$

While each of the equations (1) of problem 7.11 contains the displacements of the three neighbouring points (long-range interaction), (1) here contains the displacement x in a given point ξ (short-range interaction). The last term

$$-\frac{\varkappa a^2}{12\rho}\frac{\partial^4 x}{\partial \xi^4}$$

in (1) corresponds to approximate account of the small difference between the system considered and a continuous one. This leads, in particular, to the fact that the phase velocity appears to be dependent on the wavelength.

Substituting $x = a\cos(\omega t - k\xi)$ in the equation, we obtain

$$v = \frac{\omega}{k} = \sqrt{\frac{\varkappa}{\rho}\left(1 - \frac{a^2 k^2}{12}\right)}.$$

Such phenomena, which are due to the long-range interaction, are called *spatial dispersion*.

§8

Non-linear oscillations

8.1. a) We solve the equation of motion

$$\ddot{x} + \omega_0^2 x = -\beta x^3 \tag{1}$$

using the method of successive approximations (cf. problem 6.35):

$$x = x_0 + \delta x = A e^{i\omega_0 t} + A^* e^{-i\omega_0 t} + \delta x, \tag{2}$$

where $A(t)$ is the complex amplitude

$$A = \tfrac{1}{2}\, a e^{i\varphi}.$$

The "force"

$$-\beta x_0^3 = -\beta A^3 e^{3i\omega_0 t} - 3\beta A^2 A^* e^{i\omega_0 t} - 3\beta A A^{*2} e^{-i\omega_0 t} - \beta A^{*3} e^{-3i\omega_0 t}$$

contains the resonant terms

$$-3\beta A^2 A^* e^{i\omega_0 t} - 3\beta A A^{*2} e^{-i\omega_0 t} = -3\beta |A|^2 x,$$

which are more convenient to attach to the term $\omega_0^2 x$ in left-hand side of (1).
   This results in replacement

$$\omega_0^2 \to \omega^2 = \omega_0^2 + 3\beta |A|^2.$$

For δx, we obtain

$$\delta\ddot{x} + \omega^2 \delta x = -\beta(A^3 e^{3i\omega t} + \text{comp. conj.}),$$

Exploring Classical Mechanics: A Collection of 350+ Solved Problems for Students, Lecturers, and Researchers. First Edition.
Gleb L. Kotkin and Valeriy G. Serbo, Oxford University Press (2020). © Gleb L. Kotkin and Valeriy G. Serbo 2020.
DOI: 10.1093/oso/9780198853787.001.0001

from where

$$\delta x = \frac{\beta A^3}{8\omega^2} e^{3i\omega t} + \text{comp. conj.}$$

As a result,

$$x = a\cos(\omega t + \varphi) + \frac{\beta a^3}{32\omega^2} \cos(3\omega t + 3\varphi),$$

$$\omega = \omega_0 + \frac{3\beta a^2}{8\omega_0}$$

(cf. [1], § 28 and [7], § 29.1). Fig. 154 depicts the function $x(t)$.

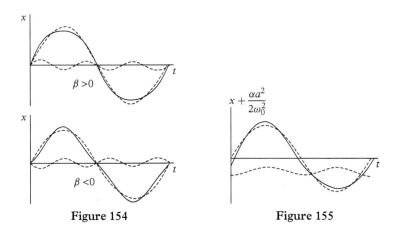

Figure 154   Figure 155

When $\beta > 0$, there is a "limitation" of the oscillations; when $\beta < 0$, the maxima become sharper. These properties of the oscillations, as well as the sign of the corrections to the frequency, can easily be considered by looking at the graph of $U(x)$. For other solution methods, see problems 1.9 and 11.26d.

b) Solving the problem in the same way as in point 8.1a, we obtain

$$\delta x = \frac{\alpha A^2}{3\omega_0^2} e^{2i\omega_0 t} - \frac{2\alpha |A|^2}{\omega_0^2} + \frac{\alpha A^{*2}}{3\omega_0^2} e^{-2i\omega_0 t};$$

that is

$$x = a\cos(\omega_0 t + \varphi) - \frac{\alpha a^2}{2\omega_0^2} + \frac{\alpha a^2}{6\omega_0^2} \cos(2\omega_0 t + 2\varphi).$$

The distortion of the oscillations is non-symmetric (Fig. 155).

In the following approximation one must take into account the term $-2\alpha x_0 \delta x$ in the "force" $-\alpha(x_0 + \delta x)^2$ which contains the resonant terms

$$-\frac{2\alpha^2}{3\omega_0^2}A^2 A^* e^{i\omega_0 t} - \frac{2\alpha^2}{3\omega_0^2}AA^{*2}e^{-i\omega_0 t} + \frac{4\alpha^2|A|^2}{\omega_0^2}x = \frac{10\alpha^2}{3\omega_0^2}|A|^2 x.$$

This results to the replacement

$$\omega_0 \to \omega_0 - \frac{5\alpha^2 a^2}{12\omega_0^3}.$$

8.2. $x = a\cos\omega t - \frac{1}{4}\gamma a^2 \cos 2\omega t + \frac{1}{4}\gamma a^2, \quad \omega = \omega_0 + \frac{1}{16}\gamma^2 a^2 \omega_0.$

8.3. $\quad \varphi = \dfrac{a\Omega^2}{g - l\Omega^2}\cos\Omega t + \dfrac{a^2\Omega^4}{2(g - l\Omega^2)(g - 4l\Omega^2)}\sin 2\Omega t$ (the notation is that of problem 5.9).

8.4. $x = x^{(0)} + x^{(1)} + \ldots,$

$$x^{(0)} = \frac{f_1 \cos\omega_1 t}{m(\omega_0^2 - \omega_1^2)} + \frac{f_2 \cos\omega_2 t}{m(\omega_0^2 - \omega_2^2)},$$

$$x^{(1)} = -\frac{\alpha f_1^2}{2m^2\omega_0^2(\omega_0^2 - \omega_1^2)^2} - \frac{\alpha f_2^2}{2m^2\omega_0^2(\omega_0^2 - \omega_2^2)^2} -$$
$$-\frac{\alpha f_1^2 \cos 2\omega_1 t}{2m^2(\omega_0^2 - 4\omega_1^2)(\omega_0^2 - \omega_1^2)^2} - \frac{\alpha f_2^2 \cos 2\omega_2 t}{2m^2(\omega_0^2 - 4\omega_2^2)(\omega_0^2 - \omega_2^2)^2} -$$
$$-\frac{\alpha f_1 f_2 \cos(\omega_1 - \omega_2)t}{m^2[\omega_0^2 - (\omega_1 - \omega_2)^2](\omega_0^2 - \omega_1^2)(\omega_0^2 - \omega_2^2)} -$$
$$-\frac{\alpha f_1 f_2 \cos(\omega_1 + \omega_2)t}{m^2[\omega_0^2 - (\omega_1 + \omega_2)^2](\omega_0^2 - \omega_1^2)(\omega_0^2 - \omega_2^2)}.$$

What combinational frequencies will occur when we take into account an anharmonic correction of the form $\delta U = \frac{1}{4}m\beta x^4$?

8.5. Let x and y be the displacements of the pendulum from the equilibrium position in the horizontal and vertical directions. We expand the Lagrangian function

$$L = \frac{1}{2}m(\dot{x}^2 + \dot{y}^2) - \frac{1}{2}k\left(\sqrt{(l-y)^2 + x^2} - l_0\right)^2 - mgy$$

in series of small x and y up to the third order inclusively

$$L = \frac{1}{2}m(\dot{x}^2 + \dot{y}^2 - \omega_1^2 x^2 - \omega_2^2 y^2 + 2\alpha x^2 y) + \ldots,$$

where

$$\omega_1^2 = \frac{g}{l}, \quad l = l_0 + \frac{mg}{k}, \quad \omega_2^2 = \frac{k}{m}, \quad \alpha = \frac{kl_0}{2ml^2}.$$

Using the same method as in problem 8.1, we obtain

$$x = a\cos(\omega_1 t + \varphi_1) - \frac{\alpha ab}{2\omega_2(2\omega_1 + \omega_2)}\cos(\omega_+ t + \varphi_+) +$$

$$+ \frac{\alpha ab}{2\omega_2(2\omega_1 - \omega_2)}\cos(\omega_- t + \varphi_-),$$

$$y = b\cos(\omega_2 t + \varphi_2) + \frac{\alpha a^2}{2\omega_2^2} + \frac{\alpha a^2}{2(\omega_2^2 - 4\omega_1^2)}\cos(2\omega_1 t + 2\varphi_1),$$

where $\omega_\pm = \omega_1 \pm \omega_2$ and $\varphi_\pm = \varphi_1 \pm \varphi_2$.

The solutions obtained are valid as long as the frequency ω_2 is not close to $2\omega_1$. When $\omega_2 = 2\omega_1$, the anharmonic corrections cease to be small and can lead to the significant pumping of the energy from the x- to y-oscillations and back. This case is considered in problem 8.10.

8.6. a) We look for a solution in the form

$$x = Ae^{i\omega t} + A^* e^{-i\omega t}.$$

Equating the coefficients of $e^{i\omega t}$, we get

$$(\omega_0^2 - \omega^2 + 2i\lambda\omega + 3\beta|A|^2)a = \tfrac{1}{2}f,$$

and, hence,

$$[(\omega_0^2 - \omega^2 + 3\beta|A|^2)^2 + 4\lambda^2\omega^2]|A|^2 = \tfrac{1}{4}f^2.$$

A study of this equation which is cubic in $|A|^2$ can be done in the same way as the study of the analogous equation (29.4) in [1].

This equation is square in ω^2, so that the dependence of $|A|^2$ on ω^2 can be easily represented graphically (see [7], § 30).

8.7. a) We look for a solution of the oscillations equation

$$\ddot{x} + 2\lambda\dot{x} + \omega_0^2(1 + h\cos 2\omega t)x + \beta x^3 = 0 \tag{1}$$

in the form

$$x = Ae^{i\omega t} + A^* e^{-i\omega t}, \tag{2}$$

and we retain only the terms containing $e^{\pm i\omega t}$.[1] Putting the coefficients of $e^{\pm i\omega t}$ equal to zero, we find

$$\begin{aligned} &\tfrac{1}{2}\, h\omega_0^2 A + \left(\omega_0^2 - \omega^2 - 2i\omega\lambda + 3\beta|A|^2\right)A^* = 0, \\ &\tfrac{1}{2}\, h\omega_0^2 A^* + \left(\omega_0^2 - \omega^2 + 2i\omega\lambda + 3\beta|A|^2\right)A = 0. \end{aligned} \tag{3}$$

We can only have a non-vanishing A if

$$\begin{vmatrix} \tfrac{1}{2}\, h\omega_0^2 & \omega_0^2 - \omega^2 - 2i\omega\lambda + 3\beta|A|^2 \\ \omega_0^2 - \omega^2 + 2i\omega\lambda + 3\beta|A|^2 & \tfrac{1}{2}\, h\omega_0^2 \end{vmatrix} = 0. \tag{4}$$

Hence,

$$|A|^2 = \frac{1}{3\beta}\left[\omega^2 - \omega_0^2 \pm \sqrt{\left(\tfrac{1}{2}\, h\omega_0^2\right)^2 - (2\omega\lambda)^2}\right]. \tag{5}$$

From (3), we get[2]

$$\begin{aligned} \sin 2\varphi &= \operatorname{Im}\frac{A}{A^*} = -\frac{4\lambda}{h\omega_0}, \\ \cos 2\varphi &= \mp\frac{2}{h\omega_0^2}\sqrt{\left(\tfrac{1}{2}\, h\omega_0^2\right)^2 - (2\omega\lambda)^2}. \end{aligned} \tag{6}$$

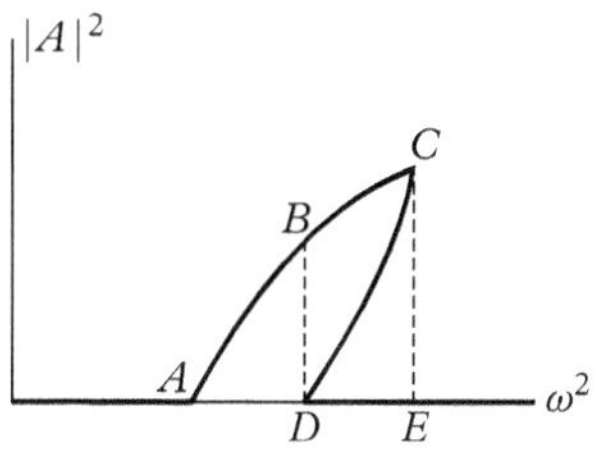

Figure 156

Thus,

$$x = a\cos(\omega t + \varphi), \tag{7}$$

where $A = \tfrac{1}{2}\, ae^{i\varphi}$.

Fig. 156 shows how $|A|^2$ depends on ω^2 (to fix the ideas, we assume $\beta > 0$). In some frequency ranges two options or three (including zero values) different amplitudes of stable oscillations are possible.

The amplitudes corresponding to the sections *AD* and *CD* are not realized in actual cases since those oscillations are unstable (for a proof of this for the section *AD*, see the next problem; for a study of the stability of the oscillations along sections *ABC*, *CD*, and *DE*, see [10]).

[1] Assume that the terms in $e^{\pm 3i\omega t}$ are appreciably smaller and will be compensated by the contribution to x from the third harmonic, as will become clear in the following discussion.

[2] Equations (6) determine the phase apart from a term $n\pi$. There is no sense in determining the phase with greater precision as a change in the phase by π corresponds simply to a shift in the time origin.

b) When we take the third harmonic into account, x has the form

$$x = Ae^{i\omega t} + A^* e^{-i\omega t} + Be^{3i\omega t} + B^* e^{-3i\omega t}. \tag{8}$$

We assume that $|B| \ll |A|$, which will be confirmed by the results. Substituting (8) into (1), we split off the terms containing $e^{3i\omega t}$; we then drop the product of B with small parameters. We find out that

$$B = \left(\tfrac{1}{16} h + \tfrac{1}{8} \beta A^2 \omega^{-2} \right) A \tag{9}$$

and, indeed, $|B| \ll |A|$.

Therefore,

$$x = a \cos(\omega t + \varphi) + b \cos(3\omega t + \psi),$$

where $b = 2|B|$, $\psi = \arg B$.

It is easy to notice that the fifth harmonic turns out to be smaller than second order ($\sim h^2 A$), the seventh $\sim h^3 A$, and so on. The even harmonics do not occur. This is the basis of the method used to evaluate the amplitudes.

8.8. a) We look for a solution to the equation of motion in the form

$$x(t) = a(t) \cos \omega t + b(t) \sin \omega t, \tag{1}$$

where $a(t)$ and $b(t)$ are slowly changing functions of the time. To determine $a(t)$ and $b(t)$, we get the following set of equations (cf. [1], § 27)

$$\dot{a} + \left(\omega - \omega_0 + \tfrac{1}{4} h \omega_0 \right) b = 0,$$
$$\dot{b} - \left(\omega - \omega_0 - \tfrac{1}{4} h \omega_0 \right) a = 0. \tag{2}$$

If $|\omega_1 - \omega_0| < \tfrac{1}{4} h \omega_0$, its solution is

$$a(t) = \alpha_1 (C_1 e^{-st} + C_2 e^{st}),$$
$$b(t) = \alpha_2 (C_1 e^{-st} - C_2 e^{st}), \tag{3}$$

where

$$s = \tfrac{1}{4} \sqrt{(h\omega_0)^2 - 16(\omega - \omega_0)^2}, \quad \alpha_{1,2} = \sqrt{h\omega_0 \pm 4(\omega - \omega_0)}.$$

Hence,

$$x = C_3 e^{st} \cos(\omega t + \varphi) + C_4 e^{-st} \cos(\omega t - \varphi), \tag{4}$$

where $\tan \varphi = \alpha_2 / \alpha_1$ (Fig. 157).

The oscillations thus increase, generally speaking, without limit. The rate of their increase which is characterized by the quantity s is, indeed, small. In actual cases, the increase in the amplitude of the oscillations is cut off, for instance, if the influence of the anharmonic terms becomes important (see problem 8.7) or the reaction of the oscillations on the device which periodically changes its frequency becomes important.

It is useful to draw attention to the analogy between the results obtained and the particular solution of the problem of the normal oscillations of a chain of particles connected by springs of different stiffness (problem 7.5b). An inhomogeneity with period $2a$ along the chain leads to a build-up along the chain of the amplitude of the stable oscillations, when the "wavelength" is equal to $4a$ (Fig. 151) in a similar way that a periodic change with time of the frequency of an oscillator will lead to an increase in the amplitudes with time.

An even more complete analogy can be observed in problem 7.7. The region of instability with respect to parametric resonance corresponds to the forbidden zone in the spectrum of oscillations of the chain.

Similar equations are obtained in quantum mechanics in the problem of the motion of a particle in a periodic field. In that problem, we also met with "forbidden bands" and "surface states".

b) If $|\omega - \omega_0| > \frac{1}{4} h\omega_0$, we have

$$x = C\beta_1 \sin(\Omega t + \psi)\cos\omega t - C\beta_2 \cos(\Omega t + \psi)\sin\omega t,$$

where $\Omega = \frac{1}{4}\sqrt{16(\omega - \omega_0)^2 - (h\omega_0)^2}$,

$$\beta_{1,2} = \begin{cases} \sqrt{4(\omega - \omega_0) \pm h\omega_0}, & \text{when} \quad \omega > \omega_0, \\ \pm\sqrt{4(\omega_0 - \omega) \mp h\omega_0}, & \text{when} \quad \omega < \omega_0. \end{cases}$$

The oscillations are beats:

$$x = C\sqrt{4|\omega - \omega_0| \mp h\omega_0 \cos(2\Omega t + 2\psi)}\cos(\omega t + \theta), \quad \text{when } \omega \gtrless \omega_0,$$

where θ is a slowly varying phase (see Fig. 158). If the frequency approaches the limit of the instability region, the depth of the modulation of the oscillations approaches the total amplitude and their period increases without limit.

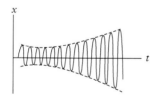

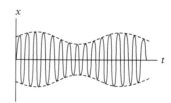

Figure 157   Figure 158

What is the form of the oscillations when $|\omega - \omega_0| = \frac{1}{4} h\omega_0$?

8.9. Let $x = e^{i\omega_1 t}$, when $0 < t < \tau$. We then have in the interval $\tau < t < 2\tau$,

$$x = ae^{i\omega_2 t} + be^{-i\omega_2 t}$$

where a and b are determined from the "matching" condition for $t = \tau$:

$$x(\tau - 0) = x(\tau + 0), \quad \dot{x}(\tau - 0) = \dot{x}(\tau + 0).$$

Hence, we have

$$a = \frac{\omega_1 + \omega_2}{2\omega_2} e^{i(\omega_1 - \omega_2)\tau},$$

$$b = \frac{\omega_2 - \omega_1}{2\omega_2} e^{i(\omega_1 + \omega_2)\tau}.$$

Similarly, we find that for $2\tau < t < 3\tau$

$$x = \alpha e^{i\omega_1 t} + \beta e^{-i\omega_1 t},$$

where

$$\alpha = e^{-i\omega_1 \tau} \left(\cos \omega_2 \tau + i \frac{\omega_1^2 + \omega_2^2}{2\omega_1 \omega_2} \sin \omega_2 \tau \right),$$

$$\beta = i \sin \omega_2 \tau \, e^{3i\omega_1 \tau} \frac{\omega_1^2 - \omega_2^2}{2\omega_1 \omega_2}.$$

It is clear that the oscillation in the form

$$Ae^{i\omega_1 t} + Be^{-i\omega_1 t} \tag{1}$$

for $0 < t < \tau$, after a period of 2τ will become

$$A(\alpha e^{i\omega_1 t} + \beta e^{-i\omega_1 t}) + B(\alpha^* e^{-i\omega_1 t} + \beta^* e^{i\omega_1 t}) =$$
$$= (\alpha A + \beta^* B)e^{i\omega_1 t} + (\beta A + \alpha^* B)e^{-i\omega_1 t}.$$

We now look for such a linear combination (1) that it retains its form—apart from a multiplying factor—after a period 2τ:

$$(\alpha A + \beta^* B)e^{i\omega_1 t} + (\beta A + \alpha^* B)e^{-i\omega_1 t} = \mu(Ae^{i\omega_1 t'} + Be^{-i\omega_1 t'}),$$

where $t' = t - 2\tau$,

$$\alpha A + \beta^* B = \mu e^{-2i\omega_1 \tau} A,$$
$$\beta a + \alpha^* B = \mu e^{2i\omega_1 \tau} B. \tag{2}$$

Set (2) has a non-trivial solution, provided

$$(\alpha - \mu e^{-2i\omega_1 \tau})(\alpha^* - \mu e^{2i\omega_1 \tau}) - \beta\beta^* = 0,$$

whence

$$\mu_{1,2} = \gamma \pm \sqrt{\gamma^2 - 1},$$

where

$$\gamma = \operatorname{Re}(\alpha e^{2i\omega_1 \tau}) = \cos\omega_1 \tau \cos\omega_2 \tau - \frac{\omega_1^2 + \omega_2^2}{2\omega_1 \omega_2} \sin\omega_1 \tau \sin\omega_2 \tau.$$

After n periods, the oscillation

$$x_{1,2} = A_{1,2}(e^{i\omega_1 t} + \lambda_{1,2} e^{-i\omega_1 t}), \quad 0 < t < \tau,$$
$$\lambda_{1,2} = \mu_{1,2} \frac{e^{-2i\omega_1 \tau} - \alpha}{\beta^*}$$

has changed to

$$x_{1,2}(t) = \mu_{1,2}^n A_{1,2}(e^{i\omega_1 t'} + \lambda_{1,2} e^{-i\omega_1 t'}), \quad 0 < t' = t - 2n\tau < \tau.$$

Any oscillation is a superposition of oscillations such as $x_{1,2}$; in particular, the real oscillation (which is the only one which has a direct physical meaning)

$$x(t) = A e^{i\omega_1 t} + A^* e^{-i\omega_1 t}, \quad 0 < t < \tau,$$

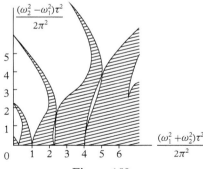

Figure 159

is the sum of $x_1(t) + x_2(t)$ with

$$A_1 = \frac{A^* - \lambda_2 A}{\lambda_1 - \lambda_2}, \quad A_2 = \frac{\lambda_1 A - a^*}{\lambda_1 - \lambda_2}.$$

If $\gamma < 1$, $|\mu_{1,2}| = 1$ and the oscillations $x_{1,2}(t)$ (and at the same time $x(t)$) remain bounded.

If, however, $\gamma > 1$, we have $\mu_1 > 1$, and the amplitude of the oscillations increases without bound. This is the case of the onset of parametric resonance. We can easily verify that if the frequency difference is small, $|\omega_1 - \omega_2| \ll \omega_1$, this condition is satisfied if the frequencies lie close to $n\pi/\tau$:

$$|(\omega_1 + \omega_2)\tau - 2\pi n| < \frac{(\omega_1 - \omega_2)^2 \tau}{\omega_1 + \omega_2}.$$

We show in Fig. 159 (taken from [17]) the regions of instability against parametric resonance.

8.10. The equations of motion are

$$\ddot{x} + \omega^2 x - 2\alpha xy = 0,$$
$$\ddot{y} + 4\omega^2 y - \alpha x^2 = 0.$$

We look for a solution in the form

$$x = A e^{i\omega t} + A^* e^{-i\omega t} + \delta x,$$
$$y = B e^{2i\omega t} + B^* e^{-2i\omega t} + \delta y,$$

assuming that A and B are the slowly varying amplitudes of the oscillations, while more rapidly oscillating terms δx and δy can be neglected: $|\ddot{A}| \ll \omega|\dot{A}| \ll \omega^2|A|$, $|\ddot{B}| \ll \omega|\dot{B}| \ll \omega^2|B|$, $\delta x \sim \delta y \ll |A|$.

Leaving only the terms with $e^{i\omega t}$ ($e^{2i\omega t}$, respectively) and neglecting $|\ddot{A}|$, $|\ddot{B}|$, we get

$$\omega \dot{A} + i\alpha B A^* = 0,$$
$$4\omega \dot{B} + i\alpha A^2 = 0. \tag{1}$$

Clearly, from (1) follows

$$|A|^2 + 4|B|^2 = C = \text{const} \tag{2}$$

(this is the law of conservation of energy) and

$$A^{*2} B + A^2 B^* = D = \text{const}. \tag{3}$$

Using (1), we find

$$\omega \frac{d}{dt}|A|^2 = -i\alpha(A^{*2}B - A^2B^*).$$ (4)

Squaring (4) and taking into account (2) and (3), we get

$$\left(\frac{d}{dt}|A|^2\right)^2 = -\frac{\alpha^2}{\omega^2}\left[(A^{*2}B + A^2B^*)^2 - 4|A|^4|B|^2\right] =$$

$$= \frac{\alpha^2}{\omega^2}\left[|A|^4(C - |A|^2) - D^2\right].$$ (5)

This equation is similar to the law of conservation of energy for the problem about the one-dimensional motion of a particle with a coordinate $|A|^2$. It is helpful to study this equation using the graph of the "potential energy" $U(|A|^2) = (|A|^2 - C)|A|^4$ (see Fig. 160).

Fig. 160 shows that the amplitude $|A|$ experiences oscillations which lead to the beats. The dependence of the amplitudes $|A|$ and $|B|$ on the time can be expressed in the elliptic functions (we will not do this here).

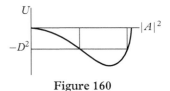

Figure 160

Note that, in this case, not only the depth of beating, but also the period will depend on the initial amplitudes and phases. This is in direct contrast to the vibrations of oscillators with linear coupling (see problem 6.9).

This problem is relevant, for instance, to the coupling of the longitudinal and flexural oscillations of a molecule CO_2 (the so-called *Fermi resonance*; see [18]) and to the *doubling and division* of the light frequency in nonlinear optics (see [19]).

8.11. $\omega^2 = \frac{a^2\gamma^2}{2l^2} + \frac{g}{l}$. When $\frac{a^2\gamma^2}{2l^2} > \frac{g}{l}$, a second stable equilibrium position appears in the form the straight upward vertical; the oscillation frequency around it is equal to $\omega^2 = \frac{a^2\gamma^2}{2l^2} - \frac{g}{l}$ (see [1], § 30, problem 1).

8.12. a)

$$U_{\text{eff}} = \frac{\alpha^2}{m\omega^2}\left[\frac{a^2}{r^6} + \frac{3(\mathbf{ar})^2}{r^8}\right],$$ (1)

Notice that the dependence $U_{\text{eff}} \propto r^{-6}$ is characteristic of intermolecular forces. If we substitute in (1) the values[1] $\alpha \sim e^2 \sim (5 \cdot 10^{-10}\,\text{ESU})^2$, $a \sim 10^{-8}$ cm, and $\omega \sim 10^{16}\,\text{s}^{-1}$, which are typical for atoms, and as a mass we choose the electron mass, $m \sim 10^{-27}$ g,

---

[1] In the CI system: $\alpha \sim \frac{e^2}{4\pi\varepsilon_0} \sim \frac{(10^{-19}\,\text{K})^2}{10^{-11}\,(\text{F/m})\cdot 10}$, $m \sim 10^{-30}$ kg, $a \sim 10^{-10}$ m, $U_{\text{eff}} \sim 10^{-18}\,(a/r)^6$ J.

we get $U_{\mathrm{eff}} \sim 10^{-40}$ erg $\cdot$ cm$^6/r^6$, which is close to the correct value for van der Waals interaction, as far as order of magnitude is concerned. This result may serve as an indication of the physical nature of this interaction. A complete calculation of the van der Waals forces is only possible using quantum mechanics.

b)

$$U_{\mathrm{eff}} = \frac{\alpha^2}{m(\omega^2 - \omega_0^2)} \left[\frac{a^2}{r^6} + \frac{3(\mathbf{a r})^2}{r^8} \right],$$

where ω_0 is the eigen-frequency of the oscillator.

8.13. The motion along the z-axis is nearly uniform, $z = vt$. In the xy-plane, the particle is acted upon by a fast oscillating force

$$f_x = 2Ax \sin kvt, \quad f_y = 2Ay \sin kvt.$$

The corresponding effective potential energy is $U_{\mathrm{eff}} = \frac{1}{2}m\Omega^2(x^2 + y^2)$, where $\Omega = \sqrt{2}A/(mkv)$.

According to the initial conditions, we have for the frequency of the force oscillation $kv \gg \Omega$ so that the force is, indeed, fast oscillating. Thus, in the xy-plane, the particle performs a harmonic oscillation with frequency Ω around the z-axis.

This problem illustrates the principle of strong focus of particle beams in accelerators.

8.14. The equations of motion are

$$m\ddot{x} = \frac{e}{c}B(x)\dot{y},$$

$$m\ddot{y} = -\frac{e}{c}B(x)\dot{x}.$$

We look for the law of motion in the form

$$x = X + \xi, \qquad y = Y + \eta, \tag{1}$$

where the terms ξ and η describe the fast motion over the almost circular orbit, while X and Y are the slow displacements of its centre (cf. [1], § 30). Substituting (1) into the equations of motion, we expand $B(X + \xi)$ in powers of ξ:

$$\ddot{X} + \ddot{\xi} = \omega\dot{Y} + \omega\dot{\eta} + \frac{e}{mc}B'(X)\xi(\dot{Y} + \dot{\eta}),$$

$$\ddot{Y} + \ddot{\eta} = -\omega\dot{X} - \omega\dot{\xi} - \frac{e}{mc}B'(X)\xi(\dot{x} + \dot{\xi}),$$

and then consider separately the fast oscillating and slowly changing terms. For the oscillating terms, we obtain

$$\ddot{\xi} = \omega\dot{\eta}, \quad \ddot{\eta} = -\omega\dot{\xi}, \quad \omega = \frac{e}{mc}B(X),$$

where

$$\xi = r\cos\omega t, \quad \eta = -r\sin\omega t.$$

For the slowly changing terms, we have

$$\ddot{X} = \omega\dot{Y} + \frac{e}{mc}B'(X)\langle\xi\dot{\eta}\rangle,$$

$$\ddot{Y} = -\omega\dot{X} - \frac{e}{mc}B'(X)\langle\xi\dot{\xi}\rangle,$$

(2)

where

$$\langle\xi\dot{\eta}\rangle = -r^2\omega\langle\cos^2\omega t\rangle = -\tfrac{1}{2}r^2\omega, \quad \langle\xi\dot{\xi}\rangle = 0.$$

Since $\ddot{X}, \ddot{Y} \sim \varepsilon\omega\dot{x}, \varepsilon\omega\dot{Y}$, the left parts in (2) can be put equal to zero.

As a result,

$$\dot{Y} = \frac{er^2}{2mc}B'(X) = \tfrac{1}{2}\varepsilon v, \quad \dot{x} = 0.$$

The rate of the displacement of the orbit centre (the drift velocity) in a more general case is considered in [2], § 22, problem 3 and in [10], § 25.

8.15. The equation of motion of the ball is

$$m\ddot{y} = -\frac{dU(y)}{dy} + f(t).$$

The proper motion of the ball under the action of a spring is described by the "low-frequency" displacement $x = y - y_0\cos\gamma t$, for which

$$m\ddot{x} = -\frac{dU(x + y_0\cos\gamma t)}{dx}.$$

If we average this proper motion over a period of $2\pi/\gamma$ of the high-frequency motion using

$$\langle\cos^{2n+1}\gamma t\rangle = 0, \quad \langle\cos^2\gamma t\rangle = \tfrac{1}{2},$$

we get the effective force and the corresponding effective potential energy

$$U_{\text{eff}}(x) = Ax^2 + Bx^4, \quad A = -C + 3By_0^2.$$

The graph of the function $U_{\text{eff}}(x)$ is shown in Fig. 161.

When $A > 0$ or

$$T = y_0^2 > T_c = \frac{C}{3B},$$

the ball oscillates near the point $x = 0$ with the frequency

$$\omega = \sqrt{2A/m} \propto \sqrt{T - T_c}.$$

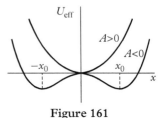

Figure 161

When $A < 0$ or $T < T_c$, the minima of $U_{\text{eff}}(x)$ are located at the points

$$\pm x_0 = \pm\sqrt{\frac{-A}{2B}},$$

and the ball oscillates near one of the points with the frequency

$$\omega = \sqrt{\frac{-4A}{m}} \propto \sqrt{T_c - T}.$$

The emerging picture is very close to the picture of the phase transitions of the second kind, described by the phenomenological theory of Landau [20]. The fast forced oscillations are analogous to the thermal motion (corresponding to the optical modes of oscillations of the system which is not connected with the transition), and the value $T = y_0^2$ is the analog of temperature. When T is large, the system oscillates around the equilibrium position $x = 0$. In this case, there is the symmetry with respect to the replacement of $x \to -x$. When the temperature decreases reaching a value $T < T_c$, the ball begins to oscillate around one of the new equilibrium positions: x_0 or $-x_0$. In this case, the symmetry $x \to -x$ obviously vanishes. Moreover, the value T_c is an analog of the temperature of the phase transition of the second kind. In the neighbourhood of $T = T_c$, the value x_0 is small, $x_0 \propto \sqrt{T_c - T}$, and frequency ω of the eigen-oscillations is small.

§9

Rigid-body motion. Non-inertial coordinate systems

9.1. Let us introduce the generalized coordinates as follows: the angle of the string deflection from the vertical, φ, and the angle between the vertical and the radius of the ring connecting the point of its attachment to the centre of the ring, ψ. The kinetic energy of the ring is the sum of the energy of its rotation around the ring centre $\frac{1}{2} mR^2 \dot{\psi}^2$ and the energy of the translational motion $\frac{1}{2} m(R\dot{\varphi} + R\dot{\psi})^2$ (since the angles φ and ψ are small, we can consider velocities of the suspension point and the centre of the ring as parallel). The potential energy is determined by the position of the ring centre

$$U = mgR\left[(1 - \cos\varphi) + (1 - \cos\psi)\right] \approx \tfrac{1}{2} mgR(\varphi^2 + \psi^2).$$

Using Lagrangian function

$$L = \tfrac{1}{2} mR^2 \left[(\dot{\varphi} + \dot{\psi})^2 + \dot{\psi}^2 - \frac{g}{R}\left(\varphi^2 + \psi^2\right)\right],$$

we get the equations of motion

$$\ddot{\varphi} + \ddot{\psi} + \frac{g}{R}\varphi = 0, \quad \ddot{\varphi} + 2\ddot{\psi} + \frac{g}{R}\varphi = 0.$$

The solution of these equations has the form

$$\begin{pmatrix} \varphi \\ \psi \end{pmatrix} = \begin{pmatrix} -1 + \sqrt{5} \\ 2 \end{pmatrix} A_1 \cos(\omega_1 t + \chi_1) + \begin{pmatrix} -1 - \sqrt{5} \\ 2 \end{pmatrix} A_2 \cos(\omega_2 t + \chi_2),$$

where the normal frequencies are equal to

$$\omega_{1,2} = \sqrt{\frac{3 \pm \sqrt{5}}{2} \frac{g}{R}}.$$

Exploring Classical Mechanics: A Collection of 350+ Solved Problems for Students, Lecturers, and Researchers. First Edition.
Gleb L. Kotkin and Valeriy G. Serbo, Oxford University Press (2020). © Gleb L. Kotkin and Valeriy G. Serbo 2020.
DOI: 10.1093/oso/9780198853787.001.0001

9.2. a)

$$\begin{pmatrix} 2a^2(m+M) & 2a^2(m-M) & 0 \\ 2a^2(m-M) & 2a^2(m+M) & 0 \\ 0 & 0 & 4a^2(m+M) \end{pmatrix},$$

b)

$$\begin{pmatrix} 4a^2M & 0 & 0 \\ 0 & 4a^2m & 0 \\ 0 & 0 & 4a^2(m+M) \end{pmatrix}.$$

9.3. For both cases, one of the principal axes is the axis perpendicular to the plane of the figure going through the centre of mass of the figure (z-axis). The principal x-axis is orientated at an angle φ with respect to $O'x'$ of each of two figures. The principal y-axis is perpendicular to the x-axis. Both these axes pass through the centre of mass of the figures.

a) The coordinates of the centre of mass in the $O'x'y'$ system ($x' = b, y' = a$) are

$$I_{zz} = 2(a^2 + b^2)(M + m),$$
$$I_{xx} = (a^2 + b^2)(M + m) \mp \sqrt{(b^2 - a^2)^2(M+m)^2 + 4a^2b^2(M-m)^2},$$
$$I_{yy} = (a^2 + b^2)(M + m) \pm \sqrt{(b^2 - a^2)^2(M+m)^2 + 4a^2b^2(M-m)^2},$$

when $a \geqslant b$, $\varphi = \frac{1}{2} \arctan \dfrac{2ab(M - m)}{(a^2 - b^2)(M + m)}$.

b) The coordinates of the centre of mass O in the $O'x'y'z'$ system (Fig. 162) are $x' = y' = a$, $z' = 0$. In the $Ox''y''z''$ coordinate system with the axes which are parallel to the axes $x'y'z'$, the inertia tensor is

$$I_{ik}'' = 4ma^2 \begin{pmatrix} 3 & 1 & 0 \\ 1 & 1 & 0 \\ 0 & 0 & 4 \end{pmatrix}.$$

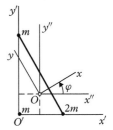

Figure 162

In the transition to the $Oxyz$ system, which is rotated over an angle φ around z''-axis, the coordinates are converted to:

$$x = x'' \cos \varphi + y'' \sin \varphi, \quad y = -x'' \sin \varphi + y'' \cos \varphi, \quad z = z'',$$

and the components of the inertia tensor, as products of the coordinates, are

$$I_{xx} = I''_{xx} \cos^2 \varphi + 2I''_{xy} \sin \varphi \cos \varphi + I''_{yy} \sin^2 \varphi =$$
$$= 4ma^2 (3 \cos^2 \varphi + \sin 2\varphi + \sin^2 \varphi),$$
$$I_{yy} = 4ma^2 (\cos^2 \varphi - \sin 2\varphi + 3 \sin^2 \varphi),$$
$$I_{zz} = 16ma^2,$$
$$I_{xy} = 4ma^2 (-\sin 2\varphi + \cos 2\varphi),$$
$$I_{xz} = I_{yz} = 0.$$

We take the angle φ such that the condition $I_{xy} = 0$ will satisfy; for instance, $\varphi = \pi/8$. Then

$$I_{xx} = 4ma^2 (2 + \sqrt{2}), \quad I_{yy} = 4ma^2 (2 - \sqrt{2}).$$

9.4. $I_\mathbf{n} = \displaystyle\sum_{i,k} I_{ik} n_i n_k.$

9.5. First, we express the kinetic and potential energy for the solid disc and then make the corrections by subtracting the contribution of the hole.

Moments of inertia of the disc relative to its centre are

$$I_3^\mathrm{d} = \tfrac{1}{2} mR^2, \quad I_1^\mathrm{d} = \tfrac{1}{4} mR^2,$$

where m is the mass of the disc. With respect to the suspension point A, the moments of inertia are

$$I_{Ai}^\mathrm{d} = I_i^\mathrm{d} + mR^2, \quad I_{A3}^\mathrm{d} = \tfrac{3}{2} mR^2, \quad I_{A1}^\mathrm{d} = \tfrac{5}{4} mR^2.$$

The contribution of the hole can be obtained by making the substitution $m \to -m/4$, $R \to R/2$. As a result, the moments of inertia for the tag with respect to the suspension point are equal to

$$I_3 = \frac{45}{32} mR^2, \quad I_1 = \frac{45}{64} mR^2.$$

Kinetic energy of the tag at oscillations in the plane of the disc or across are $\frac{1}{2} I_3 \dot{\varphi}^2$ and $\frac{1}{2} I_1 \dot{\varphi}^2$, respectively.

Potential energy of the disc at an angle φ (in the plane of the disc or across) is $U^\mathrm{d}(\varphi) = mgR(1 - \cos \varphi)$. After the same substitutions (that is, taking the hole into account), the potential energy of the tag equals $U(\varphi) = \frac{7}{8} mgR(1 - \cos \varphi) \approx \frac{7}{16} mgR\varphi^2$.

As a result, the frequency of small oscillations for the motion in the plane of the tag is

$$\omega_3 = \sqrt{\frac{28}{45}\frac{g}{R}},$$

while for the motion across the plane of the tag, it is

$$\omega_1 = \sqrt{\frac{56}{75}\frac{g}{R}}.$$

9.6. The centre of mass is a point on the axis of symmetry at a distance

$$\frac{(R-r)r^3}{R^3-r^3}$$

to the left of the centre of the ball. The body is a symmetric top. The moment of inertia with respect to the axis of symmetry is

$$I_3 = \frac{m}{R^3-r^3}\frac{2}{5}(R^5-r^5),$$

and with respect to any two perpendicular axes passing through the centre of mass, the moment of inertia is

$$I_1 = I_2 = \frac{m}{R^3-r^3}\left[\frac{2}{5}(R^5-r^5) - \frac{(R-r)^2 r^3 R^3}{(R^3-r^3)}\right].$$

9.7. $D_{ik} = \left(\sum_n I_{nn}\right)\delta_{ik} - 3I_{ik}$ (see [2], § 99).

9.8. The centre of mass lies on the axis of the hemisphere at a distance $\frac{3}{8}R$ from the centre of the ball, where R is the radius of the ball. The moment of inertia around any axis perpendicular to the axis of symmetry is

$$I = \frac{2}{5}mR^2 - m\left(\frac{3}{8}R\right)^2 = \frac{83}{320}mR^2,$$

where m is the mass of the hemisphere. The centre of mass can only move along the vertical. Let φ be the angle over which the hemisphere is turned, and z the height of the centre of mass above the plane so that

$$z = R - \frac{3}{8}R\cos\varphi.$$

The Lagrangian function of the system is

$$L = \frac{1}{2}I\dot{\varphi}^2 + \frac{1}{2}m\dot{z}^2 - mgz,$$

or when φ is small

$$L = \tfrac{1}{2} I \dot{\varphi}^2 - \tfrac{3}{16} mgR\varphi^2.$$

The frequency of the small oscillations is thus given by

$$\omega = \sqrt{\frac{120}{83} \frac{g}{R}}.$$

9.9. The current distances from the centre of mass of the Earth–Moon system to the Earth and the Moon are equal to $\dfrac{m}{M+m}R$ and $\dfrac{M}{M+m}R$, respectively, and the angular momentum of the system is

$$m \left(\frac{MR}{M+m} \right)^2 \Omega_{\mathrm{M}} + M \left(\frac{mR}{M+m} \right)^2 \Omega_{\mathrm{M}} + I\Omega_{\mathrm{E}} = \mathcal{J}\Omega_{\mathrm{M}} + I\Omega_{\mathrm{E}},$$

$$\mathcal{J} = \frac{Mm}{M+m}R^2, \tag{1}$$

where Ω_{M} and Ω_{E} are the angular velocities of the Moon around the Earth and the Earth around its own axis ($\Omega_{\mathrm{E}}/\Omega_{\mathrm{M}} \approx 28$). In (1), we considered the Moon as a material point, and for the Earth we took into account the rotation around the centre of mass and around its own axis (with the moment of inertia $I = \tfrac{2}{5} Ma^2$).

At the moment when a day becomes equal to a month, the angular velocity of the Earth's rotation ω will coincide with the angular velocity of the Moon. Simultaneously the distance from the Earth to the Moon (according to the Kepler's third law) will become equal to $R(\Omega_{\mathrm{M}}/\omega)^{2/3}$, and the angular momentum

$$\left[\mathcal{J} \left(\frac{\Omega_{\mathrm{M}}}{\omega} \right)^{4/3} + I \right] \omega. \tag{2}$$

From (1) and (2), we find for the dimensionless variable $x = \omega/\Omega_{\mathrm{E}}$ the equation

$$x(1 + k - x)^3 = k^3 \frac{\Omega_{\mathrm{M}}}{\Omega_{\mathrm{E}}}, \tag{3}$$

where

$$k = \frac{\mathcal{J}\Omega_{\mathrm{M}}}{I\Omega_{\mathrm{E}}} = \frac{5}{2} \left(\frac{R}{a} \right)^2 \frac{m}{M+m} \frac{\Omega_{\mathrm{M}}}{\Omega_{\mathrm{E}}} \approx 3.8.$$

Equation (3), or $x(4.8 - x)^3 = 2.01$, has two real roots: $x_1 \approx 1/55$ and $x_2 \approx 4$. The first root corresponds to the future, and the second root, to the past. Accordingly, in the first case, the month will be equal to 55 current days, whereas in the second case, the month equalled 6 hours. The distance from the Earth to the Moon will be $1.6R$, and it used to be $2.6a$ (in this model!).

For more realistic (than the one considered here) models of evolution of the Earth–Moon system see, for instance, [21], Ch. 2.

9.10. a) The body rotates with the angular velocity $\frac{2}{7}\omega$; the fraction $\frac{5}{7}$ of the initial kinetic energy transfers into the heat.

b) The line of the centres rotates with the angular velocity of $\frac{\sqrt{2}}{7}\omega$ around the direction of the angular momentum $\mathbf{M}$, which has an angle of $45°$ with respect to this line. The body rotates around a line of the centres with the angular velocity $\frac{5}{14}\omega$; the fraction $\frac{19}{28}$ of the initial kinetic energy transfers into the heat.

9.11. $f_{1,2} = m\left(g \pm \frac{v^2}{14r}\right)$.

9.12. It is helpful to use the moving coordinate system with the origin at centre of mass and axes x_1, x_2, x_3 parallel to the edges $AB = a$, $AD = b$, $AA' = c$ of the parallelepiped. In this system, the angular velocity is

$$\mathbf{\Omega} = \Omega\mathbf{n}, \quad \text{where} \quad \mathbf{n} = \mathbf{l}/l, \quad \mathbf{l} = (a, b, c),$$

and the angular momentum of the parallelepiped

$$\mathbf{M} = (I_1\Omega_1, I_2\Omega_2, I_3\Omega_3),$$

where

$$I_1 = \tfrac{2}{3}m(b^2 + c^2) = \tfrac{2}{3}m(l^2 - a^2), \quad I_2 = \tfrac{2}{3}m(l^2 - b^2), \quad I_3 = \tfrac{2}{3}m(l^2 - c^2)$$

are its principal moments of inertia. Thus,

$$\mathbf{M} = \tfrac{2}{3}m\Omega l^2\,\mathbf{n} - \frac{2\Omega m}{3l}\cdot(a^3, b^3, c^3).$$

The vector $\mathbf{M}$ is fixed with respect to the system $x_1x_2x_3$; that is, it rotates in the laboratory system with the angular velocity $\mathbf{\Omega}$, and therefore, $\dot{\mathbf{M}} = [\mathbf{\Omega}, \mathbf{M}]$.

Let the forces acting upon the parallelepiped at the points A and C' be equal to $-\mathbf{f}$ and $\mathbf{f}$ (we do not take into account gravity). The moment of these forces $\mathbf{K} = [\mathbf{l}, \mathbf{f}]$. The equation of motion $\dot{\mathbf{M}} = \mathbf{K}$ leads to the equality

$$\Omega[\mathbf{n}, \mathbf{M}] = l[\mathbf{n}, \mathbf{f}],$$

which allows us to determine the component $\mathbf{f}_\perp$ of the force $\mathbf{f}$ perpendicular to the vector $\mathbf{n}$:

$$\mathbf{f}_\perp = \frac{\Omega}{l}\mathbf{M}_\perp = \frac{\Omega}{l}\{\mathbf{M} - \mathbf{n}(\mathbf{Mn})\} =$$

$$= \frac{2m\Omega^2}{3l^4}(a^4 + b^4 + c^4)\cdot(a, b, c) - \frac{2m\Omega^2}{3l^2}\cdot(a^3, b^3, c^3).$$

The component of the force $\mathbf{f}$ parallel to the diagonal AC' cannot be found in the model which considers the parallelepiped (and hinges) as a non-deformable rigid body. It is easy to see that if we act upon the parallelepiped at the points A and C' by forces $N\mathbf{n}$ and $-N\mathbf{n}$, we will not affect its motion.

As a result, the forces applied to the A and C' hinges are equal to $\mathbf{f}$ and $-\mathbf{f}$,

$$\mathbf{f} = -\frac{2m\Omega^2}{3l^2}(a^3, b^3, c^3) + N'(a, b, c),$$

where N' is an indefinable value. We have introduced

$$N' = N - \frac{2m\Omega^2}{l^5}(a^4, b^4, c^4).$$

In a laboratory system, the vector $\mathbf{f}_\perp$ rotates with the angular velocity $\mathbf{\Omega}$.

9.13. The moment of inertia of the ellipsoid with respect to the axis of symmetry is $I_3 = \frac{2}{5}Ma^2$; with respect to any perpendicular to that axis and passing through the centre of mass, we have $I_1 = \frac{1}{5}M(a^2 + c^2)$, where M is the mass of the ellipsoid.

The impinging particle of mass $m \ll M$ transfers to the ellipsoid a momentum $\mathbf{p} = (p_x, p_y, p_z) = mv(0, -1, 0)$ and an angular momentum $\mathbf{M} = mv(\rho_1, 0, -\rho_2)$.

In the system of reference moving with the velocity $\mathbf{p}/(M + m)$, we find (cf. [1], §33 and [7], §47) that the ellipsoid will rotate around the c-semi-axis with an angular velocity

$$\Omega_3 = \frac{M_z}{I_3} = -\frac{5mv\rho_2}{2Ma^2},$$

while at the same time it is precessing around the direction of $\mathbf{M}$ with an angular velocity

$$\Omega = \frac{|\mathbf{M}|}{I_1} = \frac{5mv\sqrt{\rho_1^2 + \rho_2^2}}{M(a^2 + c^2)}.$$

9.14. We denote the angular velocity of rotation of the disc around its axis as $\dot\psi$ and the angle between this axis and the direction on the north as φ. The angular velocity of the disc in the inertial system is $\boldsymbol\omega = \mathbf{\Omega} + \dot{\boldsymbol\varphi} + \dot{\boldsymbol\psi}$, while its projections are

$\omega_3 = \dot{\psi} + \Omega\cos\alpha\cos\varphi$ on the disc axis, $\omega_1 = \dot{\varphi} + \Omega\sin\alpha$ on the vertical, and $\omega_2 = \Omega\cos\alpha\sin\varphi$ on the horizontal axis perpendicular to the disc axis. The Lagrangian function equals the kinetic energy (we take into account that $I_1 = I_2$):

$$L = \tfrac{1}{2}I_1(\dot{\varphi} + \Omega\sin\alpha)^2 + \tfrac{1}{2}I_1\Omega^2\cos^2\alpha\sin^2\varphi + \tfrac{1}{2}I_3(\dot{\psi} + \Omega\cos\alpha\cos\varphi)^2.$$

It is helpful to study the motion using the integrals of motion p_ψ and $E = p_\psi\dot{\psi} + p_\varphi\dot{\varphi} - L$:

$$p_\psi = I_3(\dot{\psi} + \Omega\cos\alpha\cos\varphi),$$
$$E = \tfrac{1}{2}I_1(\dot{\varphi}^2 - \Omega^2\sin^2\alpha) - \tfrac{1}{2}I_1\Omega^2\cos^2\alpha\sin^2\varphi +$$
$$+ \tfrac{1}{2}I_3(\dot{\psi}^2 - \Omega^2\cos^2\alpha\cos^2\varphi).$$

Excluding $\dot{\psi}$, we find

$$E = \tfrac{1}{2}I_1\dot{\varphi}^2 + U_{\text{eff}}(\varphi) + \text{const},$$

where

$$U_{\text{eff}}(\varphi) = -p_\psi\Omega\cos\alpha\cos\varphi + \tfrac{1}{2}I_1\Omega^2\cos^2\alpha\cos^2\varphi.$$

Let us consider in detail the important case of $\dot{\psi} \gg \Omega$ only. Then $p_\psi \approx I_3\dot{\psi}$ and

$$U_{\text{eff}}(\varphi) \approx -p_\psi\Omega\cos\alpha\cos\varphi.$$

The function $U_{\text{eff}}(\varphi)$ has a minimum at $\varphi = 0$ (i.e., in the north direction). The axis of the gyrocompass oscillates around this direction. For small oscillations, we have

$$U_{\text{eff}}(\varphi) = \tfrac{1}{2}I_3\dot{\psi}\Omega\cos\alpha \cdot \varphi^2 + \text{const},$$

and the oscillation frequency of the axis is

$$\sqrt{\frac{I_3}{I_1}\Omega\dot{\psi}\cos\alpha}.$$

For example, for a gyroscope making about ten thousand turns per minute, the oscillation period equals approximately half a minute $\left(\text{for } \frac{I_3}{I_1}\cos\alpha \sim 1\right)$.

How can we take the moment of inertia of the gyrocompass' frame into account?

9.15. It is helpful to use the moving coordinate system with the vertical z-axis and the x-axis which passes through the following points: the O and the point of contact of the disc with the table surface. This system of reference rotates around the z-axis with the angular velocity $\dot{\boldsymbol{\varphi}}$. Here we have the equation for the angular momentum projections on the moving axes

$$\left(\frac{d\mathbf{M}}{dt}\right)_i + [\dot{\boldsymbol{\varphi}}, \mathbf{M}]_i = \mathbf{K}_i, \tag{1}$$

where $\mathbf{M}$ is the angular momentum of the top with respect to the stationary point O, and $\mathbf{K}$ is the moment of the forces acting upon the top (see [1], §36 and [7], §45). The projection of (1) onto the x-axis is

$$\dot{M}_x - \dot{\varphi}M_y = K_x.$$

It is evident that $M_y = 0$ and $K_x = 0$; therefore, $M_x = \text{const.}$

Let $I_1 = I_2 \neq I_3$ be the principal moments of inertia of the top with respect to point O. In the initial moment, we have

$$M = \Omega I_3, \quad M_x = \Omega I_3 \cos\theta.$$

When the slipping stops, the x-axis becomes the instant axis of rotation; that is, the angular velocity of the top $\boldsymbol{\omega}$ will be directed along the x-axes, and $M_x = I_3\omega\cos^2\theta + I_1\omega\sin^2\theta$. Thus,

$$\omega = \frac{\Omega I_3 \cos\theta}{I_3 \cos^2\theta + I_1 \sin^2\theta}.$$

Note that $\dot{\varphi} = -\omega\tan\theta$ when the slipping has stopped.

9.16. Let $a = b \neq c$ be the semi-axes of the ellipsoid, R, Θ, and Φ, the spherical coordinates of the centre of mass of the ellipsoid, and θ, φ, and ψ the Euler angles, and let the x_3-axis of the moving system of reference be along the c-semi-axis. The kinetic energy of the body is (see [1], §35)

$$T = \tfrac{1}{2}m(\dot{R}^2 + R^2\dot{\Theta}^2 + R^2\dot{\Phi}^2\sin^2\Theta) +$$
$$+ \tfrac{1}{2}I_1(\dot{\varphi}^2\sin^2\theta + \dot{\theta}^2) + \tfrac{1}{2}I_3(\dot{\varphi}\cos\theta + \dot{\psi})^2, \tag{1}$$

where $I_1 = I_2 = \tfrac{1}{5}m(a^2 + c^2)$ and $I_3 = \tfrac{2}{5}ma^2$ are the moments of inertia of the ellipsoid with respect to the x_1-, x_2-, and x_3-axes.

The given potential energy for the interaction of the ellipsoid with the gravitational centre can be transformed into the form

$$U = -\frac{GmM}{R} - \frac{GMD}{4} \cdot \frac{3\cos^2\alpha - 1}{R^3}, \tag{2}$$

where $D = 2(I_1 - I_3)$ and α is the angle between the radius vector $\mathbf{R}$ and the x_3-axis.

The unit vector $\mathbf{e}_3$, which determines the direction of the x_3-axis, has the components

$$\mathbf{e}_3 = (\sin\theta\sin\varphi, \, -\sin\theta\cos\varphi, \, \cos\theta).$$

Hence, we have

$$\cos\alpha = \frac{\mathbf{Re}_3}{R} = \cos\theta\cos\Theta + \sin\theta\sin\Theta\sin(\varphi - \Phi). \tag{3}$$

From (1), (2), and (3), we finally get the expression for the Lagrangian function $L = T - U$.

9.17. Let us first consider the Sun's influence alone without the Moon's influence. The origin of the coordinate system will coincide with the Sun; the X_3-axis will be directed perpendicular to the plane of the Earth's orbit, and the x_3-axis, to the north (see notations of the preceding problem).

The angular velocity of the Earth's axis precession $\dot\varphi$ is evidently small compared to the daily $\dot\psi$ and the annual $\dot\Phi$ velocities of the Earth's rotation. Therefore, in the Lagrangian function we save only the first-order terms in $\dot\varphi$. Moreover, we put R, $\Theta = \pi/2$, and $\dot\Phi$ as the constants; we also average $\cos^2\alpha$ over year: $\langle\cos^2\alpha\rangle = \frac{1}{2}\sin^2\theta$. As a result, we obtain

$$L = \tfrac{1}{2}I_1\dot\theta^2 + \tfrac{1}{2}I_3(\dot\psi^2 + 2\dot\psi\dot\varphi\cos\theta) - \frac{3GMm(a^2 - c^2)}{20R^3}\sin^2\theta.$$

Since $p_\psi = I_3(\dot\psi + \dot\varphi\cos\theta)$ and $p_\varphi = I_3\dot\psi\cos\theta$ are conserved, it follows that $\dot\psi$ and $\theta = 23°$ are conserved as well (up to values of the order of $\dot\varphi$). The equation of motion for the angle θ

$$I_1\ddot\theta + I_3\dot\varphi\dot\psi\sin\theta + \frac{3GMm(a^2 - c^2)}{10R^3}\sin\theta\cos\theta = 0$$

(taking into account that $\ddot\theta \sim \dot\theta^2 \sim \dot\varphi^2$) leads to the equation

$$\dot\varphi = -\frac{3GM}{4R^3\dot\psi}\frac{a^2 - c^2}{a^2}\cos\theta.$$

Substituting

$$\frac{a^2 - c^2}{a^2} \approx \frac{2(a - c)}{a}, \quad \frac{GM}{R^3} = \dot\Phi^2,$$

where $2\pi/\dot{\Phi} = 1$ year, we get

$$\dot{\varphi} \approx -\frac{3}{2}\frac{a-c}{a}\frac{\dot{\Phi}^2}{\dot{\psi}}\cos\theta \approx -16''\,\text{per year}.$$

The angular velocity of the precession caused by the Moon is obtained from (2) by replacement the mass of the Sun M with the mass of the Moon and R with the distance from the Earth to the Moon. This angular velocity turns out to be $-31''$ per year. The total angular velocity is $\dot{\varphi} = -48''$ per year. The observed value $\dot{\varphi}_{\exp} = -50.2''$ per year (see [21], Ch. 2).

Thus, the Earth's axis rotates around the x_3-axis with a period of about 26 thousand years in the opposite direction with respect to the rotation of the Earth around the Sun (the so-called precedence of the equinoxes).

9.18.

$$\dot{M}_1 + \left(\frac{1}{I_2} - \frac{1}{I_3}\right)M_2 M_3 = K_1,$$

$$\dot{M}_2 + \left(\frac{1}{I_3} - \frac{1}{I_1}\right)M_3 M_1 = K_2,$$

$$\dot{M}_3 + \left(\frac{1}{I_1} - \frac{1}{I_2}\right)M_1 M_2 = K_3.$$

When $I_1 = I_2$ and $K_{1,2,3} = 0$, we get

$$M_1 = B\cos(\omega t + \alpha), \quad M_2 = B\sin(\omega t + \alpha), \quad M_3 = \text{const},$$

where $\omega = \left(\frac{1}{I_1} - \frac{1}{I_3}\right)M_3$ (see [1] § 36; [7] § 47, and compare problem 10.21).

9.19. Let us consider the motion around an axis which lies close to the principal x_1-axis. From the Euler equation (see [1], equation (36.5))

$$\dot{\Omega}_1 + \frac{I_3 - I_1}{I_1}\Omega_2\Omega_3 = 0,$$

we get $\Omega_1 = \text{const}$, neglecting terms proportional to $\Omega_{2,3}/\Omega_1 \ll 1$. The two other equations become linear in Ω_2 and Ω_3. Assuming that

$$\Omega_{2,3} \propto e^{st}, \tag{1}$$

we get for s the equation

$$s^2 = \frac{(I_1 - I_3)(I_2 - I_1)}{I_2 I_3}\Omega_1^2. \tag{2}$$

When $I_2 < I_1 < I_3$ or $I_3 < I_1 < I_2$, (2) has real roots so that according to (1) the rotation around the x_1-axis is unstable.

If, however, I_1 is either the largest or the smallest moment of inertia, (2) has imaginary roots; that is, the change in Ω_2 and Ω_3 are in the nature of oscillations and rotation around the x_1-axis is stable.

9.20. The motion of the ball is determined by the equations

$$m\dot{\mathbf{v}} = m\mathbf{g} + \mathbf{f}, \tag{1}$$

$$I\dot{\boldsymbol{\omega}} = [\mathbf{a}, \mathbf{f}], \tag{2}$$

$$\mathbf{v} + [\boldsymbol{\omega}, \mathbf{a}] = 0, \tag{3}$$

where m is the mass of the ball; $I = \frac{2}{5}ma^2$, its moment of inertia; $\mathbf{a}$, radius of the ball directed to the point of its touch with by cylinder; $\mathbf{f}$, the force applied upon the ball at this point (i.e. the sum of the reaction and friction forces); $\mathbf{v}$, the velocity; and $\boldsymbol{\omega}$, the angular velocity of the ball.

It is helpful to use the cylindrical coordinates with the z-axis directed along the axis of the cylinder. It is necessary to take into account that projection of the derivatives of any vector $\mathbf{A}$ on the axes of the moving system of reference is determined by the equations

$$\begin{aligned}(\dot{\mathbf{A}})_\varphi &= \dot{A}_\varphi + [\dot{\boldsymbol{\varphi}}, \mathbf{a}]_\varphi = \dot{A}_\varphi + \dot{\varphi} A_r, \\ (\dot{\mathbf{A}})_r &= \dot{A}_r + [\dot{\boldsymbol{\varphi}}, \mathbf{A}]_r = \dot{A}_r - \dot{\varphi} A_\varphi,\end{aligned} \tag{4}$$

where $r = b - a$, φ, and z are the coordinates of the centre of the ball and $\dot{\varphi} = v_\varphi/r$. From the equations

$$m\dot{v}_\varphi = f_\varphi, \quad I\dot{\omega}_z = af_\varphi, \quad v_\varphi + a\omega_z = 0$$

we obtain

$$v_\varphi = \text{const}, \quad \omega_z = \text{const}, \quad f_\varphi = 0;$$

while from the equations

$$m\dot{v}_z = -mg + f_z, \quad v_z - a\omega_\varphi = 0,$$
$$I(\dot{\omega}_\varphi + \dot{\varphi}\omega_r) = -af_z, \quad I(\dot{\omega}_r - \dot{\varphi}\omega_\varphi) = 0$$

it follows that

$$(I + ma^2)\ddot{\omega}_\varphi + I\dot{\varphi}^2\omega_\varphi = 0,$$

and hence

$$\omega_\varphi = C\cos(\Omega t + \alpha), \quad \Omega = \sqrt{\frac{I}{I + ma^2}} \quad \dot\varphi = \sqrt{\frac{2}{7}}\frac{v_\varphi}{r},$$

$$\omega_r = -\frac{5gr}{2av_\varphi} + \sqrt{\frac{7}{2}}C\sin(\Omega t + \alpha),$$

$$z = z_0 + a\frac{C}{\Omega}\sin(\Omega t + \alpha).$$

Thus, the ball makes the harmonic oscillations in height, and the radial component of the angular velocity is varying at the same time with these oscillations.

9.21. a) We use as our generalized coordinates the X- and Y-coordinates of the centre of the disc and the Euler angles φ, θ, and ψ (see [1], §35). We take the Z-axis to be vertical and the (moving) x_3-axis along the axis of the disc. The angle φ is between the intersection of the plane of the disc with the XY-plane and the X-axis (the line of nodes[1]). The Lagrangian function is

$$L = \tfrac{1}{2}m(\dot X^2 + \dot Y^2 + a^2\dot\theta^2\cos^2\theta) + \tfrac{1}{2}I_1(\dot\theta^2 + \dot\varphi^2\sin^2\theta) +$$
$$+ \tfrac{1}{2}I_3(\dot\varphi\cos\theta + \dot\psi)^2 - mga\sin\theta,$$

where $I_1 = I_2$ and I_3 are the principal moments of inertia of the disc and the x_1-, x_2-, and x_3-axes are along its principal axes; a is the radius of the disc and m, its mass; the centre of the disc is at a distance $Z = a\sin\theta$ from the XY-plane. The generalized momenta

$$m\dot X = p_X, \quad m\dot Y = p_Y,$$
$$I_1\dot\varphi\sin^2\theta + I_3\cos\theta\,(\dot\varphi\cos\theta + \dot\psi) = p_\varphi \equiv M_Z, \qquad (1)$$
$$I_3(\dot\varphi\cos\theta + \dot\psi) = p_\psi \equiv M_3$$

together with the energy are integrals of motion. In the coordinate system moving with the constant velocity $(\dot X, \dot Y, 0)$, the centre of the disc moves only vertically. From (1), we have

$$\dot\varphi = \frac{M_Z - M_3\cos\theta}{I_1\sin^2\theta}, \quad \dot\psi = \frac{M_3}{I_3} - \frac{M_Z - M_3\cos\theta}{I_1\sin^2\theta}\cos\theta, \qquad (2)$$

---

[1] In the following discussion, we will use the ξ- and η-axes lying in the horizontal plane: the ξ-axis is along the line of nodes and the η-axis is at right angles to the ξ-axis; that is, $\mathbf{e}_\eta = [\mathbf{e}_Z, \mathbf{e}_\xi]$.

and substituting this into the expression for energy, we obtain

$$E = \frac{1}{2}(I_1 + ma^2 \cos^2 \theta)\dot{\theta}^2 + \frac{M_3^2}{2I_3} + \frac{(M_Z - M_3 \cos \theta)^2}{2I_1 \sin^2 \theta} + mga \sin \theta. \tag{3}$$

The dependence $\theta(t)$ is determined from this equation in the form of quadratures, after which one can get $\varphi(t)$ and $\psi(t)$ from (2). The dip angle of the disc oscillates, and the precession velocity $\dot{\varphi}$ and the angular velocity of rotation around the axis of the disc $\dot{\psi}$ change at the same time. Equations (2) and (3) are similar to those for the motion of a heavy symmetric top (e.g. see, equations (4)–(7) from [1], § 35, problem 1).

The rolling of the disc is stable if $\Omega_3^2 > mgaI_1/I_3^2$. The rotation is stable if $\Omega_Z^2 > mga/I_1$.

b) If the rolling disk is not slipping, in addition to the force of gravity mg and the vertical reaction force $\mathbf{Q}$, there is also the horizontal friction force $\mathbf{f}$ acting on the disc. It is helpful to write the equation of motion

$$\dot{\mathbf{M}} = [\mathbf{a}, \mathbf{Q}] + [\mathbf{a}, \mathbf{f}]$$

and its components along the Z-, x_3-, and ξ-axes[1]

$$\dot{M}_Z = f_\xi a \cos \theta, \quad \dot{M}_3 = f_\xi a, \tag{4}$$

$$\frac{d}{dt} \frac{\partial L}{\partial \dot{\theta}} - \frac{\partial L}{\partial \theta} = f_\eta a \sin \theta. \tag{5}$$

Here $\mathbf{a}$ is the vector from the centre of the disc to the point of its contact with the plane.

If we take the components of the equation

$$m\dot{\mathbf{V}} = \mathbf{f} + \mathbf{Q} + m\mathbf{g}$$

along the ξ- and η-axes (see equation (4) of the preceding problem), we have the following equations for f_ξ and f_η:

$$f_\xi = m(\dot{\mathbf{V}})_\xi = m(\dot{V}_\xi - \dot{\varphi} V_\eta),$$
$$f_\eta = m(\dot{\mathbf{V}})_\eta = m(\dot{V}_\eta + \dot{\varphi} V_\xi). \tag{6}$$

The condition of rolling without slipping $\mathbf{V} + [\boldsymbol{\Omega}, \mathbf{a}] = 0$ leads to

$$V_\xi = -a(\dot{\psi} + \dot{\varphi} \cos \theta), \quad V_\eta = -a\dot{\theta} \sin \theta. \tag{7}$$

---

[1] These equations are the Lagrangian equations for the Euler angles when there is no friction force.

Substituting (1), (6), and (7) in (4) and (5), we obtain a set of equations for the Euler angles. The disc can move along the plane without slipping and without leaving the plane, provided

$$|f| \leqslant \mu m(g + \ddot{Z}), \quad g + \ddot{Z} \geqslant 0,$$

where μ is the friction coefficient.

If we put $\dot{\theta} = 0$, we get $\ddot{\varphi} = \ddot{\psi} = 0$. Moreover, the following relation exists between θ, $\dot{\varphi}$, and $\dot{\psi}$:

$$I_3'\dot{\varphi}(\dot{\varphi}\cos\theta + \dot{\psi})\sin\theta - I_1\dot{\varphi}^2\sin\theta\cos\theta + mga\cos\theta = 0 \tag{8}$$

(from now on, we use $I_{1,3}' = I_{1,3} + ma^2$). The centre of the disc moves with a velocity, which has a constant absolute magnitude

$$V = a|\Omega_3| = a|\dot{\psi} + \dot{\varphi}\cos\theta|,$$

along a circle with radius $R = V/|\dot{\varphi}|$.

Condition (8) can also be expressed as

$$I_3'RV^2 = |I_1 aV^2\cos\theta - mga^2R^2\cot\theta|.$$

In particular, if the mass of the disc is concentrated in its centre ($I_1 = I_3 = 0$), we get the elementary relation: $V^2 = gR|\cot\theta|$.

Gyroscopic effects which appear when $I_{1,3}$ differs from zero may turn out to be important. For example, when $R \gg a$, we have $\frac{3}{2}V^2 = gR|\cot\theta|$ for a uniform disc ($2I_1 = I_3 = \frac{1}{2}ma^2$), but $2V^2 = gR|\cot\theta|$ for a hoop ($2I_1 = I_3 = ma^2$).

If the disc is rolling vertically, we have

$$\theta = \pi/2, \quad \dot{\varphi} = 0, \quad \dot{\psi} = \Omega_3 = \text{const}. \tag{9}$$

To study the stability of this motion, we put

$$\theta = (\pi/2) - \beta, \quad \beta \ll 1, \quad \dot{\varphi} \sim \dot{\beta} \sim \beta\Omega_3 \ll \Omega_3, \quad \ddot{\varphi} \sim \ddot{\beta} \sim \dot{\Omega}_3 \ll \dot{\varphi}\Omega_3$$

into (1) and (4)–(7), retaining only the first-order terms. We then get

$$M_Z = I_1\dot{\varphi} + I_3\Omega_3\beta = \text{const}, \quad \Omega_3 = \text{const},$$
$$I_1'\ddot{\beta} - I_3'\Omega_3\dot{\varphi} - mga\beta = 0, \tag{10}$$

and hence

$$I_1'\ddot{\beta} + \left(\frac{I_3 I_3'}{I_1}\Omega_3^2 - mga\right)\beta = \frac{I_3'}{I_1}M_Z\Omega_3.$$

If

$$\Omega_3^2 > \frac{I_1 mga}{I_3 I_3'},\tag{11}$$

small oscillations in β and $\dot{\varphi}$ occur with the angle θ differing from $\pi/2$:

$$\beta = \frac{I_3'\Omega_3 M_Z}{I_1 I_1'\omega^2} + \beta_0\cos(\omega t + \delta),$$

$$\dot{\varphi} = -\frac{mgaM_Z}{I_1 I_1'\omega^2} - \frac{I_3}{I_1}\Omega_3\beta_0\cos(\omega t + \delta),$$

where

$$\omega^2 = \frac{I_3 I_3'}{I_1 I_1'}\Omega_3^2 - \frac{mga}{I_1'}.$$

The direction of motion of the disc also executes small vibrations; the disc's motion is not about a straight line, but about the circle of radius $a\Omega_3 I_1^2\omega^2/(mgaM_Z)$.

Therefore, if there are small deviations in the initial conditions from (9), we may have either small oscillations near an "equilibrium" motion along a straight line—if $M_z = 0$, $\beta_0 \neq 0$—or a new "equilibrium" motion—if $M_z \neq 0$, $\beta_0 = 0$.

If the inequality (11) is not satisfied, the motion is not stable.

One can say that the motion with $\theta = \text{const}$ can occur, if in the θ-, $\dot{\varphi}$-, $\dot{\psi}$-"space" the representative point lies on the surface determined by (8). It can easily be seen that, if condition (11) holds, the motion is stable with respect to perturbations which remove the representative (θ-, $\dot{\varphi}$-, $\dot{\psi}$)-point from the surface (8) while it is neutral with respect to perturbation shifting the point along the surface. A similar situation holds for the stability of the motion of the disc along the smooth plane; we must merely replace $I_{1,3}'$ by $I_{1,3}$.

Rotation of the disc around its vertical diameter is stable, if $\Omega_Z^2 > mga/I_1'$.

c) If there can be no rotation around a vertical axis, the following condition must be satisfied:

$$\Omega_Z = \dot{\varphi} + \dot{\psi}\cos\theta = 0.\tag{12}$$

In this case, there is an additional, "frictional torque" $\mathbf{N}$, which is directed along the vertical, acting on the disc. Instead of (4), we then have

$$\dot{M}_Z = f_\xi a \cos\theta + N, \quad \dot{M}_3 = f_\xi a + N\cos\theta. \tag{13}$$

The integration of the equations of motion can easily be reduce to quadratures (in contrast to the equations under problem 9.21b).

Motion with a constant angle of inclination is possible if conditions (8) is satisfied as well as condition (12), which means that $\dot{\varphi}$ and $\dot{\psi}$ are completely determined by the angle θ. In this case,

$$R = a|\sin\theta\tan\theta|, \quad V^2 = \frac{mga^3\sin\theta}{I_3' + I_1\cot^2\theta}.$$

Rolling of the disc in a vertical position is stable if $\Omega_3^2 > mga/I_3'$.

d) If there is a small inclination, the term $\delta L = -mga X$ must be added to the Lagrangian function; the Y-axis lies in the plane of the disc and is horizontal, the X-axis lies in the plane at right angles to the Y-axis, pointing upwards, and the Z-axis is at right angles to the plane. In this case, an interesting effect can be observed: although the additional force is directed against the X-axis, the disc moves in the opposite direction to the Y-axis; that is, the disc moves without losing height.

Substituting

$$\theta = (\pi/2) - \beta, \quad \beta \ll 1, \quad \dot{\psi} \sim \dot{\beta} \ll \Omega_Z, \quad \ddot{\psi} \sim \ddot{\beta} \sim \dot{\Omega}_Z \ll \dot{\psi}\Omega_Z$$

into in (1) and (4) to (7) and adding the contribution from δL, we get

$$I_1'\ddot{\beta} + (I_1 - I_3')\Omega_Z^2\beta - I_3'\Omega_Z\dot{\psi} - mga\beta = -mga\alpha\cos\Omega_Z t,$$
$$I_3'\ddot{\psi} + (I_3 + 2ma^2)\Omega_Z\dot{\beta} = mga\alpha\sin\Omega_Z t,$$

and hence

$$\psi = -\left(2 + \frac{2ma^2}{I_3'} + \frac{mga}{I_3'\Omega_Z^2}\right)\alpha\sin\Omega_Z t, \quad \beta = -2\alpha\cos\Omega_Z t.$$

Substituting (7) and $\theta, \varphi = \Omega_Z t$, and ψ into

$$\dot{X} = -V_\xi\sin\varphi - V_\eta\cos\varphi, \quad \dot{Y} = V_\xi\cos\varphi - V_\eta\sin\varphi$$

and averaging over one period of rotation, we find

$$\langle\dot{X}\rangle = 0, \quad \langle\dot{Y}\rangle = \left(1 + \frac{ma^2}{I_3'} + \frac{mga}{2I_3'\Omega_Z^2}\right)\alpha a\Omega_Z;$$

that is, the disc is displaced without losing height.

9.22. a) The position of the ball is determined by the coordinates of its centre of mass X, Y, Z and the Euler angles θ, φ, ψ (see [1], § 35), which determine the orientation of the principal axes of inertia. Let the Z-axis be vertical and the x_3-axis be directed from the centre of mass to the geometric centre of the ball; let $x_3 = b$ be the position of the geometrical centre of the ball and a be its radius. The analysis of the motion of the ball is analogous to the analysis in the preceding problem.

If $M_Z \neq M_3$, there is a minimum $U_{\mathrm{eff}}(\theta)$ when $\theta_0 \neq 0, \pi$, and if the energy $E = U_{\mathrm{eff}}(\theta_0)$, we can have a steady rotation of the ball with a constant angle $\theta = \theta_0$ and $Z = a - b\cos\theta_0$. The instantaneous point of contact of ball and plane has the velocity $v = (b\dot\varphi + a\dot\psi)\sin\theta_0$ along the direction of the line of nodes.

b) Let

$$\mathbf{R}_{\mathrm{c}} = (X + b\sin\theta\sin\varphi,\ Y - b\sin\theta\cos\varphi,\ a)$$

be the coordinates of the geometric centre of the ball, and let

$$\mathbf{\Omega} = (\Omega_X, \Omega_Y, \Omega_Z) = (\dot\theta\cos\varphi + \dot\psi\sin\theta\sin\varphi,\ \dot\theta\sin\varphi - \dot\psi\sin\theta\cos\varphi,\ \dot\varphi + \dot\psi\cos\theta)$$

be the angular rotational velocity of the ball, while $\mathbf{a} = (0, 0, -a)$ is the radius vector from the geometric centre of the ball to the point where the ball is in contact with the plane. The condition that the ball rolls without slipping

$$\dot{\mathbf{R}}_{\mathrm{c}} + [\mathbf{\Omega}, \mathbf{a}] = 0$$

is a non-holonomic constraint

$$
\begin{aligned}
\dot X &= \dot\theta(a - b\cos\theta)\sin\varphi - (a\dot\psi + b\dot\varphi)\sin\theta\cos\varphi, \\
\dot Y &= -\dot\theta(a - b\cos\theta)\cos\varphi - (a\dot\psi + b\dot\varphi)\sin\theta\sin\varphi.
\end{aligned}
\tag{4}
$$

The equation of motion are

$$m\ddot X = \lambda_1, \quad m\ddot Y = \lambda_2, \tag{5}$$

$$\dot M_Z = (\lambda_1\cos\varphi + \lambda_2\sin\varphi)b\sin\theta, \tag{6}$$

$$\dot m_3 = (\lambda_1\cos\varphi + \lambda_2\sin\varphi)a\sin\theta, \tag{7}$$

$$\frac{d}{dt}\frac{\partial L}{\partial\dot\theta} - \frac{\partial L}{\partial\theta} = (-\lambda_1\sin\varphi + \lambda_2\cos\varphi)(a - b\cos\theta); \tag{8}$$

they contain the friction forces and the torques produced by these forces on the right-hand sides. Using the constraint conditions (4), we can express the Lagrangian multipliers λ_1 and λ_2 in terms of the Euler angles. Indeed, the components of the friction force along the line of nodes $f_\parallel$ and at right angles to it, $f_\perp$, are given by

$$f_\parallel = m\dot{\theta}\dot{\varphi}(a - b\cos\theta) - m\frac{d}{dt}[(a\dot{\psi} + b\dot{\varphi})\sin\theta], \tag{9}$$

$$f_\perp = -m\frac{d}{dt}[\dot{\theta}(a - b\cos\theta)] - m\dot{\varphi}(a\dot{\psi} + b\dot{\varphi})\sin\theta. \tag{10}$$

Note that (6) and (7) are the same as (1) if one substitutes into (1) the friction force $f_\parallel$ (9) instead of the dry friction force $\mp f$. The quantity $M_z a - M_3 b = C$ is the integral of motion (2), as discussed before.

9.23. To solve the problem, it is necessary to take into account that the height of the particle above the Earth, h, is small compared to the radius R of the Earth and that the centrifugal acceleration, $R\Omega^2$ (where Ω is the angular velocity of the Earth), is small compared to the acceleration of free fall, g, towards the surface of the Earth. There are thus two small parameters in the problem:

$$\varepsilon_1 = \frac{h}{R} \lesssim \varepsilon_2 = \frac{R\Omega^2}{g} \sim 0.003.$$

Let $\mathbf{R} + \mathbf{r}$ be the coordinate of the particle reckoned from the centre of the Earth. The equation of motion has the following form if we take into account the dependence of $\mathbf{g}$ on the distance r from the surface of the Earth (up to the second order in ε_2):

$$\ddot{\mathbf{r}} = -GM\frac{\mathbf{R} + \mathbf{r}}{|\mathbf{R} + \mathbf{r}|^3} + 2[\mathbf{v}, \boldsymbol{\Omega}] + [\boldsymbol{\Omega}, [\mathbf{R} + \mathbf{r}, \boldsymbol{\Omega}]], \tag{1}$$

where G is the gravitational constant and M is the mass of the Earth. We expand the first term into series in small parameter $r/R \lesssim \varepsilon_1$:

$$\ddot{\mathbf{r}} = \mathbf{g} + 2[\mathbf{v}, \boldsymbol{\Omega}] + \mathbf{g}_1 + [\boldsymbol{\Omega}, [\mathbf{r}, \boldsymbol{\Omega}]], \quad \mathbf{r}(0) = \mathbf{h}, \quad \mathbf{v}(0) = 0, \tag{2}$$

$$\mathbf{g} = -GM\frac{\mathbf{R}}{R^3} + [\boldsymbol{\Omega}, [\mathbf{R}, \boldsymbol{\Omega}]], \quad \mathbf{g}_1 = \frac{GM}{R^2}\left(3\mathbf{R}\frac{(\mathbf{r}\mathbf{R})}{R^3} - \frac{\mathbf{r}}{R}\right)[1 + O(\varepsilon_1)].$$

The Coriolis acceleration $2[\mathbf{v}, \boldsymbol{\Omega}] \sim gt\Omega \sim \sqrt{\varepsilon_1\varepsilon_2}g$ and $g_1 \sim \varepsilon_1 g$ have the first order of smallness, while $[\boldsymbol{\Omega}, [\mathbf{r}, \boldsymbol{\Omega}]] \sim \varepsilon_1\varepsilon_2 g$ has the second order.

The vertical $\mathbf{h}$ is anti-parallel to the vector $\mathbf{g}$ and, it makes a small angle $\alpha = \varepsilon_2 \sin\lambda\cos\lambda$ with the vector $\mathbf{R}$ (here λ is the northern [geocentric] latitude; i.e. the angle between the plane of the equator and the vector $\mathbf{R}$). We choose the z-axis along the vertical upwards, the x-axis along the meridian towards to the south, and the y-axis along the latitude to the east, then

$$\mathbf{g} = (0, 0, -g), \quad \mathbf{R} = R(\sin\alpha, 0, \cos\alpha), \quad \boldsymbol{\Omega} = \Omega(-\cos(\alpha + \lambda), 0, \sin(\alpha + \lambda)).$$

In zero approximation, the particle moves with acceleration $-g$ along the z-axis. In the first approximation, the Coriolis acceleration results in the deviation to the east, and the magnitude of the displacement is $y \sim \sqrt{\varepsilon_1 \varepsilon_2}\, gt^2 \sim \sqrt{\varepsilon_1 \varepsilon_2}\, h$. The acceleration $\mathbf{g}_1$ has the component $\sim \varepsilon_1 g$ along z, and components along x and y only of the second order; therefore, in the first approximation, the effect of $\mathbf{g}_1$ will only lead to an increase of the time of the fall from a height of h by $\sim \varepsilon_1 \sqrt{2h/g}$. The deviation to the south thus occurs only in the second order. Then we write (2) in projections into the chosen axes and take the terms of the zero, first order, and second order for the z-, y-, and x-component, respectively:

$$\ddot{z} = -g, \quad \ddot{y} = -2\dot{z}\Omega \cos \lambda,$$

$$\ddot{x} = 2\dot{y}\Omega \sin \lambda + \frac{g}{R}(3z \sin \alpha - x) + \Omega^2 z \sin \lambda \cos \lambda.$$

Using the method of successive approximations, we obtain

$$z = h - \tfrac{1}{2}gt^2, \quad y = \tfrac{1}{3}gt^3 \Omega \cos \lambda, \quad x = 2ht^2\Omega^2 \sin \lambda \cos \lambda.$$

Substituting the time of the fall $t = \sqrt{2h/g}$, we find deviations to the east and south

$$y = \tfrac{1}{3}\sqrt{8\varepsilon_1 \varepsilon_2}\, h \cos \lambda, \qquad x = 2\varepsilon_1 \varepsilon_2 h \sin 2\lambda.$$

9.24. We use the reference system, which is rotating with a vessel; the x-axis is directed along AB, the origin of the coordinates is placed in a fixed point, that is, the intersection of the AB- and CD-axes. The angular velocity of the system is

$$\boldsymbol{\Omega} = \boldsymbol{\omega}_1 + \boldsymbol{\omega}_2 = (\omega_2, \, \omega_1 \cos \omega_2 t, \, -\omega_1 \sin \omega_2 t).$$

Besides the force of gravity $m\mathbf{g}$, the inertia forces act upon each particle of fluid of mass m in this system of reference (see [1], § 39):

the Coriolis force $2m[\mathbf{v}, \boldsymbol{\Omega}]$,

the centrifugal force $\dfrac{\partial}{\partial \mathbf{r}} \dfrac{1}{2} m[\boldsymbol{\Omega}, \mathbf{r}]^2$,

and the force $m[\mathbf{r}, \dot{\boldsymbol{\Omega}}]$, where $\dot{\boldsymbol{\Omega}}$ is the rate of change of the vector $\boldsymbol{\Omega}$ in the inertial system.

When the resin hardens, the particle velocities (relative to the vessel) turn to zero, and the Coriolis force vanishes. The other forces need to be averaged over the periods of rotation:

$$\langle m\mathbf{g} \rangle = -mg\left(\langle \cos \omega_1 t \rangle, \, \langle \sin \omega_1 t \sin \omega_2 t \rangle, \, \langle \sin \omega_1 t \cos \omega_2 t \rangle \right) = 0,$$

if $\omega_1 \neq \omega_2$, and

$$\langle m[\mathbf{r}, \dot{\boldsymbol{\Omega}}] \rangle = m[\mathbf{r}, \langle \dot{\boldsymbol{\Omega}} \rangle] = 0,$$

since $\langle \dot{\boldsymbol{\Omega}} \rangle = \langle [\boldsymbol{\omega}_1, \boldsymbol{\omega}_2] \rangle = [\langle \boldsymbol{\omega}_1 \rangle, \boldsymbol{\omega}_2] = 0$. Finally,

$$\left\langle \frac{\partial}{\partial \mathbf{r}} \frac{1}{2} m [\boldsymbol{\Omega}, \mathbf{r}]^2 \right\rangle = \frac{1}{2} m \frac{\partial U(\mathbf{r})}{\partial \mathbf{r}},$$

where

$$U(\mathbf{r}) = \left\langle [\boldsymbol{\Omega}, \mathbf{r}]^2 \right\rangle = \left\langle \Omega^2 r^2 - (\boldsymbol{\Omega} \mathbf{r})^2 \right\rangle =$$
$$= \left(\omega_1^2 + \omega_2^2 \right) r^2 - \left\langle (x\omega_2 + y\omega_1 \cos \omega_2 t - z\omega_1 \sin \omega_2 t)^2 \right\rangle = \qquad (1)$$
$$= \omega_1^2 x^2 + \left(\tfrac{1}{2} \omega_1^2 + \omega_2^2 \right) (y^2 + z^2).$$

The surface of the liquid will be located along the level $U(\mathbf{r}) = \text{const}$. The resin will harden in the form of an ellipsoid of rotation.

What will change in this result when $\omega_1 = \omega_2$?

When $g = 0$ and $\omega_2 \to 0$, we get an obviously wrong answer from (1). Why?

9.25.

$$t = \sqrt{\frac{m}{2}} \int \frac{dr}{\sqrt{E + M\Omega - U_{\text{eff}}}},$$

$$\varphi = \sqrt{\frac{m}{2}} \int \frac{\left(\dfrac{M}{mr^2} - \Omega \right) dr}{\sqrt{E + M\Omega - U_{\text{eff}}}},$$

where E is the energy, M is the angular momentum in the rotating system of reference, and $U_{\text{eff}}(r) = U(r) + M^2/(2mr^2)$. Bear in mind that $E = E_0 - M\Omega$ and $M = M_0$, where E_0 and M_0 are the energy and angular momentum, respectively, in the inertial system.

It is interesting to note that the centrifugal potential energy $-\frac{1}{2} m\Omega^2 r^2$ does not enter into $U_{\text{eff}}(r)$.

9.26. In the system of reference, which is fixed in the frame, the Lagrangian function of the problem is the same as the one considered in problem 6.37, with $z = 0$ and with the parameters

$$\omega_B = -2\Omega, \qquad \omega_{1,2}^2 = \frac{2}{m} \left(k_{1,2} + \frac{f_{2,1}}{l} \right) - \Omega^2.$$

When $\omega_{1,2}^2 > 0$, the motion of the particles is the same as that of an anisotropic oscillator in the magnetic field $B = -2mc\Omega/e$. The orbit of the particle for the case $\omega_1 = \omega_2$ is shown in Fig. 109 of problem 2.36. In particular, if $\omega_1 = \omega_2 = 0$, the motion of the particle is the same as the motion of a free particle in a magnetic field:

$$x = x_0 + a\cos\omega_B t, \quad y = y_0 - a\sin\omega_B t;$$

that is, the particle moves uniformly along a circle of radius a and centre (x_0, y_0). It is interesting to investigate what the motion of the particle is in the rest system, which corresponds to the latter case, especially for $a = 0$ or for $x_0 = y_0 = 0$.

If the centrifugal force is larger than the restoring force from the two springs, $\omega_{1,2}^2 < 0$, then the particle still executes small oscillations. Although the potential energy has a maximum at $x = y = 0$, the equilibrium position is stable because of the Coriolis force.

If ω_1^2 and ω_2^2 have opposite signs, so that the point $x = y = 0$ is a saddle point of the potential energy, the equilibrium position is unstable.

It is interesting to compare these results with those of problem 5.4. One sees that in a system rotating with an angular velocity Ω the point r_0, φ_0 lies on the ridge of the "potential circus"; the potential energy

$$U(r) = -\frac{\alpha}{r^n} - \frac{1}{2}m\Omega^2 r^2$$

has a maximum in the direction of the radius vector, but it does not change in the direction of the azimuth. In this case, one of the eigen-frequency is ω, but the other one vanishes: the equilibrium position is neutral with respect to some perturbations (e.g. a changes in φ_0).

9.27. The Lagrangian function of a particle, which is expressed in its coordinates and velocity in the system of reference rotating with the paraboloid, has the form

$$L = \frac{m}{2}(\mathbf{v} + [\boldsymbol{\omega}, \mathbf{r}])^2 + m g r =$$
$$= \frac{m}{2}\left[(\dot{x} - \omega y)^2 + (\dot{y} + \omega x)^2 + \dot{z}^2 - g\left(\frac{x^2}{a} + \frac{y^2}{b}\right)\right].$$

The quantity $\dot{z}$ can be omitted for the small oscillations, and then the equations of motion are

$$\ddot{x} - 2\omega\dot{y} + \left(\frac{g}{a} - \omega^2\right)x = 0,$$
$$\ddot{y} + 2\omega\dot{x} + \left(\frac{g}{b} - \omega^2\right)y = 0. \tag{1}$$

We look for a solution in the form

$$x = Ae^{i\Omega t}, \qquad y = Be^{i\Omega t};$$

and for Ω^2 we get

$$\Omega^4 - \left(\frac{g}{a} + \frac{g}{b} + 2\omega^2\right)\Omega^2 + \left(\frac{g}{a} - \omega^2\right)\left(\frac{g}{b} - \omega^2\right) = 0.$$

It is easy to prove that its roots are real. However, when

$$\left(\frac{g}{a} - \omega^2\right)\left(\frac{g}{b} - \omega^2\right) < 0$$

one of the roots becomes negative, $\Omega_1^2 < 0$, so that the corresponding motion

$$x = A_1 e^{|\Omega_1|t} + A_2 e^{-|\Omega_1|t},$$
$$y = B_1 e^{|\Omega_1|t} + B_2 e^{-|\Omega_1|t}$$

leads to the departure of the particle from the origin. This means that the lower position of the particle becomes unstable. Assuming for definiteness that $a > b$, we obtain the instability region:

$$\frac{g}{a} < \omega^2 < \frac{g}{b}.$$

Note that if $\omega^2 > g/b$, the motion is stable although the potential energy in the rotating system

$$U(x,y) = -\frac{1}{2} m \left(\omega^2 - \frac{g}{a}\right) x^2 - \frac{1}{2} m \left(\omega^2 - \frac{g}{b}\right) y^2$$

is not a potential pit, but a potential hump. Stability in this case is provided by the action of the Coriolis force (see also [7], § 23).

9.28. a) The potential energy (including the centrifugal one) is

$$U(x) = k(x - a)^2 - \frac{1}{2} m \gamma^2 x^2.$$

The equilibrium position x_0 is determined by the condition $U'(x_0) = 0$; it is equal to

$$x_0 = \frac{2ka}{2k - m\gamma^2}.$$

Note that when the rotation frequency γ becomes greater than the frequency of the eigen-oscillations of the particle $\sqrt{2k/m}$, we have $x_0 < 0$; at the same time, when $m\gamma^2 \gg 2k$, the equilibrium position is close to the axis of rotation.

The sign of $U''(x_0) = 2k - m\gamma^2$ defines the stability of the equilibrium position: when $m\gamma^2 < 2k$, the equilibrium is stable; when $m\gamma^2 > 2k$, it is not.

b) Obviously, the equilibrium position is the same as in the preceding point 9.28a. To study the stability of the oscillations, we will consider the potential energy of the system at the small displacements from the equilibrium position:

$$U(x, y, z) = U(x_0, 0, 0) + \frac{1}{2} k_1 (x - x_0)^2 + \frac{1}{2} k_2 y^2 + \frac{1}{2} k_3 z^2,$$

where

$$k_1 = 2k - m\gamma^2,$$

$$k_2 = \frac{f + k(x_0 - a)}{l + x_0 - a} + \frac{f - k(x_0 - a)}{l - x_0 + a} - m\gamma^2 = k_3 - m\gamma^2.$$

The sign of $\frac{\partial^2 U(x_0, 0, z)}{\partial z^2}\big|_{z=0} = k_3$ defines the stability of the equilibrium position with respect to the small deviations along the z-axis. When $k_3 > 0$, the z-oscillations are stable. When $k_3 < 0$, they are unstable since the deviation of the particle from the point $(x_0, 0, 0)$ increases in time (until the interaction with the walls of the frame, which we do not consider here, becomes significant). Upon comparing this results with problem 9.27, we can conclude that the equilibrium position with respect to the small deviations in the xy-plane is stable when $k_1 k_2 > 0$, and it is unstable when $k_1 k_2 < 0$.

Let us consider in detail the case of fast rotation of the frame when

$$m\gamma^2 \gg 2k. \tag{1}$$

In that case,

$$k_3 \approx 2\frac{fl - ka^2}{l^2 - a^2} = 2k\frac{(l - l_0)l - a^2}{l^2 - a^2}, \tag{2}$$

where we introduce the length of a free string l_0 in accordance with the equation $f = k(l - l_0)$. From (2), it follows that $0 < k_3 < 2k$ when $fl > ka^2$, while $k_3 < 0$ when $fl < ka^2$. Besides, we find that in the limit (1) both $k_1 < 0$ and $k_2 < 0$. This means that when $m\gamma^2 \gg 2k$ and $fl > ka^2$, the equilibrium position is stable, in contrast to the results of 9.28a.

9.29. We use the Cartesian coordinates in the rotating system of reference with the origin of this system places in the centre of mass. The system itself rotates with an angular velocity ω around the z-axis, while the stars are located along the x-axis (Fig. 163).

Let the x coordinates of the stars be

$$a_{1,2} = \mp\frac{m_{2,1}}{m_1 + m_2}a,$$

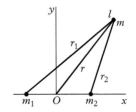

Figure 163

where $m_{1,2}$ are the masses of the stars and a is the distance between them. From equality

$$\frac{m_1 m_2}{m_1 + m_2}a\omega^2 = \frac{Gm_1 m_2}{a^2}$$

(where G is the gravitational constant), we obtain

$$\omega^2 = G\frac{m_1 + m_2}{a^3}.$$

The potential energy of a body of mass m (including the centrifugal potential energy) is

$$U(x, y, z) = -\frac{Gmm_1}{r_1} - \frac{Gmm_2}{r_2} - \tfrac{1}{2}m\omega^2(x^2 + y^2),$$

where $r_{1,2}$ are the distances to the stars and $\mathbf{r} = (x, y, z)$ is the radius vector of the body. The equilibrium position of the body is determined by the condition $\dfrac{\partial U}{\partial \mathbf{r}} = 0$, or

$$\frac{\partial U}{\partial x} = Gm\left(\frac{m_1}{r_1^3} + \frac{m_2}{r_2^3} - \frac{m_1 + m_2}{a^3}\right)x - Gm\left(\frac{m_1 a_1}{r_1^3} + \frac{m_2 a_2}{r_2^3}\right) = 0,$$

$$\frac{\partial U}{\partial y} = Gm\left(\frac{m_1}{r_1^3} + \frac{m_2}{r_2^3} - \frac{m_1 + m_2}{a^3}\right)y = 0,$$

$$\frac{\partial U}{\partial z} = Gm\left(\frac{m_1}{r_1^3} + \frac{m_2}{r_2^3}\right)z = 0.$$

Hence, $z = 0$ and (when $y \neq 0$) $r_1 = r_2 = a$.

Thus, the stars and the equilibrium point are at the vertices of a right triangle. There are two such points (the so-called *Lagrange points*):

$$x_0 - a_1 = a_2 - x_0 = \tfrac{1}{2}a, \quad y_0 = \pm\tfrac{\sqrt{3}}{2}a, \quad z_0 = 0.$$

The potential energy near the Lagrange points has the form

$$U(x_0 + x_1, y_0 + y_1, z_0 + z_1) =$$
$$= U(x_0, y_0, z_0) - \tfrac{3}{8}m\omega^2 x_1^2 - m\alpha x_1 y_1 - \tfrac{9}{8}m\omega^2 y_1^2 + \tfrac{1}{2}m\omega^2 z_1^2,$$
$$\alpha = \pm\frac{3\sqrt{3}}{4a^3}G(m_1 - m_2).$$

The motion along the z-axis is obviously stable. The equations of motion in the xy-plane read

$$\ddot{x}_1 - \tfrac{3}{4}\omega^2 x_1 - \alpha y_1 - 2\omega\dot{y}_1 = 0,$$
$$\ddot{y}_1 - \tfrac{9}{4}\omega^2 y_1 - \alpha x_1 + 2\omega\dot{x}_1 = 0.$$

The substitutions $x = Ae^{i\Omega t}, y = Be^{i\Omega t}$ lead to the equation for Ω:

$$\Omega^4 - \omega^2\Omega^2 + \tfrac{27}{16}\omega^4 - \alpha^2 = 0.$$

Its roots are real for $16\alpha^2 \geqslant 23\omega^4$, that is, when

$$27(m_1 - m_2)^2 \geqslant 23(m_1 + m_2)^2.$$

This condition is satisfied if the mass of one of the stars (say, the first one) satisfies the condition $m_1/m_2 \geqslant \tfrac{1}{2}(25 + 3\sqrt{69}) \approx 25$. In this case, the motion of bodies in the neighbourhood of Lagrange points is stable. Stability of motion is provided by the Coriolis force (cf. problem 9.28).

On the x-axis, there are three more points where $\dfrac{\partial U}{\partial \mathbf{r}} = 0$, but the motion near them is unstable.

Note that for the Sun–Jupiter system, asteroids are observed just near the Lagrange points (for more detail see [5], § 3.12).

§10

The Hamiltonian equations of motion. Poisson brackets

10.1. Let $\boldsymbol{\varepsilon}$ be a vector of infinitesimal displacement; we then have

$$\mathbf{r}_a \to \mathbf{r}'_a = \mathbf{r}_a + \boldsymbol{\varepsilon}, \quad \mathbf{p}_a \to \mathbf{p}'_a = \mathbf{p}_a,$$
$$H(\mathbf{r}_a, \mathbf{p}_a) = H(\mathbf{r}'_a, \mathbf{p}'_a).$$

Hence, $\sum_a \dfrac{\partial H}{\partial \mathbf{r}_a} = 0$. Using the Hamiltonian equation, we get

$$\dot{\mathbf{P}} = \sum_a \dot{\mathbf{p}}_a = -\sum_a \frac{\partial H}{\partial \mathbf{r}_a} = 0, \qquad \mathbf{P} = \text{const.}$$

For an infinitesimal rotation $\delta\boldsymbol{\varphi}$, we obtain

$$\mathbf{r}_a \to \mathbf{r}'_a = \mathbf{r}_a + [\delta\boldsymbol{\varphi}, \mathbf{r}_a], \quad \mathbf{p}_a \to \mathbf{p}'_a = \mathbf{p}_a + [\delta\boldsymbol{\varphi}, \mathbf{p}_a],$$
$$H(\mathbf{r}_a, \mathbf{p}_a) = H(\mathbf{r}'_a, \mathbf{p}'_a), \quad \sum_a \left\{ \frac{\partial H}{\partial \mathbf{r}_a} [\delta\boldsymbol{\varphi}, \mathbf{r}_a] + \frac{\partial H}{\partial \mathbf{p}_a} [\delta\boldsymbol{\varphi}, \mathbf{p}_a] \right\} = 0 =$$
$$= \sum_a \{ -\dot{\mathbf{p}}_a [\delta\boldsymbol{\varphi}, \mathbf{r}_a] + \dot{\mathbf{r}} [\delta\boldsymbol{\varphi}, \mathbf{p}_a] \} = -\delta\boldsymbol{\varphi} \sum_a \frac{d}{dt} [\mathbf{r}_a, \mathbf{p}_a],$$

or

$$\mathbf{M} = \sum_a [\mathbf{r}_a, \mathbf{p}_a] = \text{const.}$$

10.2. $H = \dfrac{p_\theta^2}{2I_1} + \dfrac{(p_\varphi - p_\psi \cos\theta)^2}{2I_1 \sin^2\theta} + \dfrac{p_\psi^2}{2I_3}.$

Exploring Classical Mechanics: A Collection of 350+ Solved Problems for Students, Lecturers, and Researchers. First Edition.
Gleb L. Kotkin and Valeriy G. Serbo, Oxford University Press (2020). © Gleb L. Kotkin and Valeriy G. Serbo 2020.
DOI: 10.1093/oso/9780198853787.001.0001

10.3. $H = \frac{p^2}{2(1+2\beta x)} + \frac{1}{2}\omega^2 x^2 + \alpha x^3$. In particular, for small oscillations ($|\alpha x| \ll \omega^2, |\beta x| \ll 1$)

$$H = \frac{1}{2}p^2 + \frac{1}{2}\omega^2 x^2 + \alpha x^3 - \beta x p^2 + 2\beta^2 x^2 p^2 - \ldots,$$

and up to and including terms linear in α and β the extra term in the Hamiltonian function of the harmonic oscillator is connected with the extra term in the Lagrangian function through the equation $\delta H = -\delta L$ (see [1], § 40).

10.4. $x = a\cos(\omega t + \varphi), p = -\omega_0 a\sin(\omega t + \varphi)$, where

$$\omega = (1 + 2\lambda E_0)\omega_0, \quad E_0 = \frac{1}{2}\omega_0^2 a^2.$$

10.5. $p = p_0 + Ft, \; x = x_0 + \frac{A}{F}(\sqrt{p_0 + Ft} - \sqrt{p_0})$. The given Hamiltonian function describes the motion of a charged vortex ring in liquid helium in the presence of a homogeneous electric field along the x-axis [26]. The characteristic features of such motion are the following: the momentum of the vortex increases with the time, while velocity of its motion drops $\dot{x} = a/(2\sqrt{p_0 + Ft})$.

10.6. $\dot{\mathbf{r}} = \frac{c\mathbf{p}}{np}, \quad \dot{\mathbf{p}} = \frac{cp}{n^2}\frac{\partial n}{\partial \mathbf{r}}, \quad p = |\mathbf{p}|$.

The given Hamiltonian function describes the propagation of light in a transparent medium with refractive index $n(\mathbf{r})$ in the geometric optics approximation (see [3], § 65). The "particle" is a wave packet, and $\mathbf{r}(t)$ gives the law of its motion; $\dot{\mathbf{r}}$ is its group velocity, while the vector $\mathbf{p}$, which is perpendicular to the wave front, determines the wave vector.

When $n(\mathbf{r}) = ax$, the trajectory is

$$x = C_1 \cosh\left(\frac{y}{C_1} + C_2\right),$$

where the C_1 and C_2 are determined by the initial and final points of the trajectory, respectively.

10.7. a) $L = \frac{1}{2}m(\mathbf{v} - \mathbf{a})^2$;

b) $L = 0$: such "particles" cannot be described using a Lagrangian function (see [2], § 53).

10.8. The given vector potential determines a magnetic field $\mathbf{B}$ in the z-direction.
The Hamiltonian function is

$$H(x, y, z, p_x, p_y, p_z) = \frac{p_x^2 + p_z^2}{2m} + \frac{1}{2m}\left(p_y - \frac{e}{c}Bx\right)^2.$$

Since H depend neither on y nor on z, we have

$$p_y = \text{const}, \quad p_z = \text{const}.$$

If we present H in the form

$$H = \frac{p_x^2}{2m} + \tfrac{1}{2} m\omega^2 (x - x_0)^2 + \frac{p_z^2}{2m},$$

where

$$\omega = \frac{eB}{mc}, \quad x_0 = \frac{cp_y}{eB},$$

we see that x and p_x are obtained from the same Hamiltonian function as a harmonic oscillator. Therefore, we have

$$x = a\cos(\omega t + \varphi) + x_0, \quad p_x = -m\omega a \sin(\omega t + \varphi).$$

To determine y and z, we use the equations

$$\dot{y} = \frac{\partial H}{\partial p_y} = \frac{1}{m}\left(p_y - \frac{e}{c}Bx\right) = -\omega a\cos(\omega t + \varphi), \quad \dot{z} = \frac{p_z}{m},$$

whence

$$y = -a\sin(\omega t + \varphi) + y_0, \quad z = \frac{p_z}{m}t + z_0.$$

The particle moves along a spiral with its axis parallel to **B**. The generalized momentum p_y determines the distance of that axis from the yz-plane.

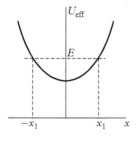

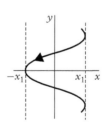

Figure 164   Figure 165

10.9. The magnetic field is directed along the z-axis and equals $2hx$. The motion along the z-axis is uniform. Here we consider only the motion in the xy-plane. The Hamiltonian function

$$H = \frac{p_x^2}{2m} + \frac{1}{2m}\left(p_y - \frac{eh}{c}x^2\right)^2$$

depends neither on y nor on t. Therefore, the integrals of motion are the generalized momentum p_y and energy E:

$$p_y = m\dot{y} + \frac{eh}{c}x^2,$$

$$E = \tfrac{1}{2}m\dot{x}^2 + U_{\text{eff}}(x), \quad U_{\text{eff}}(x) = \frac{1}{2m}\left(p_y - \frac{eh}{c}x^2\right)^2.$$

When $p_y \leqslant 0$, the function $U_{\text{eff}}(x)$ is shown in Fig. 164, while an approximate view of the orbits can be found in Fig. 165. Note that the velocity

$$\dot{y} = -\frac{|p_y|}{m} - \frac{eh}{mc}x^2$$

is negative everywhere and oscillates near the value p_y/m.

When $p_y > 0$, the function

$$U_{\text{eff}}(x) = \frac{e^2 h^2}{2mc^2}(x^2 - x_0^2)^2, \quad x_0 = \sqrt{\frac{p_y c}{eh}}$$

is shown in Fig. 166. Now the velocity

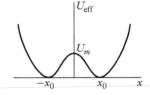

$$\dot{y} = \frac{eh}{mc}(x_0^2 - x^2)$$

Figure 166

has both positive and negative values for any value of E. The approximate view of the orbits is given in Fig. 167; Fig. 167a–e corresponds to the decreasing values of energies.

At high energies, $E \gg U_m = p_y^2/(2m)$, the amplitude of the oscillations along the x-axis is large; the average over the period value $\langle x^2 \rangle$ is larger than x_0^2. Therefore, the averaged value

$$\langle \dot{y} \rangle = \frac{eh}{mc}\left(x_0^2 - \langle x^2 \rangle\right)$$

is negative (see Fig. 167a). When the energy decreases, the value $\langle \dot{y} \rangle$ increases to zero (Fig. 167b) and then becomes positive (Fig. 167c). When energy $E = U_m$, the particle,

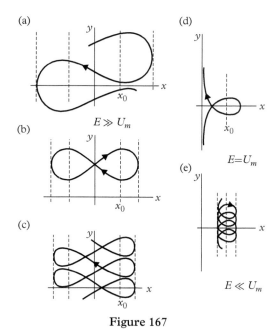

Figure 167

having in the initial moment $x > x_0$ and $\dot{x} < 0$, asymptotically approaches the y-axis (Fig. 167d).

Finally, when $E < U_m$, the particle moves either in the region near $(-x_0)$ or in the region near x_0 (Fig. 167e). That $\langle \dot{y} \rangle > 0$ is easily demonstrable. When $|x - x_0| \ll x_0$, the particle moves in a circle, the centre of which slowly drifts along the y-axis. To find the velocity of the drift, it is necessary to take into account the first anharmonic corrections (see [1], § 28 and [7], § 29)

$$x = x_0 + a \cos \omega t - \frac{a^2}{4x_0}(3 - \cos 2\omega t),$$

when calculating $\langle x^2 \rangle$. That leads to

$$\langle \dot{y} \rangle = \frac{cE}{2hx_0^2 e}$$

(cf. problem 8.14).

10.10. Introducing the coordinates of the centre of mass **R** and the relative motion **r** (cf. problems 2.29 and 2.30), we present the Lagrangian function of the system in the form

$$L = \tfrac{1}{2} M \dot{\mathbf{R}}^2 + \frac{e}{2c} [\mathbf{B}, \mathbf{r}] \dot{\mathbf{R}} + L_1(\mathbf{r}, \dot{\mathbf{r}}) + \frac{e}{2c} [\mathbf{B}, \mathbf{R}] \dot{\mathbf{r}}, \tag{1}$$

where $M = m_1 + m_2$,

$$L_1(\mathbf{r}, \dot{\mathbf{r}}) = \tfrac{1}{2} m \dot{\mathbf{r}}^2 + \frac{e}{2c} [\mathbf{B}', \mathbf{r}] \dot{\mathbf{r}} + \frac{e^2}{r},$$

$$\mathbf{B}' = \frac{m_2 - m_1}{m_2 + m_1} \mathbf{B}$$

and m is the reduced mass.

We rewrite the last term in (1) as

$$\frac{e}{2c} [\mathbf{B}, \mathbf{R}] \dot{\mathbf{r}} = \frac{e}{2c} [\mathbf{B}, \mathbf{r}] \dot{\mathbf{R}} + \frac{d}{dt} \frac{e}{2c} [\mathbf{B}, \mathbf{R}] \mathbf{r}.$$

Omitting the total derivative with respect to time, we have

$$L = \tfrac{1}{2} M \dot{\mathbf{R}}^2 + \frac{e}{c} [\mathbf{B}, \mathbf{r}] \dot{\mathbf{R}} + L_1(\mathbf{r}, \dot{\mathbf{r}}).$$

The Hamiltonian function of the system now has the form

$$H = \frac{1}{2M} \left(\mathbf{P} - \frac{e}{c} [\mathbf{B}, \mathbf{r}] \right)^2 + \frac{1}{2m} \left(\mathbf{p} - \frac{e}{2c} [\mathbf{B}', \mathbf{r}] \right)^2 - \frac{e^2}{r}.$$

This function has no evident dependence on $\mathbf{R}$ and t; therefore, the generalized momentum

$$\mathbf{P} = \frac{\partial L}{\partial \dot{\mathbf{R}}} = M \dot{\mathbf{R}} + \frac{e}{c} [\mathbf{B}, \mathbf{r}] \tag{2}$$

and the total energy of the system

$$E = \tfrac{1}{2} M \dot{\mathbf{R}}^2 + \tfrac{1}{2} m \dot{\mathbf{r}}^2 - \frac{e^2}{r}$$

are conserved. The last equation can be rewritten in the form

$$E = \tfrac{1}{2} m \dot{\mathbf{r}}^2 + U_{\text{eff}}(\mathbf{r}), \quad U_{\text{eff}}(\mathbf{r}) = -\frac{e^2}{r} + \frac{1}{2M} \left(\mathbf{P} - \frac{e}{c} [\mathbf{B}, \mathbf{r}] \right)^2.$$

From here follows (with the assumption $\mathbf{P} = \mathbf{const}$) that a particle of mass m moves in the field with the effective potential energy $U_{\mathrm{eff}}(\mathbf{r})$ and in the constant uniform magnetic field $\mathbf{B}'$. If the z-axis is directed along $\mathbf{B}$, then

$$U_{\mathrm{eff}}(\mathbf{r}) = -\frac{e^2}{r} + \tfrac{1}{2} m\omega^2 [(x-a)^2 + (y-b)^2] + \mathrm{const},$$

$$\omega = \frac{eB}{c\sqrt{m_1 m_2}}, \quad b = -\frac{cP_x}{eB}, \quad a = \frac{cP_y}{eB}.$$

After finding $\mathbf{r}(t)$, the law of motion of the centre of mass can be determined from (2):

$$\mathbf{R}(t) = \frac{\mathbf{P}t}{M} - \frac{e}{Mc} \left[\mathbf{B}, \int_0^t \mathbf{r}(t)\, dt \right] + \mathbf{R}_0.$$

10.11.

$$\mathbf{p} = \mathbf{p}_0 + e\mathbf{E}t, \quad \varepsilon(\mathbf{p}) - e\mathbf{E}\mathbf{r} = \varepsilon_0,$$
$$(\mathbf{r} - \mathbf{r}_0)e\mathbf{E} = \varepsilon(\mathbf{p}_0 + e\mathbf{E}t) - \varepsilon(\mathbf{p}_0).$$

Here $\mathbf{r}_0$, $\mathbf{p}_0$, and ε_0 are the constants.

10.12.

$$\dot{\mathbf{p}} = e\mathbf{E} + \frac{e}{c} [\mathbf{v}, \mathbf{B}].^1$$

10.13. a) $\varepsilon(\mathbf{p}) = E$, $p_B = \mathrm{const}$, where p_B is the component of the momentum along the magnetic field $\mathbf{B}$. The trajectory in the momentum space is determined by the intersection of the two surfaces: $\varepsilon(\mathbf{p}) = E$ and $p_B = \mathrm{const}$.

b) From the equation of motion $\dot{\mathbf{p}} = \frac{e}{c} [\dot{\mathbf{r}}, \mathbf{B}]$, it is clear that the projection of the electron orbit on a plane perpendicular to the magnetic field $\mathbf{B}$ is obtained from the orbit in momentum space by a rotation over $\pi/2$ around $\mathbf{B}$ and by changing the scale by a factor $\frac{c}{eB}$.

10.14.

$$S = \int_{E_{\min}}^{E} dE \oint \frac{dp}{|\mathbf{v}_\perp|}, \quad T = \frac{c}{eB} \oint \frac{dp}{|\mathbf{v}_\perp|} = \frac{c}{eB} \frac{\partial S}{\partial E},$$

where $\mathbf{v}_\perp$ is the component of the vector $\dfrac{\partial \varepsilon}{\partial \mathbf{p}}$, which is orthogonal to $\mathbf{B}$.

---

[1] For detail about the motion of electrons in a metal (problems 10.10–10.14) see, for istance, [16].

10.15. a) $-\sum_k e_{ijk} x_k;^1 \quad -\sum_k e_{ijk} p_k; \quad -\sum_k e_{ijk} M_k.$

b) $\mathbf{ab};$

$$\{\mathbf{aM}, \mathbf{br}\} = \left\{\sum_i a_i M_i, \sum_j b_j x_j\right\} = \sum_{ij} a_i b_j \{M_i, x_j\} =$$

$$= -\sum_{ijk} a_i b_j e_{ijk} x_k = -[\mathbf{a}, \mathbf{b}]\,\mathbf{r}; \quad -[\mathbf{a}, \mathbf{b}]\mathbf{M}.$$

c) $0; \ n\mathbf{r}\,r^{n-2}; \ 2\mathbf{a}\,(\mathbf{ar}).$

10.16. $\{A_i, A_j\} = -\sum_k e_{ijk} A_k, \{A_i, A_4\} = 0$, where i, j, and k take on the values 1, 2, and 3, respectively (cf. problem 10.15a).

10.17.

$$\{M_i, \Lambda_{jk}\} = -\sum_l e_{ijl} \Lambda_{lk} - \sum_l e_{ikl} \Lambda_{lj};$$

$$\{\Lambda_{jk}, \Lambda_{il}\} = \delta_{ij} M_{lk} + \delta_{ik} M_{lj} + \delta_{jl} M_{ik} + \delta_{kl} M_{ij},$$

where $M_{kl} = p_k x_l - p_l x_k.$

10.18. When the system as a whole is rotated around the z-axis over an infinitesimal angle ε, the change $\delta\varphi$ in any function of the coordinates and momenta is in the first order in ε given by

$$\delta\varphi = \varphi(x - \varepsilon y, y + \varepsilon x, z, p_x - \varepsilon p_y, p_y + \varepsilon p_x, p_z) - \varphi(x, y, z, p_x, p_y, p_z) =$$

$$= \varepsilon\left(-\frac{\partial\varphi}{\partial x}y + \frac{\partial\varphi}{\partial y}x - \frac{\partial\varphi}{\partial p_x}p_y + \frac{\partial\varphi}{\partial p_y}p_x\right) = \varepsilon\{M_z, \varphi\}.$$

If φ is a scalar, this change under rotation must vanish, and thus $\{\varphi, M_z\} = 0$. If $\varphi = f_x$ is the component of a vector function, its change under rotation is $\delta f_x = -\varepsilon f_y$, and thus

$$\{M_z, f_x\} = -f_y \quad \text{or} \quad \{M_z, \mathbf{f}\} = [\mathbf{n}, \mathbf{f}]$$

(cf. [1], § 42, problems 3 and 4).

---

[1] e_{ijk} is the completely antisymmetric tensor:

$$e_{123} = e_{231} = e_{321} = 1, \quad e_{132} = e_{321} = e_{213} = -1;$$

all other components of e_{ijk} vanish.

What is the value of the Poisson brackets $\{M_z, T_{xx}\}$, where T_{xx} is a component of a tensor function?

10.19. $[\mathbf{f}, a\mathbf{M}] = [\mathbf{f}, \mathbf{a}]; \quad \{\mathbf{f}\mathbf{M}, \mathbf{l}\mathbf{M}\} = [\mathbf{f}, \mathbf{l}]\,\mathbf{M} + \sum_{ik} M_i M_k \{f_i, l_k\}.$

10.20. Substituting into the second formula of the preceding problem $\mathbf{f} = \mathbf{e}_\zeta$ and $\mathbf{l} = \mathbf{e}_\xi$, where $\mathbf{e}_\zeta$ and $\mathbf{e}_\xi$ are unit vectors along the ζ- and ξ-axes of the moving system of reference, we have

$$\{M_\zeta, M_\xi\} = +M_\eta. \tag{1}$$

This equation differs in the sign of the right-hand side from the analogous relation for the components of the angular momentum along the axis of the fixed system of coordinates,

$$\{M_z, M_x\} = -M_y. \tag{2}$$

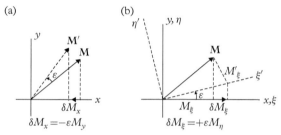

Figure 168

As was shown in problem 10.18 (see also [5], §9.7), the Poisson brackets (2) characterizes the change of the component M_x when the coordinate system is rotated as a whole over an infinitesimal angle ε (Fig. 168a)

$$\delta M_x = \varepsilon\{M_z, M_x\} = -\varepsilon M_y.$$

The Poisson brackets in (1), which are equal to

$$\{\mathbf{e}_\zeta \mathbf{M}, \mathbf{e}_\xi \mathbf{M}\} = \{M_\zeta, \mathbf{e}_\xi\}\mathbf{M},$$

characterizes the change in the components of the fixed vector $\mathbf{M}$ along the $\mathbf{e}_\xi$-axis when the moving coordinate system is rotated over an infinitesimal angle around the ζ-axis (Fig. 168b; in the figure the ξ, η, ζ-axis are the same as as x, y, z-axes before the rotation).

10.21. $\dot{M}_\alpha = \sum_{\beta\gamma\delta} e_{\alpha\beta\gamma}(I^{-1})_{\gamma\delta}M_\beta M_\delta$. In particular, if we choose the moving system of coordinates such that the tensor of inertia $I_{\alpha\beta}$ is diagonal, we obtain the Euler equation (see [1], §36) using the relation $M_\alpha = I_\alpha \Omega_\alpha$).

10.22. The equation of motion is

$$\dot{M}_i = \{H, M_i\} = \gamma \sum_{jk} e_{ijk} M_j B_k, \quad \text{or} \quad \dot{\mathbf{M}} = -\gamma \, [\mathbf{B}, \mathbf{M}];$$

that is, the vector $\mathbf{M}$ processes with an angular velocity $-\gamma \mathbf{B}$.

a) The vector $\mathbf{M}$ precesses around the direction of $\mathbf{B}$:

$$M_x = M_x(0)\cos(\gamma B_0 t) + M_y(0)\sin(\gamma B_0 t),$$
$$M_y = -M_x(0)\sin(\gamma B_0 t) + M_y(0)\cos(\gamma B_0 t),$$
$$M_z = M_z(0).$$

b) The vector $\mathbf{M}$ rotates with an angular velocity $-\gamma \mathbf{B}$, which in turn rotates around the z-axis with an angular velocity ω. It is helpful to use the rotating system of coordinates, in which the vector $\mathbf{B}$ is fixed. In that system, the components of the angular velocity of the vector $\mathbf{M}$ are equal to

$$\omega'_x = -\gamma B_1, \quad \omega'_y = 0, \quad \omega'_z = -\gamma B_0 - \omega \equiv \varepsilon.$$

At a given initial condition, the components $\mathbf{M}$ in the rotating system of reference are

$$M'_x = -a\frac{\varepsilon}{\lambda} M_0 (1 - \cos \lambda t),$$
$$M'_y = a M_0 \sin \lambda t,$$
$$M'_z = \left(\frac{\varepsilon^2}{\lambda^2} + a^2 \cos \lambda t \right) M_0,$$

where

$$\lambda = \sqrt{\varepsilon^2 + \gamma^2 B_1^2}, \quad a = \frac{\gamma B_1}{\sqrt{\varepsilon^2 + \gamma^2 B_1^2}}.$$

In the fixed system, we have

$$M_x = M'_x \cos \omega t - M'_y \sin \omega t,$$
$$M_y = M'_x \sin \omega t + M'_y \cos \omega t,$$
$$M_z = M'_z.$$

When $B_1 \ll B_0$, the dependence of amplitudes $M_{x,y}$ on ω is resonant: generally speaking, these amplitudes are small $\sim M_0 B_1 / B_0$, but for $|\varepsilon| = |\omega + \gamma B_1| \lesssim \gamma B_1$, they increase dramatically, reaching values $\sim M_0$. In particular, for $\omega = -\gamma B_0$, we get

$$M_x = M_0 \sin\gamma B_1 t \sin\gamma B_0 t,$$
$$M_y = M_0 \sin\gamma B_1 t \cos\gamma B_0 t,$$
$$M_z = M_0 \cos\gamma B_1 t.$$

10.23. $\{v_i, v_j\} = -\dfrac{e}{m^2 c}\sum_k e_{ijk}B_k.$

10.24. a) $\mathbf{p}(t) = \mathbf{p} + \mathbf{F}t,\ \ \mathbf{r}(t) = \mathbf{r} + \dfrac{\mathbf{p}t}{m} + \dfrac{\mathbf{F}t^2}{2m};$
 b) $p(t) = p\cos\omega t - m\omega q \sin\omega t,$

$$q(t) = q\cos\omega t + \dfrac{p}{m\omega}\sin\omega t.$$

Of course, these quantities can more simply be evaluated without using Poisson brackets. However, this method can easily be taken over in quantum mechanics.

10.25. a) According to the preceding problem, we have

$$\frac{df}{dt} = \{h, f\} = \frac{\partial H}{\partial f}\{f, f\} = 0.$$

b) The Hamiltonian function is

$$H = \frac{p_r^2}{2m} + \frac{1}{2mr^2}f(\theta, \varphi, p_\theta, p_\varphi),$$

where

$$f(\theta, \varphi, p_\theta, p_\varphi) = p_\theta^2 + \frac{p_\varphi^2}{\sin^2\theta} + 2ma\cos\theta.$$

Integrals of motion are E, p_φ, and, according to the preceding problem, f.

10.26. a)

$$\{A_i, A_j\} = \frac{2H}{m}\sum_{k=1}^{3}\varepsilon_{ijk}M_k,$$

$$\{A_i, M_j\} = -\sum_{k=1}^{3}\varepsilon_{ijk}A_k;$$

b) $\{H, \mathbf{J}_{1,2}\} = 0$, $\{\mathcal{J}_{1i}, \mathcal{J}_{2j}\} = 0$,

$$\{\mathcal{J}_{1i}, \mathcal{J}_{1j}\} = -\sum_{k=1}^{3} \varepsilon_{ijk} \mathcal{J}_{1k}, \quad \{\mathcal{J}_{2i}, \mathcal{J}_{2j}\} = -\sum_{k=1}^{3} \varepsilon_{ijk} \mathcal{J}_{2k},$$

$$H = -\frac{m\alpha^2}{4\left(\mathbf{J}_1^2 + \mathbf{J}_2^2\right)}.$$

The vectors $\mathbf{J}_1$ and $\mathbf{J}_2$ are the independent integrals of motion. Each has the same Poisson brackets for their components as an usual angular momentum. The presence of two such "momenta" is closely related to the so-called hidden symmetry of the hydrogen atom (see [22], Ch. I, § 5).

§11

Canonical transformations

11.1. a)
$$q = \sqrt{\frac{2P}{m\omega}}\sin Q, \qquad p = \sqrt{2m\omega P}\cos Q,$$

$$\dot{Q} = \omega + \frac{\dot{\omega}}{2\omega}\sin 2Q, \qquad \dot{p} = -P\frac{\dot{\omega}}{\omega}\cos 2Q.$$

In this case, P and Q are action and angle variables. These variables are more helpful than p and q for solving the problem by the method of the perturbation theory, if the frequency $\omega(t)$ changes slowly, $|\dot{\omega}| \ll \omega^2$ (see problem 13.11).

b)
$$q = \frac{F}{m\omega^2} + \sqrt{\frac{2P}{m\omega}}\sin Q, \qquad p = \sqrt{2m\omega P}\cos Q,$$

$$\dot{Q} = \omega - \frac{\dot{F}}{\omega}\frac{\cos Q}{\sqrt{2m\omega P}}, \qquad \dot{P} = -\frac{\dot{F}}{\omega}\sqrt{\frac{2P}{m\omega}}\sin Q.$$

11.2. $\Psi(p, Q) = -Q\left(1 + \ln\frac{p^2}{4Q}\right).$

11.3. The function $\Phi(q_1, q_2, \ldots, q_s, P_1, P_2, \ldots, P_s)$ determines a canonical transformation, provided

$$\det\frac{\partial^2\Phi}{\partial q_i \partial P_k} \neq 0.$$

11.4. Let

$$Q = q\cos\alpha - p\sin\alpha, \qquad P = q\sin\alpha + p\cos\alpha.$$

Then we have

$$\{P, Q\}_{p,q} = -\{q, p\}_{p,q}\sin^2\alpha + \{p, q\}_{p,q}\cos^2\alpha = 1.$$

Exploring Classical Mechanics: A Collection of 350+ Solved Problems for Students, Lecturers, and Researchers. First Edition.
Gleb L. Kotkin and Valeriy G. Serbo, Oxford University Press (2020). © Gleb L. Kotkin and Valeriy G. Serbo 2020.
DOI: 10.1093/oso/9780198853787.001.0001

For a system with one degree of freedom, this is sufficient for the transformation to be canonical.

11.5. One sees easily (and subsequent calculations verify) that the canonical transformation must be close to the identity transformation and that the terms $ax^2 P$ and bP^3 in the generating function are small. To solve the equations

$$p = P + 2axP, \quad Q = x + ax^2 + 3bP^2,$$

which determine the canonical transformation, for x and p, we replace x by Q and p by P in the small terms:

$$p = P + 2aQP, \quad x = Q - aQ^2 - 3bP^2. \tag{1}$$

We proceed in the similar fashion when we express the Hamiltonian function in terms of the new variables:

$$H'(Q, P) = \tfrac{1}{2}(P^2 + \omega^2 Q^2) + \alpha Q^3 + \beta QP^2 + 2aQP^2 - a\omega^2 Q^3 - $$
$$- 3b\omega^2 QP^2 + \text{terms of fourth degree in } Q \text{ and } P.$$

Putting

$$\alpha - a\omega^2 = 0, \quad \beta + 2a - 3b\omega^2 = 0,$$

the third-order terms vanish. In the approximation indicated in the problem, we have

$$Q = A\cos\omega t, \quad P = -\omega A\sin\omega t,$$

and from (1) (cf. [1], § 28)

$$x = A\cos\omega t - \frac{\alpha A^2}{\omega^2} - \left(\beta + \frac{\alpha}{\omega^2}\right) A^2 \sin^2 \omega t.$$

11.6. Reducing the Hamiltonian function to the form considered in problem [10.4], we get

$$x = Q - \frac{5\beta}{8\omega_0^2} Q^3 - \frac{9\beta}{8\omega_0^4} QP^2,$$

where

$$Q = A\cos\omega t, \quad P = -\omega_0 A\sin\omega t, \quad \omega = \omega_0 + \frac{3\beta}{2\omega_0} A^2$$

(cf. [1], § 28).

11.7. Because the terms $xP_X + yP_Y$ in the generating function correspond to the identical transformation, it is clear in advance that the parameters a, b, and c will be proportional to α. Therefore, the canonical transformation for p_x with required precision has the following form

$$p_x = \frac{\partial \Phi}{\partial x} = P_X + ayP_X + 2cxP_Y \approx P_X + aYP_X + 2cXP_Y.$$

Acting similarly, we get

$$p_y \approx P_Y + aXP_X, \quad x \approx X - aXY - 2bP_XP_Y, \quad y \approx Y - bP_X^2 - cX^2.$$

Substituting these expressions into a new Hamiltonian function

$$H' = H + \frac{\partial \Phi}{\partial t} = H$$

on the condition that this function corresponds (with the necessary accuracy) to the two independent oscillators,

$$H' = \frac{1}{2m}\left[P_X^2 + (m\omega_1 X)^2 + P_Y^2 + (m\omega_2 Y)^2\right],$$

we find

$$a = -\frac{2\alpha}{4\omega_1^2 - \omega_2^2}, \quad b = -\frac{2\alpha}{(4\omega_1^2 - \omega_2^2)m^2\omega_2^2}, \quad c = -\frac{\alpha(2\omega_1^2 - \omega_2^2)}{(4\omega_1^2 - \omega_2^2)\omega_2^2}.$$

The new canonical variables correspond to the two free oscillators

$$X = A\cos(\omega_1 t + \varphi_1), \quad Y = B\cos(\omega_2 t + \varphi_2),$$

$$P_X = -m\omega_1 A\sin(\omega_1 t + \varphi_1), \quad P_Y = -m\omega_2 B\sin(\omega_2 t + \varphi_2).$$

As a result, with the required accuracy, we find

$$x(t) = A\cos(\omega_1 t + \varphi_1) - \frac{\alpha AB}{2\omega_2(2\omega_1 + \omega_2)}\cos(\omega_+ t + \varphi_+) +$$

$$+\frac{\alpha AB}{2\omega_2(2\omega_1 - \omega_2)}\cos(\omega_- t + \varphi_-),$$

$$y(t) = B\cos(\omega_2 t + \varphi_2) + \frac{\alpha A^2}{2\omega_2^2} - \frac{\alpha A^2}{2(4\omega_1^2 - \omega_2^2)}\cos(2\omega_1 t + 2\varphi_1),$$

where $\omega_\pm = \omega_1 \pm \omega_2$ and $\varphi_\pm = \varphi_1 \pm \varphi_2$.

The obtained solutions are valid until the frequency ω_2 is close to $2\omega_1$. For $\omega_2 = 2\omega_1$, anharmonic corrections stop being small and can lead to significant transferring energy from x to y oscillations and back (see problem 8.10).

11.8. $H'(P, Q) = H(P, Q)$; when $X = A\sin(\omega t + \varphi)$, $Y = 0$ the oscillator perform a motion along an ellipse:

$$x = A\cos\lambda\,\sin(\omega t + \varphi),$$
$$y = A\sin\lambda\,\cos(\omega t + \varphi).$$

11.9. To make the notation less cumbersome, it is helpful for the time being to put $m = \omega = e = c = 1$. One can easily reintroduce these factors in the final expressions. The transformation of problem 11.8 is a rotation in the xp_y- and yp_x-planes, which therefore leaves the form of that part of the Hamiltonian function, which is equal to

$$\tfrac{1}{2}(x^2 + y^2 + p_x^2 + p_y^2),$$

invariant. On the other hand, the correction due to the terms $\tfrac{1}{2}B^2x^2 - Bxp_y$ is equal to

$$\tfrac{1}{2}B^2(X^2\cos^2\lambda + P_Y^2\sin^2\lambda + 2XP_Y\sin\lambda\cos\lambda) +$$
$$+ B(X^2 - P_Y^2)\sin\lambda\cos\lambda - B(\cos^2\lambda - \sin^2\lambda)XP_Y.$$

The off-diagonal term XP_Y vanishes if we put

$$\sin^2\lambda - \cos^2\lambda + B\sin\lambda\cos\lambda = 0, \quad \text{that is,} \quad \tan 2\lambda = \frac{2}{B}.$$

After a few simple transformations, the Hamiltonian function is reduced to the form

$$H = \frac{1}{2m}\left(P_X^2 + P_Y^2\tan^2\lambda\right) + \tfrac{1}{2}m\omega^2\left(X^2\cot^2\lambda + Y^2\right). \tag{1}$$

The variables X and Y thus perform harmonic oscillations with frequencies which are, respectively, equal to

$$\omega_1 = \omega\cot\lambda = \sqrt{\omega^2 + \left(\frac{eB}{2mc}\right)^2} + \frac{eB}{2mc},$$

$$\omega_2 = \omega\tan\lambda = \sqrt{\omega^2 + \left(\frac{eB}{2mc}\right)^2} - \frac{eB}{2mc}$$

(cf. [2], § 21, problem). Each of the coordinates X and Y corresponds to a motion along an ellipse; an arbitrary oscillation is a superposition of two such motions (cf. problems 6.37 and 11.8).

It is interesting to note that when $B \to 0$, $\lambda = \pi/4$ (and not $\lambda = 0$). This means that even for very weak field B the "normal" oscillations is "circularly polarized". On the other hand, oscillations corresponding to the coordinates X or Y with $\lambda = 0$, which if there were no field B would be linear polarized, slowly change their direction of polarization, as soon as there is a field B present.

If the magnetic field is variable, we must add to the Hamiltonian function (1) the partial derivative with respect to the time of the generating function

$$\Phi = - \left(m\omega xy + \frac{P_X P_Y}{m\omega} \right) \tan\lambda + \frac{x P_X + y P_Y}{\cos\lambda}$$

(expressing it in terms of X, Y, P_X, and P_Y; see also the footnote to problem 13.26).

11.10. Putting into the canonical transformation of the preceding problem

$$\omega = \omega_2, \quad \tan 2\lambda = \frac{2\omega_B \omega_2}{\omega_B^2 + \omega_1^2 - \omega_2^2}, \quad \omega_B = \frac{eB}{mc},$$

we get

$$H' = \frac{1}{2m} \left(P_X^2 + \frac{\Omega_2^2}{\omega_2^2} P_y^2 + p_z^2 \right) + \tfrac{1}{2} m \left(\Omega_1^2 X^2 + \omega_2^2 Y^2 + \omega_3^2 z^2 \right),$$

where $\Omega_{1,2}$ was defined in problem 6.37.

11.11. The transformation ($\lambda = \pi/4$)

$$q_{s1} = \frac{1}{\sqrt{2}} \left(X_s + \frac{P_{Ys}}{Nm\omega_s} \right), \quad q_{s2} = \frac{1}{\sqrt{2}} \left(Y_s + \frac{P_{Xs}}{Nm\omega_s} \right)$$

leaves the form of the Hamiltonian function

$$H = \frac{p_0^2}{2Nm} + \sum_{s=1}^{R} \left[\frac{p_{s1}^2 + p_{s2}^2}{2Nm} + \tfrac{1}{2} Nm\omega_s^2 (q_{s1}^2 + q_{s2}^2) \right] = $$

$$= \frac{p_0^2}{2Nm} + \sum_{s=1}^{R} \left[\frac{P_{Xs}^2 + P_{Ys}^2}{2Nm} + \tfrac{1}{2} Nm\omega_s^2 (X_s^2 + Y_s^2) \right].$$

invariant (cf. problem 11.8). The oscillation corresponding to $X_s = A\cos(\omega_s t + \beta)$ is

$$x_n = \frac{A}{\sqrt{2}}\sin(\omega_s t + n\varphi_s + \beta),$$

and the oscillation corresponding to $Y_s = B\cos(\omega_s t + \beta)$ is

$$x_n = \frac{B}{\sqrt{2}}\sin(-\omega_s t + n\varphi_s - \beta).$$

11.12. The new Hamiltonian function is $H' = \omega P_1$, and in the new variables, the equations of motion have the form

$$\dot{P}_1 = \dot{P}_2 = \dot{Q}_2 = 0, \quad \dot{Q}_1 = \omega.$$

What is the change in the Hamiltonian function H', if $\mathbf{B}$ depends on the time?

11.13. The transformation is $p = \alpha P$, $r = Q/\alpha$, which is a similarity transformation.

11.14. The gauge transformation

$$\mathbf{A}' = \mathbf{A} + \nabla f(\mathbf{r}, t), \quad \varphi' = \varphi - \frac{1}{c}\frac{\partial f(\mathbf{r}, t)}{\partial t}$$

can be written as a canonical transformation,

$$\mathbf{R} = \mathbf{r}, \quad \mathbf{P} = \mathbf{p} + \frac{e}{c}\nabla f, \quad H' = H - \frac{e}{c}\frac{\partial f}{\partial t},$$

if one uses the generating function

$$\Phi(\mathbf{r}, \mathbf{P}, t) = \mathbf{r}\mathbf{P} - \frac{e}{c}f(\mathbf{r}, t).$$

11.15. $\Phi(q, P) = qP - F(q, t)$.

11.16. b) $F_\tau(q, Q) = -\frac{1}{2}F\tau(q + Q) - \frac{m}{2\tau}(q - Q)^2$;

c) $F_\tau(q, Q) = \frac{m\omega}{2\sin\omega\tau}[2qQ - (q^2 + Q^2)\cos\omega\tau]$.

11.17. a) $\mathbf{Q} = \mathbf{r} + \delta\mathbf{a}$, $\mathbf{P} = \mathbf{p}$: a shift of the systems as a whole over $\delta\mathbf{a}$ (or a shift of the coordinate system over $-\delta\mathbf{a}$).

b) Up to and including first-order terms, we have

$$\mathbf{Q} = \mathbf{r} + [\delta\boldsymbol{\varphi}, \mathbf{r}], \quad \mathbf{P} = \mathbf{p} + [\delta\boldsymbol{\varphi}, \mathbf{p}].$$

The transformation is a rotation of the coordinate system over an angle $-\delta\boldsymbol{\varphi}$.

c) $Q(t) = q(t + \delta\tau)$, $P(t) = p(t + \delta\tau)$, $H'(P, Q, t) = H(p, q, t + \delta\tau)$. The transformation is a shift in the time by $\delta\tau$ (cf. [1], § 45, [7] § 36.2).

d) $\mathbf{Q} = \mathbf{r} + 2\mathbf{p}\,\delta\alpha$, $\mathbf{P} = \mathbf{p} - 2\mathbf{r}\,\delta\alpha$.

The transformation is a rotation over an angle $2\delta\alpha$ in each of the $x_i p_i$-planes
$(i = 1, 2, 3)$ in phase space.

11.19. a) $\Phi(\mathbf{r}, \mathbf{P}) = \mathbf{r}\mathbf{P} + \mathbf{n}\mathbf{P}\delta a + \mathbf{n}\,[\mathbf{r}, \mathbf{P}]\,\delta\varphi$, where δa is the displacement along the
direction of $\mathbf{n}$ while $\delta\varphi = \frac{2\pi}{h}\,\delta a$ is the angle of rotation around $\mathbf{n}$; (h is the pitch of the
screw);

b) $\Phi(\mathbf{r}, \mathbf{P}, t) = \mathbf{r}\mathbf{P} - \mathbf{V}\mathbf{P}t + m\mathbf{r}\mathbf{V}$;

c) $\Phi(\mathbf{r}, \mathbf{P}, t) = \mathbf{r}\mathbf{P} - t\delta\boldsymbol{\Omega}\,[\mathbf{r}, \mathbf{P}]$.

11.20. $\delta f(q, p) = \lambda\{W, f\}_{p,q}$. Indeed, substituting the values of the new variables,

$$P = p - \lambda\frac{\partial W}{\partial q}, \quad Q = q + \lambda\frac{\partial W}{\partial P},$$

into $f(Q, P)$ and expanding the expression obtained in powers of λ, we get up to and
including first-order terms

$$\delta f(q, p) = \lambda\frac{\partial f}{\partial q}\frac{\partial W(q, p)}{\partial p} - \lambda\frac{\partial f}{\partial p}\frac{\partial W(q, p)}{\partial q}.$$

11.21. Putting $\Phi = \mathbf{r}\mathbf{P} + \lambda\mathbf{r}\mathbf{p}$ in the preceding problem, we get a similarity transformation with $\alpha = 1 + \lambda$ (see problem 11.13). The given Hamiltonian function is such that

$$H'(\mathbf{P}, \mathbf{Q}) = \alpha^{-2}H(\mathbf{P}, \mathbf{Q}),$$

and therefore

$$\lambda\{H, \mathbf{r}\mathbf{p}\} = H' - H = -2\lambda H(\lambda \to 0).$$

On the other hand, $\{H, \mathbf{r}\mathbf{p}\} = \frac{d}{dt}(\mathbf{r}\mathbf{p})$, and hence

$$\mathbf{r}\mathbf{p} - 2Et = \text{const}$$

(cf. problem 4.15b).

11.23. Let $\delta_1 q$ and $\delta_1 p$ be the changes in the coordinates and momenta connected with
the transformation defined by Φ_1.[1] Then we have

---

[1] Let us as an example indicate the change in the momentum up to second-order terms:

$$\delta_1 p = P - p = -\lambda_1\frac{\partial W_1(q, P)}{\partial q} = -\lambda_1\frac{\partial W_1(q, p)}{\partial q} + \lambda_1^2\frac{\partial^2 W_1(q, p)}{\partial p\partial q}\frac{\partial W_1(q, p)}{\partial q}.$$

$$f(q + \delta_1 q, p + \delta_1 p) = f(q, p) + \lambda_1 \{W_1(q, p), f(q, p)\} + \lambda_1^2 \varphi_1(q, p). \tag{1}$$

We now apply to each of the terms of the right-hand side of (1) another transformation, defined by the function Φ_2:

$$\begin{aligned} f(q + \delta_{21} q, p + \delta_{21} p) &= f + \lambda_2 \{W_2, f\} + \lambda_1 \{W_1, f\} + \\ &+ \lambda_1 \lambda_2 \{W_2, \{W_1, f\}\} + \lambda_1^2 \varphi_1 + \lambda_2^2 \varphi_2. \end{aligned} \tag{2}$$

The transformation of $\lambda_1^2 \varphi_1(q, p)$ gives a correction of higher than the second order. If we apply these transformations in the reverse order, the result is

$$\begin{aligned} f(q + \delta_{12} q, p + \delta_{12} p) &= f + \lambda_1 \{W_1, f\} + \lambda_2 \{W_2, f\} + \\ &+ \lambda_1 \lambda_2 \{W_1, \{W_2, f\}\} + \lambda_1^2 \varphi_1 + \lambda_2^2 \varphi_2 \end{aligned} \tag{3}$$

The difference between (2) and (3) lies only in the second-order terms which are proportional to $\lambda_1 \lambda_2$. Subtracting (3) from (2), we obtain

$$\lambda_1 \lambda_2 (\{W_2, \{W_1, f\}\} - \{W_1, \{W_2, f\}\}) = \lambda_1 \lambda_2 \{f, \{W_1, W_2\}\}.$$

Therefore, we see that, in particular, shifts $\lambda W = \delta \mathbf{aP}$ (see problem 11.17) commute, while this is not the case for rotations around different axes, $\lambda W = \delta \boldsymbol{\varphi} \, [\mathbf{r}, \mathbf{P}]$.

Is the statement, which is inverse of the one in the problem, correct?

11.24. A canonical transformation with a variable parameter λ can be considered to be a "motion" where λ plays the role of the time and $W(q, p)$ plays role of the Hamiltonian function (cf. problem 11.1c). The equations of "motion" are

$$\frac{dQ}{d\lambda} = \frac{\partial W(Q, P)}{\partial P}, \quad \frac{dP}{d\lambda} = -\frac{\partial W(Q, P)}{\partial Q}.$$

One can also easily obtain these equations formally from the result of problem 11.20.

a) The infinitesimal change in the coordinates and momenta under the given canonical transformation has the form

$$\delta \mathbf{r} = -\frac{\lambda}{N} \{W, \mathbf{r}\} = \frac{\lambda}{N} \{\mathbf{Ma}, \mathbf{r}\} = -[\mathbf{n}, \mathbf{r}] \, \delta \varphi,$$

$$\delta \mathbf{p} = -[\mathbf{n}, \mathbf{p}] \, \delta \varphi,$$

where

$$\mathbf{M} = [\mathbf{r}, \mathbf{p}], \; \mathbf{n} = \frac{\mathbf{a}}{a}, \, \delta \varphi = -\frac{\lambda a}{N}.$$

This transformation is a rotation of the coordinate system over an angle $\delta\varphi$ around the direction of $\mathbf{n}$. If we take the z-axis along $\mathbf{a}$, we finally obtain

$$X = x\cos\varphi + y\sin\varphi, \quad Y = y\cos\varphi - x\sin\varphi, \quad Z = z$$

and similar equations for the components of the momentum.

b) The infinitesimal change in the coordinates and momenta under the canonical transformation given by A_1 has the form

$$\delta x = p_x\,\delta\varphi, \quad \delta y = -p_y\,\delta\varphi, \quad \delta p_x = -x\,\delta\varphi, \quad \delta p_y = y\,\delta\varphi,$$

where $\delta\varphi = \lambda/(2N)$. This transformation represents a rotation over an angle $+\delta\varphi$ in the xp_x-plane and over an angle $-\delta\varphi$ in the yp_y-plane. Therefore, we have

$$\begin{aligned} X &= x\cos\varphi + p_x\sin\varphi, & Y &= y\cos\varphi - p_y\sin\varphi, \\ P_X &= -x\sin\varphi + p_x\cos\varphi, & P_Y &= y\sin\varphi + p_y\cos\varphi. \end{aligned}$$

Similarly, A_2 and A_3 represent the rotation over an angle φ or $-\varphi$ in the xp_y- and yp_x-planes and in the xy- and p_xp_y-planes, respectively, while A_4 represents the rotation over an angle 4φ in the xp_x- and yp_y-planes.

Not every rotation in phase space is the canonical transformation. For example, rotation in the xp_y-plane is not a canonical transformation.

It is interesting to compare the motion of a two-dimensional isotropic harmonic oscillator (its Hamiltonian function is $H = \frac{1}{2}A_4$) and the motion of a particle in the xy-plane in the arbitrary potential field having an axial symmetry, $U(x^2 + y^2)$. In both cases, the integrals of motion are the angular momentum $2A_3$, the conservation of which is due to invariance of the system with respect to rotations around the z-axis. For the oscillator, in addition, there are integrals of motion A_1 and A_2, the conservation of which is connected with a "hidden" symmetry—with the invariance of the Hamiltonian function with respect to certain rotations in phase space. In this sense, the oscillator is similar to a particle in three-dimensional central field, for which there are three integral of motion $M_{x,y,z}$.

The presence of additional integrals of motion for the two-dimensional oscillator leads to fact that point (x, y, p_x, p_y) in phase space moves along a closed line, while for a particle in the arbitrary field $U(x^2 + y^2)$, the phase trajectory "fills" two-dimensional surface (see [1], § 52, [7] § 44).

11.25. a) The volume specified in momentum and in phase space does not change with time, but in coordinate space the volume is spread out. Thus, if at $t = 0$ the state of the system is represented by the rectangle $ABCD$ (Fig. 169), it goes over to the parallelogram $A'B'C'D'$ $(AD = A'D')$ after a time t, and the distance in the x-direction between the points A' and C' is equal to $\Delta x = \Delta x_0 + \frac{\Delta p_0}{m}t$. As time goes on, this parallelogram degenerates into a narrow, very long strip.

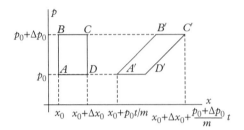

Figure 169

b) If there is a wall at the point $x = L$, the state of the system can no longer be represented by the parallelogram $A'B'C'D'$, but must have the form shown in Fig. 170a. When time marches on the initial phase volume, $ABCD$, changes into a number of very narrow parallel strips which are almost uniformly distributed inside two rectangles $0 \leqslant x \leqslant L, p_0 \leqslant p \leqslant p_0 + \Delta p_0$ and $0 \leqslant x \leqslant L, -p_0 - \Delta p_0 \leqslant p \leqslant -p_0$ (Fig. 171b).

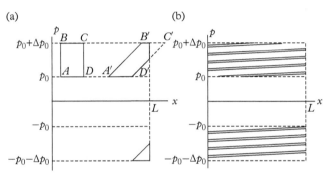

Figure 170

c) The phase orbit of an oscillator with energy E and frequency ω is the ellipse $\frac{x^2}{a^2} + \frac{p^2}{b^2} = 1$ with semi-axes $a = \sqrt{\frac{2E}{m\omega^2}}$, $b = \sqrt{\frac{2E}{m}}$. All points of the specified phase volume move along such ellipsis and return to their initial state after a period $T = 2\pi/\omega$. The dimension of the specified "volume" in coordinates space, Δx, and in momentum space, Δp, oscillate with frequency 2ω. In contrast to the case sub 11.25b there is no spreading out of the specified phase volume into the whole of the available phase space.

d) For an oscillator with friction (friction force $F_{\text{fr}} = -2m\lambda\dot{x}$), we have

$$x = ae^{-\lambda t}\cos(\omega t + \varphi),$$

$$p = m\dot{x} = -mae^{-\lambda t}[\omega\sin(\omega t + \varphi) + \lambda\cos(\omega t + \varphi)],$$

and the oscillations are damped so that the phase orbit is a spiral,

$$\frac{x^2}{a^2} + \left(\frac{p + \lambda mx}{ma\omega}\right)^2 = e^{-2\lambda t}.$$

The specified phase volume decreases until it vanishes. The non-conservation of phase volume here is connected with the fact that the system is not canonical: to describe it fully, we need know not only the Lagrangian function $L = \frac{1}{2}m\left(\dot{x}^2 - \omega_0^2 x^2\right)$ but also the dissipative function $F = \frac{1}{2}m\lambda\dot{x}^2$ (see [1], § 25).

If we choose for this system the "Lagrangian function" in the form

$$L' = \tfrac{1}{2}me^{2\lambda t}\left(\dot{x}^2 - \omega_0^2 x^2\right)$$

(cf. problem 4.20), the phase volume specified will be conserved for the appropriate canonical variables x and $p' = \dfrac{\partial L'}{\partial \dot{x}}$; however, in this case the generalized momentum $p' = m\dot{x}e^{2\lambda t}$ will not have a simple physical meaning, as before.

e) Since the period of the motion in this case depends on the energy, the phase space volume is spread out with time, "filling" the whole of the available region of phase space (cf. sub 11.25b).

Let the initial specified region be

$$x_0 < x < x_0 + \Delta x, \quad p_0 < p < p_0 + \Delta p.$$

One can easily estimate the time which is such that during that period the fastest particles have made one more oscillation than the slowest ones:

$$\tau \sim \frac{T^2}{\Delta T}, \qquad \Delta T \sim \frac{dT}{dE}\Delta E, \qquad \Delta E \sim \frac{p_0 \Delta p}{m} + \left|\frac{dU(x_0)}{dx}\right|\Delta x.$$

f) Let there be N particles such that the points in phase space which represent their state are at time $t = 0$ distributed with a density $Nw(x_0, p_0, 0)$ and that they move according to the equations

$$x = f(x_0, p_0, t),$$
$$p = \varphi(x_0, p_0, t). \tag{1}$$

Here

$$f(x_0, p_0, t) = x_0 + \frac{p_0}{m}t, \quad \varphi(x_0, p_0, t) = p_0$$

for free movement and

$$f(x_0, p_0, t) = x_0 \cos \omega t + \frac{p_0}{m\omega} \sin \omega t,$$

$$\varphi(x_0, p_0, t) = -m\omega x_0 \sin \omega t + p_0 \cos \omega t$$

for harmonic oscillators. The number of particles in the specified region of phase space, all points of which move according to the same law, remains constant; in particular, for an infinitesimal phase volume $dx\,dp$, we have

$$Nw(x, p, t)\,dx\,dp = Nw(x_0, p_0, 0)\,dx_0\,dp_0.$$

According to the Liouville theorem (see [1], §46 and [7], §40) $\dfrac{\partial(x,p)}{\partial(x_0,p_0)} = 1$, and therefore

$$w(x, p, t) = w(x_0, p_0, 0). \tag{2}$$

Using (1) to get expressions for x_0 and p_0,

$$x_0 = f(x, p, -t), \qquad p_0 = \varphi(x, p, -t)$$

and substituting this into (2), we obtain

$$w(x, p, t) = w(f(x, p, -t), \varphi(x, p, -t), 0),$$

or

$$w(x, p, t) = \frac{\exp[-\alpha(x-X)^2 - \beta(x-X)(p-P) - \gamma(p-P)^2]}{2\pi\,\Delta p_0 \Delta x_0},$$

where $X = f(X_0, P_0, t)$, $P = \varphi(X_0, P_0, t)$ while the coefficients α, β, γ for free particles are

$$\alpha = \frac{1}{2(\Delta x_0)^2}, \quad \beta = -\frac{t}{m(\Delta x_0)^2},$$

$$\gamma = \frac{1}{2(\Delta p_0)^2} + \frac{t^2}{2m^2(\Delta x_0)^2},$$

and for oscillators,

$$\alpha = \frac{\cos^2 \omega t}{2(\Delta x_0)^2} + \frac{m^2 \omega^2 \sin^2 \omega t}{2(\Delta p_0)^2},$$

$$\gamma = \frac{\cos^2 \omega t}{2(\Delta p_0)^2} + \frac{\sin^2 \omega t}{2m^2 \omega^2 (\Delta x_0)^2},$$

$$\beta = \sin \omega t \cos \omega t \left(\frac{m\omega}{(\Delta p_0)^2} - \frac{1}{m\omega (\Delta x_0)^2} \right).$$

We show in Figs. 171 and 172 how regions in phase space in which

$$2\pi \, \Delta x_0 \Delta p_0 \, w(x, p, t) \geqslant \tfrac{1}{2}$$

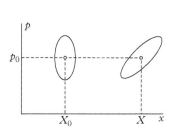

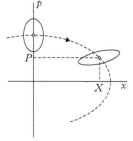

Figure 171   Figure 172

(for free particles and for harmonic oscillators, respectively) move about. These regions are ellipses which are deformed as time marches on.[1] Their centres are displaced according to the same law (1) as the particles. In the case of free particles, this ellipse is spread out without limit, but in the case of the oscillators, it only pulsates. We note that the distributions in coordinate and in momentum space are no longer independent: $w(x, p, t)$ cannot be split into two factors of the form $w_1(x, t) \cdot w_2(p, t)$.

It is interesting to consider the coordinate distribution function (independent of the values of the momenta)

$$w(x, t) = \int\limits_{-\infty}^{\infty} w(x, p, t) \, dp$$

---

[1] If the scales along the p- and x-axes in the phase space of the harmonic oscillators are chosen such that $m\omega = 1$, the phase orbits are circles, and the specified region in phase space rotates around the origin without being deformed.

or the momentum distribution function

$$\tilde{w}(p, t) = \int_{-\infty}^{\infty} w(x, p, t)\, dx.$$

These distributions turn out to be Gaussian with maxima at X and P, respectively:

$$w(x, t) = \frac{1}{\sqrt{2\pi}\,\Delta x}\, e^{-\frac{(x-X)^2}{2(\Delta x)^2}},$$

$$\tilde{w}(p, t) = \frac{1}{\sqrt{2\pi}\,\Delta p}\, e^{-\frac{(p-P)^2}{2(\Delta p)^2}},$$

where for free movement

$$(\Delta x)^2 = (\Delta x_0)^2 + \frac{(\Delta p_0)^2}{m^2} t^2, \quad (\Delta p)^2 = (\Delta p_0)^2,$$

and for the oscillators

$$(\Delta x)^2 = (\Delta x_0)^2 \cos^2 \omega t + \frac{(\Delta p_0)^2}{m^2 \omega^2} \sin^2 \omega t,$$

$$(\Delta p)^2 = (\Delta p_0)^2 \cos^2 \omega t + m^2 \omega^2 (\Delta x_0)^2 \sin^2 \omega t.$$

11.26. a) $\{a^*, a\} = -i$, $H_0 = \omega a^* a$.

b) The variables P and Q are canonical because $\{P, Q\} = 1$. The generating function is determined from the equation

$$dF = p(x, Q)\, dx - P(x, Q)\, dQ$$

and equals

$$F(x, Q, t) = \tfrac{i}{2} m\omega x^2 + \tfrac{i}{2} Q^2 e^{-2i\omega t} - i\sqrt{2m\omega}\, xQe^{-i\omega t}.$$

The new Hamiltonian function is, therefore,

$$H_0'(Q, P) = H_0 + \frac{\partial F(x, Q, t)}{\partial t} = 0.$$

c) We note that the expression $-3Q^2 P^2/(2m^2\omega^2)$ is the only term in the expression

$$x^4 = \left(\frac{Qe^{-i\omega t} - iPe^{i\omega t}}{\sqrt{2m\omega}} \right)^4$$

which does not contain the time. Whence, the average Hamiltonian function is

$$\langle H'(Q, P)\rangle = -\frac{3\beta}{8m\omega^2}Q^2P^2.$$

In the following, we will omit the brackets $\langle\rangle$ denoting averaging over time. Now it is clear that $-iQP = |Q_0|^2 = |a|^2$ is the integral of motion.

The Hamiltonian equations are

$$\dot{Q} = -i\varepsilon Q, \quad \dot{P} = i\varepsilon P, \quad \varepsilon = \frac{3\beta|Q_0|^2}{4m\omega^2},$$

where

$$Q = Q_0 e^{-i\varepsilon t}, \qquad P = iQ_0^* e^{i\varepsilon t};$$

whence

$$x = \frac{1}{\sqrt{2m\omega}}(Q_0 e^{-I\omega't} + Q_0^* e^{i\omega't}) = x_0\cos(\omega't + \varphi).$$

The influence of the additive term δU is reduced to change of frequency

$$\omega' = \omega + \frac{3\beta|Q_0|^2}{4m\omega^2} = \omega + \frac{3\beta x_0^2}{8\omega}$$

d) The new Hamiltonian function

$$H'(Q, P, t) = m^2\omega^2\alpha\left(\frac{Qe^{-i\omega t} - iPe^{i\omega t}}{\sqrt{2m\omega}}\right)^4\frac{e^{4i\omega t} + e^{-4i\omega t}}{2}$$

after averaging is reduced to

$$\langle H'(Q, P)\rangle = \tfrac{1}{8}\alpha(Q^4 + P^4).$$

For the variable $\xi = -iQP = |a|^2$, proportional to the square of the oscillation amplitude, the equation of motion is

$$\dot{\xi} = \{\langle H'\rangle, \xi\} = -\tfrac{1}{2}i\alpha(P^4 - Q^4).$$

Taking into account that

$$\tfrac{1}{4}\alpha(Q^4 + P^4) = A = \text{const},$$

we find

$$\dot{\xi}^2 = -4A^2 + \alpha^2\xi^4.$$

Thus, the variable ξ changes in the same way as the coordinate of the particle (with mass equal to one) in the field $V(\xi) = -\frac{1}{2}\alpha^2\xi^4$ and with the energy $-2A^2$ (cf. problem 1.2). Note that the amplitude increases to infinity for a finite time (the so-called explosive increase in amplitude).

Of course, the use of the average Hamiltonian function is correct only for $|\dot{\xi}| \ll \omega\xi$, that is, when $\xi \ll \omega/\alpha$ (cf. problem 8.1).

11.27. We introduce the new variables:

$$a = \frac{m\omega x + ip_x}{\sqrt{2m\omega}}e^{i\omega t}, \quad b = \frac{m\omega_2 y + ip_y}{\sqrt{2m\omega_2}}e^{i\omega_2 t}, \quad c = \frac{m\omega_3 z + ip_z}{\sqrt{2m\omega_3}}e^{i\omega_3 t},$$

($\omega = \omega_2 + \omega_3$) and canonical conjugate mometa ia^*, ib^*, ic^*. New Hamiltonian function, averaged over periods $2\pi/\omega_{2,3}$, is

$$\langle H' \rangle = \varepsilon|a|^2 + \eta(a^*bc + ab^*c^*),$$

$$\varepsilon = \omega_1 - \omega, \quad \eta = \frac{\alpha}{4\sqrt{2m\omega\omega_2\omega_3}}.$$

The equations of motion

$$\dot{a} = -i\varepsilon a - i\eta bc,$$
$$\dot{b} = -i\eta ac^*,$$
$$\dot{c} = -i\eta ab^*$$

have the integrals of motion[1]

$$\langle H' \rangle = A, \quad |a|^2 + |b|^2 = B, \quad |a|^2 + |c|^2 = C.$$

The equation

$$\frac{d}{dt}|a|^2 = i\eta(ab^*c^* - a^*bc)$$

can be presented in the form more helpful for qualitative investigation:

$$\dot{\xi}^2 + V(\xi) = 0,$$

---

[1] Integrals B and C are called *integrals of Manley–Rowe*.

where $\xi = |a|^2$ and

$$V(\xi) = (A - \varepsilon\xi)^2 - 4\xi\eta^2(B - \xi)(C - \xi).$$

In the initial moment $c = 0$, therefore $a = \varepsilon C$, $B > C$ and

$$V(\xi) = (C - \xi)^2(\varepsilon^2 - 4\xi\eta^2) - 4\xi\eta^2(B - C)(C - \xi).$$

The function $V(\xi)$ for the cases $\varepsilon^2 < 4\eta^2 C$ and $\varepsilon^2 > 4\eta^2 C$ is shown in Fig. 173.

In the first case, the variable ξ performs oscillations; that is, the system experiences beats. The energy is periodically pumped back and forth between the x-oscillator and the y and z-oscillators. In the second case (i.e., when there are small initial amplitudes and a large "detuning" ε), the y- and z-oscillations are not excited.

For more details, see [19].

11.28. a) $\langle H'\rangle = \varepsilon|a|^2 + \mu|a|^4 + \eta(a^2 + a^{*2})$, where

$$\varepsilon = \omega - \gamma, \qquad \mu = \frac{3\beta}{8m\omega^2}, \qquad \eta = \tfrac{1}{8}\hbar\omega.$$

b) The equations of motion

$$-\dot{a} = i(\varepsilon + 2\mu|a|^2)a + 2i\eta a^*,$$
$$\dot{a}^* = i(\varepsilon + 2\mu|a|^2)a^* + 2i\eta a$$

have a constant solution

$$a_0 = 0, \qquad |a_1|^2 = \frac{2\eta - \varepsilon}{2\mu}.$$

When $\xi = |a|^2$, we get

$$\dot{\xi} = -2i\eta(a^{*2} - a^2).$$

Taking into account that

$$\langle H'\rangle = \varepsilon\xi + \mu\xi^2 + \eta(a^2 + a^{*2}) = C = \text{const},$$

we obtain

$$\dot{\xi}^2 + V(\xi) = 0,$$

where

$$V(\xi) = 4\eta^2[(a^{*2} + a^2)^2 - 4|a|^4] = 4(C - \varepsilon\xi - \mu\xi^2)^2 - 16\eta^2\xi^2.$$

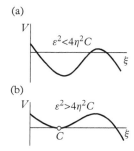

(a)

$\varepsilon^2 < 4\eta^2 C$

(b)

$\varepsilon^2 > 4\eta^2 C$

Figure 173

In the case considered, the value of C is small according to the initial conditions. In the resonance region $|\varepsilon| < 2\eta$, the graph of function $V(\xi)$ (Fig. 174) demonstrates that ξ performs the oscillations in the range from zero up to $\xi_m \approx 2|a_1|^2$.

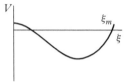

Figure 174

Thus, the transition to the stable oscillation $\xi = |a_1|^2$ (cf. problem 8.7) can be provided only by some previously unaccounted for mechanism, such as friction. This transition may be very long. Please note that this transition process has the character of beats even at zero "detuning", $\varepsilon = 0$, in contrast to the transition process in the linear oscillations (see problem 5.11).

11.29. The average Hamiltonian function is

$$\langle H'(Q, P, t)\rangle = H'(Q, P) = \tfrac{1}{2} m\omega^2 \left(\varepsilon - \tfrac{1}{4}h\right) Q^2 + \frac{1}{2m}\left(\varepsilon + \tfrac{1}{4}h\right) P^2.$$

Quantity $\sqrt{Q^2 + P^2/(m^2\omega^2)}$ represents the amplitude of the oscillation. The variables Q and P change very little over the period $2\pi/\gamma$, as can be easily seen from the Hamiltonian equations which contain small parameters ε and h.

On the Q, P-plane, a point representing the state of the system moves along the line $H'(Q, P) = C = \text{const}$. Families of such lines for area of the parametric resonance $|\varepsilon| < h/4$ and its neighbourhood $|\varepsilon| > h/4$ are shown in Fig. 175a and 175b, respectively. In the first case, the amplitude ultimately increases without limit; in the second case, it performs the bits (cf. problem 8.8).

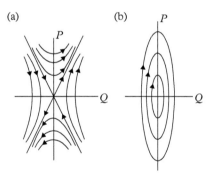

Figure 175

11.30. a) It is easy to check (cf. problem 11.4) that the given transformation is canonical.

At $V = 0$, the motion of the x-oscillator is represented by the motion of the point along the circumference in the $x, p_x/(m\omega_1)$-plane with the frequency ω_1. The radius of this circumference

$$a = \sqrt{x^2 + \frac{p_x^2}{m^2\omega_1^2}}$$

matches the amplitude of the oscillations along the x-axis. In the X, $P_X/(m\omega_1)$-plane this will be a fixed point $x = x(0)$, $P_X = p_x(0)$. Thus, the new variables are time-independent at $V = 0$ and, therefore, $H_0' = 0$.[1]

When $V \neq 0$, these variables are time-dependent, but because the new Hamiltonian function $H' = H_0' + V = V$ is small, the average motion in these variables is slow. Indeed, after averaging we have the new Hamiltonian function

$$\langle H' \rangle = -\frac{\beta}{4\omega_1\omega_2}\,(\omega_1 X P_Y - \omega_2 Y P_X),$$

and from the Hamiltonian equations

$$\dot{X} = \frac{\beta}{4\omega_1}Y, \quad \dot{Y} = -\frac{\beta}{4\omega_2}X,$$

we easily get

$$X = A\cos(\gamma t + \varphi), \quad Y = -\sqrt{\frac{\omega_1}{\omega_2}}A\sin(\gamma t + \varphi), \quad \gamma = \frac{\beta}{4\sqrt{\omega_1\omega_2}} \ll \omega_{1,2}.$$

Similarly, for the new momenta, we have

$$P_X = m\omega_1 B\cos(\gamma t + \psi), \quad P_Y = -m\sqrt{\omega_1\omega_2}B\sin(\gamma t + \psi).$$

Thus, in the $X, P_X/(m\omega_1)$-plane there occurs slow (with frequency γ) motion along an ellipse. This motion corresponds to the oscillations along the x-axis with slowly varying amplitude

$$a(t) = \sqrt{X^2 + (P_X/m\omega_1)^2} = \sqrt{A^2\cos^2(\gamma t + \varphi) + B^2\cos^2(\gamma t + \psi)},$$

that is, to the beats. Analogously, the oscillation amplitude along the y-axis is

$$b(t) = \sqrt{\frac{\omega_1}{\omega_2}}\sqrt{A^2\sin^2(\gamma t + \varphi) + B^2\sin^2(\gamma t + \psi)}.$$

---

[1] From the Hamiltonian equations for new variables (e.g. $\dot{X} = \partial H_0'/\partial P_X = 0$), it follows that H_0' is independent of them. Therefore, $H_0' = f(t)$, where $f(t)$ is an arbitrary time function, which we can put equal to zero without losing generality.

This shows that the energy of x- and y-oscillators $E_x = \frac{1}{2} m \omega_1^2 a^2(t)$ and $E_y = \frac{1}{2} m \omega_2^2 b^2(t)$, respectively, and their sum $E = E_x + E_y$ are not conserved. However, the value which can be called the total number of quanta is conserved:

$$n = \frac{E_x}{\hbar \omega_1} + \frac{E_y}{\hbar \omega_2} = \frac{m \omega_1}{2 \, hbar} C^2.$$

Here $C = \sqrt{A^2 + B^2}$, and $\hbar$ is the Planck constant.

In particular, at $\varphi = \psi = 0$ the beat amplitude reaches zero

$$x = X \cos \omega_1 t + \frac{P_X}{m \omega_1} \sin \omega_1 t = C \cos \gamma t \cos(\omega_1 t + \varphi_0),$$

$$y = -C \sqrt{\frac{\omega_1}{\omega_2}} \sin \gamma t \cos(\omega_2 t + \varphi_0), \quad \tan \varphi_0 = -\frac{B}{A},$$

and the energy oscillates with the frequency 2γ:

$$E = \frac{1}{2} m \omega_1 C^2 \left(\omega_1 \cos^2 \gamma t + \omega_2 \sin^2 \gamma t \right).$$

Note that even a weak coupling $|V| \ll H_0 = E$ leads to large changes in the energy $\Delta E \sim E$. Thus, when $\varphi = \psi = 0$ and $\omega_1 \gtrsim \omega_2$, we have

$$\Delta E = \frac{1}{2} m \omega_1 (\omega_1 - \omega_2) C^2 \sim \langle E \rangle = \frac{1}{4} m \omega_1 (\omega_1 + \omega_2) C^2.$$

Note that this problem coincides with problem 11.27 related to three interacting oscillators considered in the limit of the energy of the z-oscillator $E_z \gg E_{x,y}$ so high that the beats of the x- and y-oscillators have almost not effect on its motion

$$z = z_0 \sin \omega_3 t, \quad \omega_3 = \omega_1 - \omega_2.$$

In this case, $\beta = \frac{1}{2} \alpha z_0$, and $n \hbar$ is the same as one of the integrals of Manley–Rowe, namely, the integral B in notation of problem 11.27. The third oscillator plays the role of a large energy reservoir with which x- and y-oscillators exchange energy.

b) New canonical variables increase exponentially with time

$$X = Ae^{\gamma t} + Be^{-\gamma t}, \quad Y = \sqrt{\frac{\omega_1}{\omega_2}}\left(Ae^{\gamma t} - Be^{-\gamma t}\right),$$

$$P_X = m\omega_1\left(De^{\gamma t} + Fe^{-\gamma t}\right), \quad P_Y = -m\sqrt{\omega_1\omega_2}\left(De^{\gamma t} - Fe^{-\gamma t}\right),$$

that corresponds to the exponentially growing amplitude of the oscillations along the x- and y-axes. In this case, the difference in the number of quanta is conserved

$$\frac{E_x}{\hbar\omega_1} - \frac{E_y}{\hbar\omega_2} = \frac{2m\omega_1}{\hbar}(AB + DF).$$

§12

The Hamilton–Jacobi equation

12.2. It is clear that the trajectory is a curve in a plane. If we use polar coordinates, we can separate the variables in the Hamilton–Jacobi equation, provided we take the polar axis z along $\mathbf{a}$. The complete integral of the Hamilton–Jacobi equation is

$$S = -Et \pm \int \sqrt{\beta - 2ma\cos\theta}\, d\theta \pm \int \sqrt{2mE - \beta/r^2}\, dr. \tag{1}$$

To fix the signs in (1), we use the relations

$$p_r = m\dot{r} = \frac{\partial S}{\partial r} = \pm\sqrt{2mE - \frac{\beta}{r^2}}, \tag{2}$$

$$p_\theta = mr^2\dot{\theta} = \frac{\partial S}{\partial \theta} = \pm\sqrt{\beta - 2ma\cos\theta}. \tag{3}$$

On the initial part of the trajectory $\dot{r} < 0, \dot{\theta} > 0$ (we assume that the trajectory lies above the z-axis; see Fig. 176). We must thus take the upper sign in front of the first radical in (1) and the lower sign in front of the second radical. $\dfrac{\partial S}{\partial \beta} = B$ is the equation of the trajectory:

$$\int\limits_{0}^{\theta} \frac{d\theta}{\sqrt{\beta - 2ma\cos\theta}} + \int\limits_{\infty}^{r} \frac{dr}{r^2\sqrt{2mE - \beta/r^2}} = B. \tag{4}$$

The lower limits of the integrals can be chosen arbitrarily as long as the constant B is not determined. From our choice of lower limits and the condition that $\theta \to 0$ as $r \to \infty$, it follows that $B = 0$.

The constant β is an integral of motion of our problem, and from (3), we have

$$\beta = p_\theta^2 + 2ma\cos\theta.$$

Exploring Classical Mechanics: A Collection of 350+ Solved Problems for Students, Lecturers, and Researchers. First Edition.
Gleb L. Kotkin and Valeriy G. Serbo, Oxford University Press (2020). © Gleb L. Kotkin and Valeriy G. Serbo 2020.
DOI: 10.1093/oso/9780198853787.001.0001

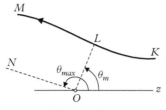

Figure 176

It can be expressed in terms of the particle parameters when $r \to \infty$ and $\theta \to 0$, that is, before the collision, when $p_\theta = mv\rho$ (ρ is the impact parameter):

$$\beta = 2m(E\rho^2 + a), \quad E = \tfrac{1}{2}.mv^2$$

When r changes from ∞ to

$$r_m = \sqrt{\frac{\beta}{2mE}} = \sqrt{\rho^2 + \frac{a}{E}},$$

which is determined by the condition $p_r = 0$, θ changes from zero to a value θ_m which is such that

$$\int\limits_{0}^{\theta_m} \frac{d\theta}{\sqrt{\beta - 2ma\cos\theta}} + \int\limits_{\infty}^{r_m} \frac{dr}{r^2\sqrt{2mE - \beta/r^2}} = 0. \tag{5}$$

A further increase in θ is accompanied by an increase in r; p_r then changes sign. The equation for the part LM of the orbit is

$$\int\limits_{\theta_m}^{\theta} \frac{d\theta}{\sqrt{\beta - 2ma\cos\theta}} - \int\limits_{r_m}^{r} \frac{dr}{r^2\sqrt{2mE - \beta/r^2}} = 0; \tag{6}$$

it is more helpful to use (5) and (6) and write it in the form:

$$\int\limits_{0}^{\theta} \frac{d\theta}{\sqrt{\beta - 2ma\cos\theta}} - \int\limits_{r_m}^{\infty} \frac{dr}{r^2\sqrt{2mE - \beta/r^2}} - \int\limits_{r_m}^{r} \frac{dr}{r^2\sqrt{2mE - \beta/r^2}} = 0. \tag{7}$$

At $r \to \infty$ the trajectory asymptotically approaches a straight line parallel to ON. The angle θ_{max} can be found from the equation[1]

$$\int_0^{\theta_{max}} \frac{d\theta}{\sqrt{\beta - 2ma\cos\theta}} = 2\int_{r_m}^{\infty} \frac{dr}{r^2\sqrt{2mE - \beta/r^2}} = \frac{\pi}{\sqrt{\beta}}. \tag{8}$$

The equation $\dfrac{\partial S}{\partial E} = A$ determines r as function of t. If we choose A such that $r(0) = r_m$, we get

$$r = \sqrt{v^2 t^2 + r_m^2} = \sqrt{\rho^2 + \frac{a}{E} + v^2 t^2}. \tag{9}$$

The integral over r in (4) and (7) can be evaluated elementarily, but those over θ reduce to elliptical integrals.

If $E\rho^2 \gg a$, we can expand the integrand in (4) and (7) in powers of $2ma/\beta \approx a/(E\rho^2)$. Up to and including first-order terms, we get

$$r\sin\theta = r_m\left(1 - \frac{a}{2E\rho^2}\cos\theta\right) \tag{10}$$

(see Fig. 177).

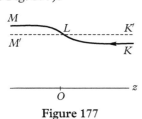

Figure 177

In this approximation, the angle over which the velocity of the particle is deflected after the scattering is zero. This can be explained by the fact that the action of the force along different sections of the orbit (which to first approximation is a straight line $K'M'$) is self-concealing.

12.3. a) To determine the angle over which the particle is deflected, we must expand in (8) of the preceding problem the powers of $a/(E\rho^2)$ up to the second order. We get the equation

$$\theta_{max} + \frac{ma}{\beta}\sin\theta_{max} + \frac{3}{4}\left(\frac{ma}{\beta}\right)^2\left(\theta_{max} + \frac{1}{2}\sin 2\theta_{max}\right) = \pi. \tag{1}$$

---

[1] We draw attention to the following method to avoid the calculation of the integral over r in (8). This integral is independent on a and must therefore equal the left-hand side of (8) also when $a = 0$. But in that case, clearly, $\theta_{max} = \pi$, and the integral over θ is trivial.

Solving this equation up to order $(ma/\beta)^2$, we find the angle of deflection[1]

$$\chi = \pi - \theta_{\max} = \frac{3}{4}\pi \left(\frac{ma}{\beta}\right)^2 = 3\pi \left(\frac{a}{4E\rho^2}\right)^2. \tag{2}$$

The scattering cross-section is

$$d\sigma = \pi |d\rho^2| = \frac{\sqrt{3\pi}\, a\, d\Omega}{16E\chi^{5/2}}. \tag{3}$$

The dependence on χ which we have obtained is the same as for small-angles scattering in the field γ/r^4, which decreases much faster than $U(\mathbf{r})$.

b) $\dfrac{d\sigma}{d\Omega} = \dfrac{\pi b}{8E\chi^3}$.

c) When $E\rho^2 \gg |b(\theta)|$, we have, instead of (10) of problems 12.2, the following expression for all θ:

$$\theta + \frac{m}{\beta}\int_0^\theta b(\theta)\,d\theta + \frac{3}{2}\frac{m^2}{\beta^2}\int_0^\theta b^2(\theta)\,d\theta = $$

$$= \begin{cases} \arcsin\frac{r_m}{r}, & \text{when} \quad 0 < \theta < \theta_m, \\ \pi - \arcsin\frac{r_m}{r}, & \text{when} \quad \theta_m < \theta < \theta_{\max}, \end{cases}$$

up to and including second-order terms. Here, the constant β is equal to $\beta = 2m[E\rho^2 + b(0)] \approx 2mE\rho^2$ and $r_m \approx \rho$.

If

$$\int_0^\pi b(\theta)\,d\theta = \pi \langle b \rangle \neq 0,$$

we can limit ourselves to the first approximation for which

$$\chi = \pi - \theta_{\max} = \frac{m}{\beta}\pi \langle b \rangle,$$

[1] We look for $\theta_{\max}$ in the form $\theta_{\max} = \theta_0 + \theta_1 + \theta_2 + \dots$, where $\theta_1 \sim (ma/\beta)\,\theta_0$. In zeroth approximation we get from (1) $\theta_0 = \pi$; in first approximation

$$\theta_1 + \left(\frac{ma}{\beta}\right)\sin\theta_0 = 0,$$

where $\theta_1 = 0$; in second approximation

$$\theta_2 + \frac{ma}{\beta}\theta_1 \cos\theta_0 + \frac{3}{4}\left(\frac{ma}{\beta}\right)^2 \left(\theta_0 + \frac{1}{2}\sin 2\theta_0\right) = 0,$$

whence follows (2).

and the small-angle scattering cross-section,

$$\frac{d\sigma}{d\Omega} = \frac{\pi \langle b \rangle}{4E\chi^3},$$

is the same as in the central field

$$U = \frac{\langle b \rangle}{r^2}.$$

If, however, $\langle b \rangle = 0$, we must take the second-order terms into account, and we get

$$\frac{d\sigma}{d\Omega} = \frac{\sqrt{3\pi}}{8E\chi^{5/2}}\sqrt{\tfrac{1}{2}\langle b^2 \rangle}.$$

12.4. a) We can separate the variables in the Hamilton–Jcobi equation if we choose the spherical coordinates with the z-axis parallel to **a**. The canonical momenta are

$$p_r = m\dot r = -\sqrt{2mE - \beta/r^2},$$
$$p_\theta = mr^2\dot\theta = \pm\sqrt{\beta - 2ma\cos\theta - p_\varphi^2/\sin^2\theta}, \qquad (1)$$
$$p_\varphi = mr^2\dot\varphi\sin^2\theta = \text{const.}$$

One finds the constant

$$\beta = p_\theta^2 + \frac{p_\varphi^2}{\sin^2\theta} + 2ma\cos\theta$$

easily by noticing that

$$p_\theta^2 + \frac{p_\varphi^2}{\sin^2\theta} = M^2,$$

where M is the total angular momentum of the particle; it is helpful to evaluate M for $r \to \infty$, $\theta \to \pi - \alpha$ (α is the angle between $\mathbf{v}_\infty$ and **a**); that is, before the collision,

$$\beta = 2m(E\rho^2 - a\cos\alpha).$$

According to (1), the particle can fall into the centre when $\beta < 0$ or

$$\rho^2 < \frac{a}{E}\cos\alpha. \qquad (2)$$

This is thus possible if $\alpha < \pi/2$ and in that case that the cross-section is

$$\sigma = \frac{\pi a}{E} \cos\alpha.$$

Averaging over all possible directions of **a** gives

$$\langle\sigma\rangle = \frac{1}{4\pi} \int\limits_{0}^{\pi/2} \frac{\pi a}{E} \cos\alpha \cdot 2\pi \sin\alpha \, d\alpha = \frac{\pi a}{4E}.$$

It is interesting that the area defined by condition (2) is a circle with centre on the axis of the particle beam, although the potential field is not symmetric with respect to that axis.
   b)

$$\sigma = \begin{cases} \dfrac{\pi a}{E} \cos\alpha - \dfrac{\pi \lambda^2}{4E^2}, & \text{when } 0 < \alpha < \alpha_m = \arccos \dfrac{\lambda^2}{4aE} \\ 0, & \text{when } \alpha_m < \alpha < \pi. \end{cases}$$

$$\langle\sigma\rangle = \frac{\pi a}{4E} \left(1 - \frac{\lambda^2}{4Ea}\right)^2.$$

   c)

$$\sigma = \begin{cases} \dfrac{\pi a}{E} \cos\alpha + 2\pi \sqrt{\dfrac{\gamma}{E}}, & \text{when } 0 < \alpha < \alpha_m = \pi - \arccos \dfrac{2\sqrt{\gamma E}}{a}, \\ 0, & \text{when } \alpha_m < \alpha < \pi. \end{cases}$$

$$\langle\sigma\rangle = \pi \left(\frac{a}{4E} + \sqrt{\frac{\gamma}{E}} + \frac{\gamma}{a}\right).$$

   d)

$$\sigma = -\frac{\pi b(\pi - \alpha)}{E},$$

provided $b(\pi - \alpha) < 0$.

12.5.

$$\sigma = \begin{cases} \pi R^2 + \dfrac{\pi a}{E} \cos\alpha, & \text{when } a\cos\alpha > -ER^2, \\ 0, & \text{when } a\cos\alpha < -ER^2, \end{cases}$$

where α is the angle between $\mathbf{v}_\infty$ and **a**.

12.6. a) We use the same notation as in problem 12.2. The equation for the initial part of the orbit $(r \to \infty, \theta \to \pi)$ is

$$\int_\theta^\pi \frac{d\theta}{\sqrt{\beta - 2ma\cos\theta}} = \int_r^\infty \frac{dr}{r^2\sqrt{2mE - \beta/r^2}}, \tag{1}$$

with

$$\beta = 2m(E\rho^2 - a). \tag{2}$$

When $\beta > 0$, the angle θ decreases when r changes from ∞ to r_m and then again increasing to ∞. The equation for the part of the orbit after r had passed through the minimum distance to the centre is

$$\int_\theta^\pi \frac{d\theta}{\sqrt{\beta - 2ma\cos\theta}} = \frac{\pi}{2\sqrt{\beta}} + \int_{r_m}^r \frac{dr}{r^2\sqrt{2mE - \beta/r^2}}. \tag{3}$$

It is clear that when $E\rho^2 \gg a$ the orbit (1) and (3) are the same as (11) in problems 12.2.

When $\beta < 0$, the particle can fall into the centre of the field (note that it follows from (2) that only values $\beta \geqslant -2ma$ are admissible.) In that case, r decreases monotonically from ∞ to 0. The angle θ decreases from π to the value θ_1 for which p_θ vanishes (section AB of the orbit; see Fig. 178). We then have $\beta - 2ma\cos\theta_1 = 0$. After that, the angle increases until it reaches the value $2\pi - \theta_1$ (section BC of the orbit)

$$\int_r^\infty \frac{dr}{r^2\sqrt{2mE - \beta r^{-2}}} = \int_{\theta_1}^\pi \frac{d\theta}{\sqrt{\beta - 2ma\cos\theta}} + \int_{\theta_1}^\theta \frac{d\theta}{\sqrt{\beta - 2ma\cos\theta}}. \tag{4}$$

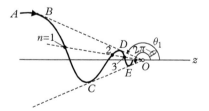

Figure 178

In point C, the momentum p_θ again changes sign, and θ decreases until it reaches the value θ_1 in point D; it then increases again, and so on.

The equation of the complete orbit can be written in the form

$$\int\limits_r^\infty \frac{dr}{r^2\sqrt{2mE-\beta r^{-2}}} = (-1)^n \int\limits_\theta^\pi \frac{d\theta}{\sqrt{\beta-2ma\cos\theta}} + 2n \int\limits_{\theta_1}^\pi \frac{d\theta}{\sqrt{\beta-2ma\cos\theta}}$$

$$(n=0,1,2,\ldots) \tag{5}$$

A single value of θ ($\theta_1 < \theta < 2\pi - \theta_1$) corresponds to an infinite number of values of r (n can take any non-negative integer value since the integral on the left-hand side of (5) increases without bound as $r \to 0$). Thus, the particle performs infinitely many oscillations between the straight lines BD and CE before it falls into the centre.

In the case of small impact parameters $E\rho^2 \ll a$, $\pi - \theta \ll 1$ so that we can write in (5) $\cos\theta \approx -1 + \frac{1}{2}(\pi-\theta)^2$. The final result is[1]

$$\theta = \pi - \rho\sqrt{\frac{2E}{a}}\sin\left[\frac{1}{\sqrt{2}}\operatorname{arsinh}\left(\frac{1}{r}\sqrt{\frac{a}{E}}\right)\right]. \tag{6}$$

The law of motion $r(t)$ is determined in the same way as in problem 12.2. When $\beta > 0$, the law is determined by (9) from problem 12.2. When $\beta < 0$, we have

$$r(t) = v\sqrt{t^2 - \tau^2}, \quad v = \sqrt{2E/m}, \quad -\infty < t < \tau = -\sqrt{|\beta|}/(2E), \tag{7}$$

and the particle falls into the centre at time τ.

b) If $\beta > 0$ ($E\rho^2 > a$), we have

$$\int\limits_\theta^\pi \frac{d\theta}{\sqrt{1 + \frac{2ma}{\beta}(1+\sin\theta)}} = \begin{cases} \arcsin\frac{r_m}{r}, & \text{when } \theta_m < \theta < \pi, \\ \pi - \arcsin\frac{r_m}{r}, & \text{when } \theta_{\min} < \theta < \theta_m. \end{cases}$$

If $\beta < 0$ ($E\rho^2 < a$), we have

$$\left(\int\limits_{\theta_1}^\pi \pm \int\limits_{\theta_1}^\theta + 2l\int\limits_{\theta_1}^{\theta_2}\right) \frac{d\theta}{\sqrt{\beta + 2ma(1+\sin\theta)}} = \int\limits_r^\infty \frac{dr}{r^2\sqrt{2mE-\beta/r^2}},$$

where l is the number of complete oscillations in angle (from θ_1 to θ_2 and back again) performed by the particle, and the ($\pm$) sign corresponds to counterclockwise (clockwise) motion (Fig. 179a), and

$$\theta_1 = -\arcsin\frac{E\rho^2}{a}, \qquad \theta_2 = \pi + \arcsin\frac{E\rho^2}{a}.$$

---

[1] $\operatorname{arsinh} x = \ln(x+\sqrt{1+x^2})$.

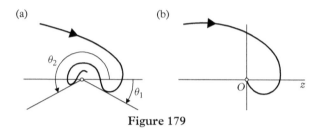

(a)   (b)

Figure 179

If $\beta = 0$ $(E\rho^2 = a)$, we have

$$r = \frac{\rho}{\sqrt{2F(\theta)}}, \quad F(\theta) = \ln \frac{\tan\left[\frac{1}{8}\pi + \frac{1}{4}(\pi - \theta)\right]}{\tan(\frac{1}{8}\pi)}, \quad -\frac{1}{2}\pi < \theta < \pi.$$

The particle moves along the orbit of Fig. 179b, and we have $r = -\sqrt{2E/m}\,t$ $(t < 0$; the particle falls into the centre at $t = 0)$.

12.7. The complete integral of the Hamilton–Jcobi equation is (see [1] § 48)

$$S = -Et + p_\varphi \varphi \pm \int \sqrt{\beta - 2ma\cos\theta - \frac{p_\varphi^2}{\sin^2\theta}}\, d\theta - \int \sqrt{2mE - \frac{\beta}{r^2}}\, dr.$$

The generalized momenta are the same as in problem 12.4a. The particle can fall into the centre if $\beta = 2m(E\rho^2 - a\cos\alpha) < 0$ (which is clearly satisfied if $\alpha^2 < 2E\rho^2/a \ll 1$). The equation for the orbit,

$$\varphi = \pm \int_{\pi - \alpha}^{\theta} \frac{p_\varphi\, d\theta}{\sin^2\theta \sqrt{\beta - 2ma\cos\theta - \dfrac{p_\varphi^2}{\sin^2\theta}}}, \tag{1}$$

$$\frac{r_0}{r} = \mp\sinh \int_{\pi - \alpha}^{\theta} \frac{\sqrt{|\beta|}\, d\theta}{\sin^2\theta \sqrt{\beta - 2ma\cos\theta - \dfrac{p_\varphi^2}{\sin^2\theta}}}, \qquad r_0^2 = \frac{|\beta|}{2mE} \tag{2}$$

can, in general, not be integrated to produce elementary functions. However, one can easily describe the motion qualitatively if one notes that (1) which gives a relation between the angles θ and φ is, apart from the notation, the same as the equation for the motion of a spherical pendulum (see [1], § 14, problem 1). The particle thus moves in such a way that the point where its radius vector intersects the surface of a sphere of radius l describes the same curve as does a spherical pendulum of length l, energy $\beta/(2ml^2)$, and angular momentum p_φ in the field of gravity $g = -a/(ml^3)$. This curve is enclosed between two "parallel" circles on the sphere corresponding to $\theta = \theta_1$ and $\theta = \theta_2$.

If $\alpha^2 < 2E\rho^2/a \ll 1$, one can easily integrate (1) and (2):

$$\theta = \pi - \sqrt{\varepsilon + \tfrac{1}{2}\alpha^2 - \left(\varepsilon - \tfrac{1}{2}\alpha^2\right)\cos\left(2\sqrt{\frac{ma}{|\beta|}}\,\text{arsinh}\,\frac{r_0}{r}\right)},$$

$$\theta = \pi - \frac{2\alpha\sqrt{\varepsilon}}{\sqrt{2\varepsilon + \alpha^2 + (2\varepsilon - \alpha^2)\cos 2\varphi}}, \quad \varepsilon = \frac{E\rho^2}{a}. \tag{3}$$

It is clear from (3) that a particle when falling into the centre moves in the region between two conical surfaces $\theta_1 \leqslant \theta \leqslant \theta_2$ rotating around the z-axis, while one complete rotation around the z-axis corresponds to two complete oscillations in the angle θ. In this approximation the orbit is closed for a spherical pendulum (it is an ellipse).

12.8. a) If the particle does not fall into the centre, the equation for a finite orbit is

$$\frac{p}{r} = 1 + e\cos f(\theta), \tag{1}$$

where

$$p = \frac{\beta}{ma}, \quad e = \sqrt{1 + \frac{2E\beta}{ma^2}}, \quad f(\theta) = \int_{\theta_1}^{\theta} \frac{d\theta}{\sqrt{1 - \frac{2ma}{\beta}\cos\theta}},$$

while the constants E and β satisfy the inequalities $-ma^2/(2\beta) < E < 0$ and $\beta > 0$.
    If $0 < \beta < 2ma$, the orbit "fills" the region $ABCDEF$ (Fig. 180),

$$r_1 \leqslant r \leqslant r_2, \quad r_{1,2} = \frac{p}{1 \pm e}, \quad \theta_1 \leqslant \theta \leqslant \theta_2, \quad \theta_1 = \arccos\frac{\beta}{2ma}, \quad \theta_2 = 2\pi - \theta_1;$$

that is, it approaches any point in this arbitrary closely.
    If $\beta = 2ma$,

$$f(\theta) = \sqrt{2}\ln\tan(\theta/4) + C_1, \tag{2}$$

and the orbit lies inside the ring $r_1 \leqslant r \leqslant r_2$ (Fig. 181).
    If $\beta > 2ma$, the orbit fills the ring $r_1 \leqslant r \leqslant r_2$. In particular, if $\beta \gg 2ma$, we have

$$f(\theta) = \theta + \zeta\sin\theta + \tfrac{3}{4}\zeta^2\theta + \tfrac{3}{8}\zeta^2\sin 2\theta + C_2, \tag{3}$$

where $\zeta = ma/\beta$. This is a slightly deformed ellipse, the nature of the deformation being determined by its orientation. Equation (3) is valid when $\arccos\zeta^{-2} \gtrsim \theta \gtrsim \arccos(-\zeta^{-2})$. It is interesting to make a comparison with the results of problem 2.28.

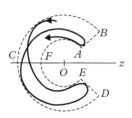

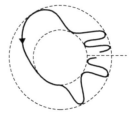

Figure 180   Figure 181

12.9. If the motion lies inside the ring $r_1 \leqslant r \leqslant r_2$, we have

$$\int_{\theta_1}^{\theta_1+2\pi} \frac{d\theta}{\sqrt{1 - \dfrac{2ma}{\beta}\cos\theta}} = 2\pi\frac{n}{l}.$$

If the motion lies in the region $r_1 \leqslant r \leqslant r_2$, $\theta_1 \leqslant \theta \leqslant \theta_2$, we have

$$\int_{\theta_1}^{\theta_2} \frac{d\theta}{\sqrt{1 - \dfrac{2ma}{\beta}\cos\theta}} = \pi\frac{n}{l}.$$

(n and l are integers.)

12.10. We can separate variables in the Hamilton–Jcobi equation, if we take the z-axis along the vector $\mathbf{a}$ (see [1],§ 48, (48.9)). The radial motion,

$$t = \sqrt{\frac{m}{2}} \int \frac{dr}{\sqrt{E + \dfrac{\alpha}{r} - \dfrac{\beta}{2mr^2}}},$$

is, when $\beta \geqslant 0$, the same as the motion of a particle in a Coulomb field $-\alpha/r$ with angular momentum β and energy E. When $\beta < 0$, the particle can fall into the centre. The equations of the orbit are $\dfrac{\partial S}{\partial p_\varphi} = $ const, $\dfrac{\partial S}{\partial \beta} = $ const. The first of these,

$$\varphi = \pm \int \frac{p_\varphi\, d\theta}{\sin^2\theta \sqrt{\beta - 2ma\cos\theta - \dfrac{p_\varphi^2}{\sin^2\theta}}}$$

is the same as the equation for the orbit of a spherical pendulum with energy $\beta/(2ml^2)$ and angular momentum $M_z = p_\varphi$ in the field of gravity $g = -a/(ml^3)$ (see [1], § 14, problem 1). The second equation connects r and θ. One can also use the analogy with a spherical pendulum for the analysis of that equation.

12.11. a) $|M_z| < \sqrt{mb/2}$.

  b) A finite orbit is possible for any value of M_z.

12.12. b) The complete integral of the Hamilton–Jacobi equation is (see [1], §48, problem 1)

$$S = -Et + p_\varphi\varphi + \int p_\xi(\xi)\,d\xi + \int p_\eta(\eta)\,d\eta,$$

where

$$p_\xi = \pm\sqrt{\tfrac{1}{2}m(E - U_\xi(\xi))}, \qquad U_\xi(\xi) = \frac{p_\varphi^2}{2m\xi^2} - \frac{m\alpha + \beta}{m\xi} - \tfrac{1}{2}F\xi,$$

$$p_\eta = \pm\sqrt{\tfrac{1}{2}m(E - U_\eta(\eta))}, \qquad U_\eta(\eta) = \frac{p_\varphi^2}{2m\eta^2} - \frac{m\alpha - \beta}{m\eta} + \tfrac{1}{2}F\eta.$$

The trajectory and the law of motion are determined by the equations

$$\frac{\partial S}{\partial \beta} = B, \quad \frac{\partial S}{\partial p_\varphi} = C, \quad \frac{\partial S}{\partial E} = A,$$

that is,

$$\int \frac{d\xi}{\xi p_\xi(\xi)} - \int \frac{d\eta}{\eta p_\eta(\eta)} = 4B, \quad \varphi - \frac{p_\varphi}{4}\int \frac{d\xi}{\xi^2 p_\xi(\xi)} - \frac{p_\varphi}{4}\int \frac{d\eta}{\eta^2 p_\eta(\eta)} = C,$$

$$-t + \frac{m}{4}\int \frac{d\xi}{p_\xi(\xi)} + \frac{m}{4}\int \frac{d\eta}{p_\eta(\eta)} = A. \tag{1}$$

(a)   (b)

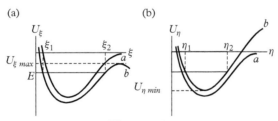

Figure 182

When studying the character of the motion, we must determine the region admissible for the values of ξ and η for given values of E, p_φ, and β. Fig. 182 gives the shape of of the effective potential energies $U_\xi(\xi)$ and $U_\eta(\eta)$.

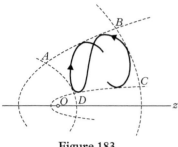

Figure 183

If $F = 0$, and when $-m\alpha < \beta < m\alpha$ (see curves 182a) and $E < 0$, the motion in both ξ and η is finite, but for $E > 0$, it is infinite. When a small force $F > 0$ appears, the curve $U_\xi(\xi)$ shows a maximum (see curves 182b); when $U_{\eta\,\min} < E < U_{\xi\,\max}$, the motion is finite, as before. The motion is restricted to the region $\xi_1 \leqslant \xi \leqslant \xi_2, \eta_1 \leqslant \eta \leqslant \eta_2$ (see Fig. 183) in the "ρz-plane", while the ρz-plane itself rotates around the z-axis with an angular velocity $\dot\varphi$. The orbit fills the region of space formed by rotating the figure $ABCD$ around the z-axis (see also problem 2.40). When $U_{\xi\,\max} < E$, the motion is infinite.

When F increases, the quantity $U_{\xi\,\max}$ decreases and $U_{\eta\,\min}$ increases. When $U_{\xi\,\max}$ becomes less than $U_{\eta\,\min}$, finite motion becomes impossible (when $\beta < -m\alpha + \frac{3}{2}(Fmp_\varphi^4)^{1/3}$, there are no $U_{\xi\,\max}$ extrema).

12.13. In elliptical coordinates, we have

$$\rho = \sigma\sqrt{(\xi^2 - 1)(1 - \eta^2)}, \quad z = \sigma\xi\eta, \quad \sigma = \sqrt{c^2 - a^2}.$$

The potential energy,

$$U = \begin{cases} \infty, & \text{when} \quad \xi > \xi_0 = c/\sigma, \\ 0, & \text{when} \quad \xi < \xi_0, \end{cases}$$

depends only on ξ, and we can separate the variables in the Hamilton–Jacobi equation (see [1], § 48).

The complete integral is

$$S = -Et + p_\varphi\varphi \pm \int \sqrt{2m\sigma^2 E + \frac{\beta - 2m\sigma^2 A(\xi)}{\xi^2 - 1} - \frac{p_\varphi^2}{(1 - \xi^2)^2}}\, d\xi \pm$$
$$\pm \int \sqrt{2m\sigma^2 E - \frac{\beta}{1 - \eta^2} - \frac{p_\varphi^2}{(1 - \eta^2)^2}}\, d\eta, \tag{1}$$

where

$$A(\xi) = (\xi^2 - \eta^2)U(\xi) = U(\xi).$$

For a particle flying through the origin, $p_\varphi = 0$. From the previous discussion, it follows that

$$p_\xi = \pm\sqrt{2m\sigma^2 E + \frac{\beta - 2m\sigma^2 A(\xi)}{\xi^2 - 1}} = \frac{m\sigma^2(\xi^2 - \eta^2)}{\xi^2 - 1}\dot{\xi}, \qquad (2)$$

$$p_\eta = \pm\sqrt{2m\sigma^2 E - \frac{\beta}{1 - \eta^2}} = \frac{m\sigma^2(\xi^2 - \eta^2)}{1 - \eta^2}\dot{\eta}. \qquad (3)$$

In the origin $(\eta = 0, \xi = 1)$, we have

$$\dot{\eta} = \pm\frac{\sqrt{2m\sigma^2 E - \beta}}{m\sigma^2}, \quad \dot{\xi} = 0, \quad \dot{z} = \sigma(\dot{\xi}\eta + \dot{\eta}\xi) = \sigma\dot{\eta}$$

and from the condition

$$\sqrt{\frac{2E}{m}}\cos\alpha = \frac{\sqrt{2m\sigma^2 E - \beta}}{m\sigma},$$

we find $\beta = 2m\sigma^2 E\sin^2\alpha$.

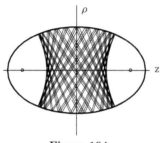

Figure 184

The region of the inaccessible values of η is determined by the condition

$$2m\sigma^2 E - \beta/(1 - \eta^2) < 0 \quad \text{or} \quad |\eta| > |\cos\alpha|.$$

The motion thus takes place in the region $|\eta| < |\cos\alpha|, 1 < \xi < \xi_0$ (see orbits in Fig. 184).

12.14. The transformation from the Cartesian coordinates to elliptical coordinates has the inverse transformation as the Jacobian of the transformation

$$\frac{\partial(x,y)}{\partial(\zeta,\varphi)} = c^2 D, \ D = \cosh^2\zeta - \cos^2\varphi = \sinh^2\zeta + \sin^2\varphi$$

vanishes only when $\zeta = 0$, $\varphi = \pm\pi$. This transformation is similar to the transition from the Cartesian coordinates to the polar ones. For a given value of ζ (and for all values of φ), the corresponding points lie on an ellipse

$$\frac{x^2}{c^2\cosh^2\zeta} + \frac{y^2}{c^2\sinh^2\zeta} = 1,$$

and for a given value φ (and for all values of ζ), the corresponding points lie on the part of the hyperbola (half of one branch)

$$\frac{x^2}{c^2\cos^2\varphi} - \frac{y^2}{c^2\sin^2\varphi} = 1.$$

The potential energy in the elliptic coordinates has a simple form

$$U(\zeta) = \begin{cases} 0, & \text{when } \zeta < \zeta_a, \\ \infty, & \text{when } \zeta > \zeta_a, \end{cases}$$

where the parameter ζ_a is related to the ellipse axes: $a = c\cosh\zeta_a$, $b = c\sinh\zeta_a$.

Taking into account that

$$\dot{y} = c(\dot\zeta\cosh\zeta\,\cos\varphi - \dot\varphi\sinh\zeta\,\sin\varphi), \quad \dot{x} = c(\dot\zeta\sinh\zeta\,\sin\varphi + \dot\varphi\cosh\zeta\,\cos\varphi),$$

$$\dot{x}^2 + \dot{y}^2 = c^2(\dot\zeta^2 + \dot\varphi^2)D,$$

we find the Lagrangian function of the system in the elliptic coordinates

$$L = \tfrac{1}{2}mc^2(\dot\zeta^2 + \dot\varphi^2)D - U(\zeta).$$

Next, we find the generalized momenta

$$p_\zeta = mc^2 D\dot\zeta, \quad p_\varphi = mc^2 D\dot\varphi$$

and the Hamiltonian function

$$H(p_\zeta, p_\varphi, \zeta, \varphi) = \frac{p_\zeta^2 + p_\varphi^2}{2mc^2 D} + U(\zeta).$$

The particle moves with a constant velocity, which changes direction after collisions with the wall indicated by the ellipse. In the collision, the angle of incidence is equal to the angle of reflection. Of course, the particle velocity and its direction of motion remain constant between the walls. However, the question remains as to which areas may be filled by the trajectory of the particle and which will be inaccessible to it (at given initial conditions). This is what we are going to find out using the Hamilton–Jacobi equation. We know in advance that these areas do not depend on the velocity value.

Since the Hamiltonian function does not have the evident dependence on time, we can use the equation for an abbreviated action $S(\zeta, \varphi, t) = -Et + S_0(\zeta, \varphi)$:

$$\frac{1}{2mc^2 D}\left(\frac{\partial S_0}{\partial \zeta}\right)^2 + \frac{1}{2mc^2 D}\left(\frac{\partial S_0}{\partial \varphi}\right)^2 + U(\zeta) = E. \tag{1}$$

Next we consider the region $\zeta < \zeta_a$ only where $U(\zeta) = 0$. In this region the variables in the Hamilton–Jacobi equation are separated:

$$S_0(\zeta, \varphi) = S_1(\zeta) + S_2(\varphi).$$

Indeed, (1) with $U(\zeta) = 0$ can be represented as

$$\left(\frac{dS_1(\zeta)}{d\zeta}\right)^2 - 2mc^2 E \cosh^2 \zeta = -\left(\frac{dS_2(\varphi)}{d\varphi}\right)^2 + 2mc^2 E \cos^2 \varphi.$$

The left-hand side of this equality is independent of φ, whereas the right-hand side is independent of ζ; therefore, each of them is a constant. Let's define such constant as $(-\beta)$. Thus,

$$S_1 = \pm \int \sqrt{2mc^2 E \cosh^2 \zeta - \beta} \, d\zeta, \quad S_2 = \pm \int \sqrt{\beta - 2mc^2 E \cos^2 \varphi} \, d\varphi.$$

The motion occurs in the region where $U = 0$, so $E > 0$; it is also clear from the form S_2 that $\beta > 0$. Thus, we obtain the complete integral of the Hamilton–Jacobie equation $S(\zeta, \varphi, t) = -Et + S_0(\zeta, \varphi, E, \beta)$, which depends on two arbitrary constants E and β. By differentiating S with respect to β and equating the result to the constant C, we get the equation for trajectory in the form $f(\zeta, \varphi, E, \beta, C) = 0$. The equation $\partial S/\partial E = t_0$ allows us to find the functions $\zeta(t)$ and $\varphi(t)$. However, we already have a clear idea of the trajectory (consisting of straight segments) and the law of motion. Generalized momenta p_ζ and p_φ can be found by differentiating S with respect to ζ and φ:

$$p_\zeta = \pm \sqrt{2mc^4 E \cosh^2 \zeta - \beta} = mc^2 D\dot{\zeta}, \tag{2}$$

$$p_\varphi = \pm \sqrt{\beta - 2mc^4 E \cos^2 \varphi} = mc^2 D\dot{\varphi}. \tag{3}$$

Constants E, β, C, t_0 are determined by the initial values of coordinates ζ, φ and velocities $\dot{\zeta}, \dot{\varphi}$.

If $\beta > 2mc^4 E$, then according to (3) p_φ will never equal zero at any value of φ. The sign of $\dot{\varphi}$ is conserved; the motion always occurs clockwise (or counterclockwise).

In accordance with (2), the condition $2mc^4 E \cosh^2 \zeta_m = \beta$ defines the minimum value ζ_m: $\zeta_m \leq \zeta \leq \zeta_a$. When ζ reaches the value ζ_a, the sign in (2) changes from positive to negative to correspond to the reflection from the wall. When ζ reaches the value ζ_m, the velocity $\dot\zeta$ becomes zero, and the negative sign in (2) is replaced by the positive one. In this case, the trajectory touches the ellipse which is limited to the area of motion (see orbits in Fig. 185)

$$\frac{x^2}{c^2 \cosh^2 \zeta_m} + \frac{y^2}{c^2 \sinh^2 \zeta_m} = 1.$$

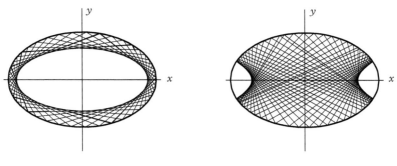

Figure 185   Figure 186

If $\beta < 2mc^4 E$, then according to (3) the function $\dot\varphi$ becomes zero for $\varphi = \varphi_m$ determined by the equation $2mc^4 E \cos^2 \varphi_m = \beta$. In this case, the sign in (3) changes. The coordinate ζ changes in the range $0 \leq \zeta \leq \zeta_a$; that is, it runs through all valid values. At the boundaries of this interval, the sign in (2) changes. The area of the particle motion is limited by both branches of the hyperbola (see orbits in Fig.186)

$$\frac{y^2}{c^2 \cos^2 \varphi_m} - \frac{x^2}{c^2 \sin^2 \varphi_m} = 1.$$

The initial values of Cartesian coordinates and velocity components allows us to find the constants

$$E = \tfrac{1}{2} mv^2, \quad \varphi_0 = 0, \quad c \cosh \zeta_0 = x_0,$$

$$\tan\alpha = \frac{v_y}{v_x} = \frac{\dot\zeta}{\dot\varphi} = \frac{p_\zeta}{p_\varphi} = \sqrt{\frac{2mc^4 E \cosh^2 \zeta_0 - \beta}{\beta}},$$

$$\beta = 2mc^4 E \cosh^2 \zeta_0 \tan^2 \alpha.$$

Each time $\dot\zeta$ or $\dot\varphi$ changes sign, the constants C and t_0 are updated, but E and β remain unchanged during the whole time of the particle motion.

Depending on whether the value $\cosh^2 \zeta_0 \tan^2 \alpha$ is larger or smaller the unit, the inaccessible area will be limited to an ellipse or a hyperbola. In the case when it equals a unit, there will be no inaccessible areas at all. It is also possible to have closed orbits.

12.15. One can separate the variables in the Hamilton–Jacobi equation using elliptical coordinates (see [1], § 48, problem 2 with $\alpha_1 = -\alpha_2 = \alpha$). For a particle coming from infinity along the z-axis, the constant $\beta = -2mE\rho^2 + 4m\alpha\sigma$, where ρ is the impact parameter.

At $\beta < 0$ the orbit is qualitatively the same as the orbit of a particle which is scattered in the field of a point dipole (see problem 12.6a).

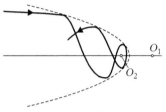

If $\beta > 0$, the particle "falls" onto the dipole (i.e., it passes in its motion through the section $O_1 O_2$) and then goes off again to infinity. Moreover, if $p_\eta(\eta_1) = 0$ when $\eta_1 < 0$, the particle moves in the region bounded by the hyperbole $\eta = \eta_1$ (see Fig. 187).

Figure 187

12.16. In the Hamilton–Jacobi equation,

$$\frac{\partial S}{\partial t} + \frac{1}{2m}\left[\left(\frac{\partial S}{\partial z}\right)^2 + \left(\frac{\partial S}{\partial r}\right)^2 + \left(\frac{1}{r}\frac{\partial S}{\partial \varphi} - \frac{e}{2c}B(z)r\right)^2\right] = 0, \tag{1}$$

we can separate the time and the angle φ:

$$S = -Et + p_\varphi \varphi + \widetilde{S}(r, z). \tag{2}$$

Considering in the following only orbits which intersect with the z-axis, we put $p_\varphi = 0$. It is not possible to separate the variables r and z, and we shall look for the integral approximately in the form of an expansion in r:

$$\widetilde{S}(r, z) = S_0(z) + r\psi(z) + \tfrac{1}{2}r^2\sigma(z) + \dots. \tag{3}$$

Since the radial momentum,

$$p_r = \frac{\partial S}{\partial r} = \psi(z) + r\sigma(z) + \dots, \tag{4}$$

for a particle flying along the z-axis (with $r = 0$) vanishes, for the particle beam considered we have $\psi(z) = 0$. Substituting (3) into (1) and comparing coefficients of the same powers of r, we obtain (cf. [2], § 56, problem 2)

$$S_0(z) = pz, \quad p = \sqrt{2mE}, \tag{5}$$

$$p\sigma'(z) + \sigma^2 + \frac{e^2}{4c^2}B^2(z) = 0. \tag{6}$$

Outside the lens (where $|z| > a$, $B(z) = 0$), we have from (6)

$$\sigma(z) = \frac{p}{z + C_1}, \quad \text{when} \quad z < -a, \tag{7}$$

$$\sigma(z) = \frac{p}{z + C_2}, \quad \text{when} \quad z > a. \tag{8}$$

The equation of the orbit,

$$\frac{\partial S}{\partial C_{1,2}} = -\frac{pr^2}{2(z + C_{1,2})^2} = B_{1,2} \tag{9}$$

is an equation of strait lines intersecting the z-axis in the points $-C_{1,2}$[1]; that is, $z_0 = -C_1$ and $z_1 = -C_2$. From (6), we get

$$p\sigma(a) - p\sigma(-a) + \int_{-a}^{a} \sigma^2 \, dz + \frac{e^2}{4c^2} \int_{-a}^{a} B^2(z) \, dz = 0. \tag{10}$$

Since $|z_{0,1}| \gg a$, it follows from (7) and (8) that

$$\sigma(\pm a) = -\frac{p}{z_{1,0}}. \tag{11}$$

Let us estimate $\int_{-a}^{a} \sigma^2 \, dz$. According to (6), $\sigma(z)$ is a monotonic function. Therefore, we have

$$\int_{-a}^{a} \sigma^2 \, dz \lesssim \frac{2ap^2}{z_{1,0}^2} \ll p\sigma(\pm a).$$

It thus follows from (10) that

$$\frac{1}{|z_0|} + \frac{1}{z_1} = \frac{e^2}{4c^2p^2} \int_{-a}^{a} B^2(z) \, dz = \frac{1}{f}. \tag{12}$$

The condition $|z_{0,1}| \gg a$ is, indeed, satisfied when $a \ll cp/(eB)$.

---

[1] When z is close to $C_{1,2}$, $\sigma \to \infty$ so that expansion (3) becomes inapplicable. However, the equations (9) for the orbits remain valid also for that region.

12.17. The whole of the calculations of the preceding problem up to (6) is applicable also in this problem. The substitution $\sigma = pf'/f$ reduces (6) to the form

$$\left(1 + \varkappa^2 z^2\right)^2 f''(z) + \frac{e^2 B^2}{4c^2 p^2} f = 0,$$

and after that the substitution

$$\varkappa z = \tan \xi, \quad -\tfrac{1}{2}\pi < \xi < \tfrac{1}{2}\pi, \quad f(z) = \frac{\eta(\xi)}{\cos \xi}$$

gives

$$\eta''(\xi) + \lambda^2 \eta(\xi) = 0,$$

where

$$\lambda^2 = 1 + \frac{e^2 B^2}{4c^2 \varkappa^2 p^2}.$$

Hence,

$$\sigma = \tfrac{1}{2} \varkappa p \sin 2\xi + \lambda \varkappa p \cos^2 \xi \cot(\lambda \xi + \alpha),$$

and the equation for the orbit becomes

$$\frac{\partial S}{\partial \alpha} = -\frac{\varkappa p r^2 \lambda \cos^2 \xi}{2 \sin^2(\lambda \xi + \alpha)} = B,$$

or

$$r \cos \xi = B' \sin(\lambda \xi + \alpha).$$

When $r = 0$,

$$\lambda \xi_n + \alpha = \pi n,$$

and thus $\alpha = -\lambda \arctan(\varkappa z_0)$ so that the points where the beam is focused are given by

$$\varkappa z_n = \tan\left(\arctan \varkappa z_0 + \frac{n\pi}{\lambda}\right).$$

Depending on the magnitude of λ, there will be one or several points of focusing.

12.18.

$$S(q, q_0, t, t_0) = f(q, \alpha(q, q_0, t, t_0), t) - f(q_0, \alpha(q, q_0, t, t_0), t_0),$$

where $f(q, \alpha, t)$ is the complete integral of the Hamilton–Jacobi equation, while the function $\alpha(q, q_0, t, t_0)$ is determined by the equation (or set of equations for the case of several degrees of freedom)

$$\frac{\partial f(q, \alpha, t)}{\partial \alpha} = \frac{\partial f(q_0, \alpha, t_0)}{\partial \alpha}.$$

§13

Adiabatic invariants

13.1. $E^2l = $ const.

Let us explain the answer. On the ring A there acts a force $\mathbf{F}$ determined by the tension T in the thread. For the small angles φ of deflection of a pendulum, we get

$$F_x = mg\varphi, \quad F_y = \tfrac{1}{2}mg\varphi^2$$

(the y-axis is directed vertically upwards; the x-axis is in the plane of oscillations.) Since the length of the thread $AB = l$ changes slowly, we can average the force over a period of the oscillation, $\varphi = \varphi_0 \cos\omega t$, $\omega = \sqrt{g/l}$, assuming the length of the thread to be constant. We get

$$F_x = 0, \quad F_y = \tfrac{1}{4}mg\varphi_0^2.$$

When the ring is displaced over a distance $dy = dl$, the energy decreases by $F_y dy = \tfrac{1}{4}mg\varphi_0^2 dl$. Since $E = \tfrac{1}{2}mgl\varphi_0^2$, we have

$$dE = -\frac{E}{2l}\,dl.$$

Hence, we have $E^2l = $ const.

13.2. After the particle has collided with both walls, its velocity v is changed by $2\dot{l}$. The condition that the change is slow means $|2\dot{l}| \ll v$.

We choose a time interval Δt such that

$$\frac{2l}{v} \ll \Delta t \ll \frac{l}{|\dot{l}|}.$$

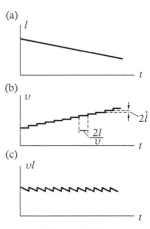

Figure 188

Exploring Classical Mechanics: A Collection of 350+ Solved Problems for Students, Lecturers, and Researchers. First Edition.
Gleb L. Kotkin and Valeriy G. Serbo, Oxford University Press (2020). © Gleb L. Kotkin and Valeriy G. Serbo 2020.
DOI: 10.1093/oso/9780198853787.001.0001

Such a Δt exists because of the slowness condition. During such a time interval there are $v\Delta t/(2l)$ pairs of collision with the walls, and the velocity is changed by

$$\Delta v = -vl\frac{\Delta t}{l}.$$

Integrating, we obtain $vl = \text{const}$ or $El^2 = \text{const}$.

It is interesting to study in somewhat more detail how the product vl changes. This is easily done by studying the functions $l(t)$ and $v(t)$ (see Fig. 188a and b). In Fig. 188c, we have drawn the function $I(t) = v(t)l(t)$. The quantity vl oscillates around the practically constant value $\langle vl \rangle$ while the amplitude of the oscillation has the relative value $\Delta I/I \sim \dot{l}/v$.

The deviation of $\langle vl \rangle$ from a constant value is of higher order of smallness:

$$\frac{d}{dt}\langle vl \rangle \sim \dot{l}^2.$$

13.3. The particle of low mass moves much faster than the piston. To determine the value of the particle velocity v, we assume that it moves between the slowly and smoothly shifting walls (bottom of the vessel and the piston). Ignoring the effect of gravity on the motion of this very fast particle, we can use the conservation of the adiabatic invariant $vX = \text{const} = C$ (see problem 13.2). To determine the law of motion of the piston, we can use the law of energy conservation

$$\tfrac{1}{2}M\dot{X}^2 + MgX + \tfrac{1}{2}mv^2 = E$$

(here we have neglected the contribution of the potential energy of a light particle). Substituting $v = C/x$, we obtain the equation for the motion of the piston:

$$\tfrac{1}{2}M\dot{X}^2 + U_{\text{eff}}(X) = E, \quad U_{\text{eff}}(X) = MgX + \frac{mC^2}{X^2},$$

Here $U_{\text{eff}}(X)$ is the effective potential energy of the piston, averaged over the period of motion of a light particle. The function $U_{\text{eff}}(X)$ has at a minimum at $X_0 = \left(\frac{2mC^2}{Mg}\right)^{1/3}$. The piston can oscillate near $X = X_0$ with frequency

$$\omega = \sqrt{\frac{U''_{\text{eff}}(X_0)}{M}} = \sqrt{\frac{3g}{X_0}}.$$

This problem demonstrates the so-called *adiabatic approximation*. We consider the motion of a system consisting of bodies that differ strongly in mass where light particles move much faster than the heavy ones. In this case, the coordinates of heavy particles can be considered as parameters defining the field in which the light particles are moving. The change of these parameters is sufficiently slow and smooth so that adiabatic invariants are conserved. Depending on the values of these parameters (i.e., on the coordinates of heavy particles), the energy of the light particles changes, and the source of these changes (positive or negative) is the kinetic energy of heavy particles. When studying the motion of heavy particles, this energy of the light particles can be added to the potential energy of the heavy particles. Of course, this implies averaging over time the motion of the light particles.

Such an approximation is used in studying the motion of atoms inside both a molecule and a crystal. Of course, in these problems the motion of particles, as well as light electrons and massive ions, is described not in terms of classical mechanics, but rather in terms of quantum mechanics. However, the essence of the adiabatic approximation is more helpful to demonstrate the approximation using a simple example from classical mechanics. The same question is dealt with in problem 13.9 where an extremely simplified model of a molecule is discussed.

13.4. If $g(t) = g - a(t)$ were constant, the motion of the ball would be described by

$$z(t) = h - \tfrac{1}{2} 2gt^2 \text{ for } -\sqrt{2h/g} < t < \sqrt{2h/g}.$$

A change in $g(t)$ by Δg leads to a change in the potential energy by $mz\Delta g$, and over a period by $m\langle z\rangle \Delta g$, where $\langle z\rangle = \tfrac{2}{3} h$ is the time average of z.

There is a slow change in the total energy, $\Delta(mgh)$, due to the changes in the potential energy. Therefore $m\Delta g \cdot \tfrac{2}{3}h = \Delta(mgh)$ or $g\Delta h + \tfrac{1}{3}h\Delta g = 0$, whence $h \propto g^{-1/3}$.

In this proof we have essentially followed by the same method which in the general case can be applied to prove that $\oint p\,dq$ is constant (see [1], §49).

Of course, we could directly have used the results of the general theory in this problem (and in the two other preceding problems).

If the plate is raised slowly (but $g(t) = $ const), then $h = $ const. This is clear when the velocity of the plate is constant (it is sufficient to change to a system of reference fixed in the plate). If the velocity changes the result cannot change as it depends, according to the general theory, only on the height of the support of the plate. It is assumed that the relative change in velocity during a time $\sqrt{2h/g}$ is small and the plate is risen smoothly.

13.5. a) $E = -A\left(1 - \dfrac{\alpha I}{\sqrt{2mA}}\right)^2$;

b) $E = -U_0\left(1 - \dfrac{\alpha I}{\sqrt{2mU_0}}\right)$;

c) $E = \alpha I\sqrt{\dfrac{2U_0}{m}} + \dfrac{\alpha^2 I^2}{2m}$;

d) $E = \left[\sqrt{\dfrac{\pi}{2m}}\,\dfrac{I}{2}\alpha^{1/n}\Gamma\left(\dfrac{1}{n} + \dfrac{3}{2}\right)\Big/\Gamma\left(\dfrac{1}{n}\right)\right]^{2n/(n+2)}.$

13.6. $h \propto (\sin\alpha)^{2/3}$.

13.7. $a \propto (\sin\alpha)^{-1/4}$.

13.8. $I = \frac{8ml\sqrt{gl}}{\pi}\left[\mathrm{E}\left(\sin\frac{\varphi_0}{2}\right) - \mathrm{K}\left(\sin\frac{\varphi_0}{2}\right)\cos^2\frac{\varphi_0}{2}\right]$, where $\mathrm{K}(k)$ and $\mathrm{E}(k)$ are the complete elliptic integrals of the first and second kind, respectively (see footnote in the solution of problem 1.7).

13.9. Let the coordinates of the particles m and M, reckoned from the point O, be x and X. The motion of the light particle can be consider approximately as the motion between two walls, one of which is moving. As the condition

$$|\dot{x}| \gg |\dot{X}| \tag{1}$$

is satisfied, the average over one period of the product $|\dot{x}|X = C$ is conserved (see problem 13.2). Eliminating $\dot{x}$ from the energy conservation law,

$$\tfrac{1}{2}m\dot{x}^2 + \tfrac{1}{2}M\dot{x} = E,$$

we find that the effect of the light particle upon the motion of the heavy particle is equivalent to the appearance of a potential energy

$$U(X) = \frac{mC^2}{2X^2}.$$

The equation

$$\tfrac{1}{2}M\dot{X}^2 + U(X) = E$$

leads to the law of motion:

$$X = \sqrt{\frac{mC^2}{2E} + \frac{2E}{M}(t - \tau)^2}.$$

The constants E, C, and τ can be determined from the initial values of X, $\dot{X}$, and $\dot{x}$ (they are independent of $x(0)$). This method of solving the problem becomes inapplicable when condition (1) is not satisfied.

13.10. We denote the coordinates of the heavy particles by $X_{1,2}$ and the light particles by x. When $X_1 < x < X_2$, the potential energy is

$$U = (x - X_1)f + (X_2 - x)f = (X_2 - X_1)f.$$

Therefore, the light particle moves freely between the heavy particles, and

$$|\dot{x}|(X_2 - X_1) = C = \mathrm{const}$$

(see the preceding problem). Taking this into account, we get the following equation for the relative motion of the heavy particles $(X = X_2 - X_1)$ from the law of energy conservation

$$\tfrac{1}{4}M\dot{X}^2 + \frac{mC^2}{2X^2} + fX = E.$$

Expanding

$$U_{\text{eff}}(X) = \frac{mC^2}{2X^2} + fX$$

near the minimum $X_0 = (mC^2/f)^{1/3}$, we find the frequency of the small oscillations of an "ion"

$$\omega^2 = \frac{2U''(X_0)}{M} = \frac{6f}{MX_0}.$$

13.11. We expand the frequency in a series in t in the equations for P and Q,

$$\dot{Q} = \omega + \frac{\dot{\omega}}{2\omega}\sin 2Q, \quad \dot{P} = -P\frac{\dot{\omega}}{\omega}\cos 2Q.$$

Restricting ourselves to the first-order corrections, we get for P and Q the equations

$$Q = (\omega_0 t + \varphi) + \tfrac{1}{2}\dot{\omega}_0 t^2 + \frac{\dot{\omega}_0}{2\omega_0}\int_0^t \sin 2Q(t)\,dt, \quad (1)$$

$$P = P_0\left(1 - \frac{\dot{\omega}_0}{\omega_0}\int_0^t \cos 2Q\,dt\right), \quad (2)$$

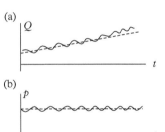

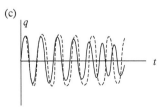

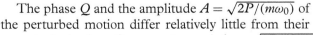

Figure 189

where ω_0 and $\dot{\omega}_0$ are the values of the frequency and its derivative at time $t_0 = 0$, and we have $\dot{\omega}_0 = \varepsilon^2\omega_0^2$ with $\varepsilon \ll 1$.

The phase Q and the amplitude $A = \sqrt{2P/(m\omega_0)}$ of the perturbed motion differ relatively little from their unperturbed values $Q_0 = \omega_0 t + \varphi$ and $A_0 = \sqrt{2P_0/(m\omega_0)}$ even for time intervals which are much longer than the period $2\pi/\omega$ of the oscillations (see Fig. 189).

Thus, for a time $t \sim 1/(\varepsilon \omega_0)$, the second term in (1) is of order unity, and the third term of order ε, and thus

$$Q = \omega_0 t + \varphi + \tfrac{1}{2} \dot{\omega}_0 t^2, \qquad P = p_0.$$

However, this change in phase leads to the fact that the perturbed motion in terms of the variables p and q,

$$q(t) = A_0 \cos\left(\omega_0 t + \varphi + \tfrac{1}{2} \dot{\omega}_0 t^2\right),$$

will differ appreciably from unperturbed motion,

$$q_0(t) = A_0 \cos(\omega_0 t + \varphi),$$

in such a way that

$$q(t) - q_0(t) \sim q_0(t).$$

When one tries to construct a perturbation theory for the variables q and p, one obtains for the first-order corrections $q_1(t)$ the equation

$$\ddot{q}_1 + \omega_0^2 q_1 = -2\omega_0 \dot{\omega}_0 t A_0 \cos(\omega_0 t + \varphi),$$

which has a resonant force which increases with time. The solution obtained in such a theory is thus applicable only for small time intervals of the order of a few periods of the oscillations

$$\frac{2\pi}{\omega_0} \ll \frac{1}{\varepsilon \omega_0}.$$

13.13. We transform the Hamiltonian function of the system

$$H(x, pt) = \frac{p^2}{2m} + \tfrac{1}{2} m \omega^2 x^2 - x F(t) = E(t) \tag{1}$$

to the form

$$\tfrac{1}{2} m \omega^2 \left(x - \frac{F}{m \omega^2}\right)^2 + \frac{p^2}{2m} - \frac{F^2}{2m\omega^2} = E(t).$$

From this, it is clear that the orbit in phase space is an ellipse which is displaced along the x-axis over a distance $F/(m\omega^2)$ with semi-axes

$$a = \sqrt{\frac{2E}{m\omega^2} + \frac{F^2}{m^2\omega^4}}, \quad b = \sqrt{2mE + \frac{F^2}{\omega^2}}.$$

Apart from the factor $1/(2\pi)$, the adiabatic invariant is the area of this ellipse

$$I = \tfrac{1}{2}ab = \frac{E + F^2/(2m\omega^2)}{\omega}. \tag{2}$$

Here the meaning of $E + F^2/(2m\omega^2)$ is that of the energy of oscillations near the displace equilibrium position (cf. problem 5.16). Substituting the value of E from (1) into (2), we can present the result in the form

$$I = \frac{m}{2\omega}\left| \dot{x} + i\omega\left(x + \frac{F}{m\omega^2}\right) \right|^2 =$$

$$= \frac{m}{2\omega}\left| \frac{1}{m}\int_0^t e^{i\omega(t-\tau)}F(\tau)d\tau + e^{i\omega t}[\dot{x}(0) + i\omega x(0)] - i\frac{F(t)}{m\omega} \right|^2.$$

Here we can use equations (22.9) and (22.10) from [1] for the quantity $\dot{x} + i\omega x$.
   Integrating by parts, we get

$$I(t) = I(0) - \frac{\dot{x}(0)}{\omega^2}\int_0^t \dot{F}(t)\sin\omega t\, dt -$$

$$- \frac{1}{\omega}\left[x(0) - \frac{F(0)}{m\omega^2}\right]\int_0^t \dot{F}(t)\cos\omega t\, dt + \frac{1}{2m\omega^3}\left| \int_0^t \dot{F}(\tau)e^{-i\omega\tau}\, d\tau \right|^2.$$

If the force changes slowly, $I(t)$ will oscillates near $I(0)$. If $F(t) \to$ const as $t \to \infty$, the total change in the adiabatic invariant, $I(\infty) - I(0)$, can be very small (see problem 5.18).

13.14. $PV^{5/3} = $ const.

13.15. a) $E = \frac{\pi^2}{2m}\left(\frac{I_1^2}{a^2} + \frac{I_2^2}{b^2} + \frac{I_3^2}{c^2}\right)$, where a, b, and c are the lengths of the edges of the parallelepiped, and $I_k = $ const.
   (b) The absolute magnitude of the components of the velocity along each of the edges are conserved.

13.16. In spherical coordinates, the variables separate. The angular momentum $\mathbf{M}$ is strictly conserved. (Moreover, M_z is an adiabatic invariant corresponding to the angle φ). The adiabatic invariant for the radial motion is

$$I_r = \frac{1}{\pi} \int_{r_{min}}^{R} \sqrt{2mE - \frac{M^2}{r^2}}\, dr. \tag{1}$$

We can find the function $E(R)$ without evaluating the integral (1). The substitution $r = Rx$ gives

$$I_r = \frac{1}{\pi} \int_{x_{min}}^{1} \sqrt{2mER^2 - \frac{M^2}{x^2}}\, dx = I_r(ER^2, M), \tag{2}$$

whence $ER^2 = \text{const.}$ We get thus for the angle of incidence α the equation

$$\sin\alpha = \frac{r_{min}}{R} = \frac{M}{\sqrt{2mER}} = \text{const.}$$

13.17. a) $E \propto \gamma^{\frac{2}{2-n}}$; b) $E \propto \gamma^{-1}$.

13.18. Equating the values of the adiabatic invariant before and after the switching on of the field,

$$\int_{r_{min}}^{r_{max}} \sqrt{E - \frac{M^2}{2mr^2} - U(r)}\, dr = \int_{r_{min}}^{r_{max}} \sqrt{E + \delta E + \frac{M^2}{2mr^2} - U - \delta U}\, dr$$

we get

$$\delta E = \langle \delta U \rangle = \frac{2}{T} \int_{r_{min}}^{r_{max}} \frac{\delta U\, dr}{\sqrt{\frac{2}{m}\left(E - \frac{M^2}{2mr^2} - U\right)}}.$$

13.19. $E = I_1\Omega_1 + I_2\Omega_2$ (in the notation of problem 6.6a). The orbit fills the rectangle

$$|Q_1| \leqslant \sqrt{I_1/\Omega_1}, \quad |Q_2| \leqslant \sqrt{I_2/\Omega_2}.$$

The conditions that the theory of adiabatic invariants is applicable is

$$|\dot{\Omega}_i| \ll \Omega_i^2, \quad |\ddot{\Omega}_i| \ll \Omega_i|\dot{\Omega}_i| \quad (i = 1, 2).$$

Outside the region of degeneracy, these conditions reduce to the same conditions for $\omega_1(t)$. In the region of degeneracy, $|\omega_1^2 - \omega_2^2| \sim \alpha$, and the second condition is more restrictive and gives $|\dot{\omega}_1| \ll \alpha$ (the region of degeneracy is traversed during a time which is considerable longer than the period of the beats).

13.20. When the coupling αxy is not present, the system splits into two independent oscillators with coordinates x and y. The corresponding adiabatic invariants are

$$I_x = \frac{E_x}{\omega_1}, \quad I_y = \frac{E_y}{\omega_2},$$

where E_x and E_y are the energies of these oscillators.

When the coupling is taken into account the system consists of two independent oscillators with coordinates Q_1 and Q_2. If the frequency changes slowly, the quantities

$$I_1 = \frac{E_1}{\Omega_1}, \qquad I_2 = \frac{E_2}{\Omega_2}.$$

are conserved.

Outside the region of degeneracy the normal oscillations are strongly localized, and when $\omega_1 < \omega_2$, then $Q_1 = x$, $Q_2 = y$, while when $\omega_1 > \omega_2$, then $Q_1 = +y$, $Q_2 = -x$. Thus, when $\omega_1 < \omega_2$, $I_x = I_1$, $I_y = I_2$, while when $\omega_2 < \omega_1$, then $I_x = I_2$, $I_y = I_1$ (Fig. 190).

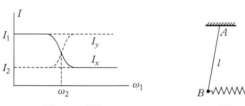

Figure 190   Figure 191

We shall illustrate this by the following example. Two pendulum, of which the length of one can be changed slowly, are coupled by a spring with small stiffness (Fig. 191). When the length l and L of the pendulums are appreciable different, the normal oscillations are practically the same as the oscillations of one or the other pendulum.

Let initially $L > l$ and the first pendulum AB oscillates with amplitude φ_0, and the pendulum CD with a very small amplitude. When L is decreased, the amplitude of the oscillations of the pendulum CD remains small until its length becomes almost equal to l. When $L \approx l$, its amplitude increases (and when $L = l$ both pendulums will oscillate with the same amplitudes, $\varphi_0/\sqrt{2}$, in anti-phase). When L is decreased further, practically all the energy transfers to the pendulum CD and its amplitude becomes $\varphi_1 = \varphi_0 (l/L)^{3/4}$, as for a separate pendulum.

If we traverse the degeneracy region relatively fast, $\dot{\omega}_1 \gg \alpha$, such a transfer of energy between the oscillators will not take place. Moreover, if $\dot{\omega}_1 \ll \omega_1^2$, $\ddot{\omega}_1 \ll \omega_1\dot{\omega}_1$, I_x and I_y are conserved.

13.21. From the equations of motion

$$\ddot{x} + \omega_1^2 x + 2\beta xy = 0, \tag{1}$$

$$\ddot{y} + \omega_2^2 y + \beta x^2 = 0, \tag{2}$$

we see easily that the coupling between the oscillators leads to a large energy transfer when $2\omega_1 \approx \omega_2$.

Let

$$x = a(t)\cos(\omega_1 t + \varphi), \quad y = b(t)\cos(\omega_2 t + \psi).$$

If $a \gg b$, the term

$$\beta x^2 = \tfrac{1}{2}\beta a^2 + \tfrac{1}{2}\beta a^2 \cos(2\omega_1 + 2\varphi)$$

in (2) will play the role of an applied force, leading to a resonance increase in y. If, however, $a \ll b$, the term

$$2\beta xy = 2\beta bx \cos(\omega_2 t + \psi)$$

in (1) leads to a parametric building-up of the oscillation in x. The detail of the study of system (1) and (2) can be found in problem 8.10.

The region of resonance interaction is

$$|2\omega_1 - \omega_2| \lesssim \frac{\beta b}{\omega_1}.$$

In general, a strong resonance interaction between the oscillators occurs when $n\omega_1 = l\omega_2$, where n and l are integers. However, the widths of the regions of frequencies in which these resonances occur are extreamly small when n and l are not too small (see [1], § 29). We can therefore neglect their influence on the motion of the oscillators for values of $\dot{\omega}_1$ that are not too small (provided they are sufficiently small that we can use the theory of adiabatic invariants).

13.22. Let the particle moving in the xy-plane at a small angle to the y-axis ($|\dot{x}| \ll |\dot{y}|$) be reflected from the x-axis and from the curve $y_0(x)$ (Fig. 192).

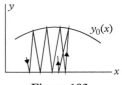

Figure 192

If we assume that we know how the particle moves in the x-direction, we can study the motion in the y-direction by taken $x(t)$ to be a slowly changing parameter. The adiabatic invariant,

$$\oint p_y \, dy = 2|p_y|y_0(x) = 2\pi I,$$

will remain constant, and that equation determines the function $p_y(x)$.

To determine $x(t)$, we can use the energy conservation law

$$m^2 \dot{x}^2 + p_y^2(x) = 2mE.$$

The minimum distance $x_{\min}$ is determined by the
condition

$$p_y^2(x_{\min}) = 2mE.$$

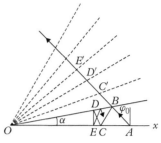

Figure 193

Substituting $y_0(x) = x \tan \alpha$ and $2\pi I = 2\sqrt{2mEl} \tan \alpha \cos \varphi_0$, we get

$$x_{\min} = l \cos \varphi_0.$$

The solution by means of the reflection method is clear from Fig. 193. This method
gives the exact solution applicable at any angle α and φ_0 but it cannot be generalized to
the cases when $y_0(x)$ is not a straight line.

13.23. $\tanh \alpha x_{\max} = \tan \varphi$, $T = \dfrac{2\pi}{\alpha v \sqrt{\cos 2\varphi}}$ (see the preceding problem).

13.24. a) The problem of the motion of a particle in a magnetic field reduces for the
given choice of vector potential to the problem of the motion of a harmonic oscillator
(see problem 10.8). The adiabatic invariant is

$$I = \frac{E - p_z^2/(2m)}{\omega} \propto \frac{v_\perp^2}{B} \propto \pi a^2 B,$$

where $v_\perp$ is the transverse to the magnetic field component of the particle velocity and
$a = cmv_\perp/(eB)$ is the radius of the particle orbit (see [2], § 21).

The relation $I \propto \pi a^2 B$ can be interpreted simply: the radius of the orbit changes in
such a way that the magnetic field flux through the area circumscribed by it remains
constant. The distance of the centre of the orbit from the yz-plane is equal to $x_0 = cp_y/(eB)$, and decreases with increasing B.

The occurrence of a drift of the orbit is connected with the appearance of an electric
field,

$$\mathbf{E} = -\frac{1}{c}\frac{\partial \mathbf{A}}{\partial t} = \left(0, -\frac{1}{c}x\dot{B}, 0\right),$$

when the magnetic field changes (cf. [2], § 22). The electric field vectors $\mathbf{E}$ and the drift
velocity $\mathbf{v}_{\mathrm{dr}}$ are shown in Fig. 194 for different orbit positions of the particle.

b) The Hamiltonian function is in cylindrical coordinates

$$H = \frac{p_z^2}{2m} + \frac{p_r^2}{2m} + \frac{1}{2mr^2}\left(p_\varphi - \frac{eB}{2c}r^2\right)^2.$$

The quantity p_z and p_φ are integrals of motion.
The adiabatic invariant for the radial motion is

$$\pi I_r = \int_{r_{min}}^{r_{max}} \sqrt{2mE_\perp - \frac{1}{r^2}\left(p_\varphi - \frac{eB}{2c}r^2\right)^2}\,dr,$$

which after the substitution $r = \zeta/\sqrt{B}$ becomes

$$\int_{\zeta_{min}}^{\zeta_{max}} \sqrt{\frac{2mE_\perp}{B} - \frac{1}{\zeta^2}\left(p_\varphi - \frac{e}{2c}\zeta^2\right)^2}\,d\zeta = \pi I_r\left(p_\varphi, \frac{E_\perp}{B}\right).[1]$$

Therefore, $E_\perp/B = $ const; that is, the energy of transverse motion, $E_\perp = \frac{1}{2}mv_\perp^2$, changes in the same way as under sub 13.23a. The distance r_0 of the centre of the orbit to the origin is

$$r_0 = \frac{r_{max} + r_{min}}{2} = \frac{\zeta_{max} + \zeta_{min}}{2\sqrt{B}} \propto \frac{1}{\sqrt{B}}.$$

When B increases, the centre of the orbit approaches the origin (Fig. 195).

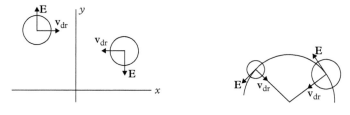

Figure 194   Figure 195

[1] It is interesting that I_r does not actually depend on p_φ when $p_\varphi > 0$. Indeed, $\frac{\partial I_r}{\partial p_\varphi}$ is the change of the angle $\Delta\varphi$ over the time of the one radial oscillation; besides, when $p_\varphi > 0$, the origin of coordinates lies outside the orbit (see Fig. 109b), and therefore $\Delta\varphi = 0$.

When B changes, there occurs an electric field

$$(\mathbf{E})_\varphi = -\frac{r}{2c}\dot{B}, \qquad (\mathbf{E})_r = (\mathbf{E})_z = 0,$$

of which the field of lines are closed circles.

In real conditions, a uniform magnetic field can exist only in a limited region of space. The electrical field occurring when the magnetic is changed depends very strongly on the shape of that region and the conditions at its boundaries (see [2], §21). For example, the field considered under sub 13.23a could occur near a conducting plane in which there was a current, while the field under sub 13.23b could be produced in a solenoid.[1]

The strong dependence of the nature of the particle motion on the weak field $\mathbf{E}$ even in the case of extremely small $\mathbf{B}$ can be explained by the presence of degeneracy (when $\mathbf{B} = \mathbf{const}$ the periods in the two coordinates x and y, or r and φ are the same).

We note that the quantity $E_\perp/B$ turns out to be an adiabatic invariant in both cases. One can prove that this result is independent of the choice of the form of $\mathbf{A}$ (see [2], §21 and [10], §25).

13.25. The adiabatic invariants are

$$I_z = \frac{E_z}{\omega}, \qquad I_\varphi = p_\varphi,$$

$$I_r = \frac{1}{\pi} \int_{\zeta_{min}}^{\zeta_{max}} \sqrt{C\zeta^2 - p_\varphi^2 - m^2\zeta^4}\,\frac{d\zeta}{\zeta} = I_r(p_\varphi, C), \tag{1}$$

where

$$C = \frac{2mE_\perp + eBp_\varphi/c}{\sqrt{\omega^2 + \left(\frac{eB}{2mc}\right)^2}}, \quad E_z = \tfrac{1}{2}mv_z^2, \quad E_\perp = \tfrac{1}{2}mv_\perp^2,$$

and $v_\perp$ is the transverse to the magnetic field component of the particle velocity. Therefore

$$E_z \propto \omega, \quad E_\perp + \frac{eB}{2mc}p_\varphi \propto \sqrt{\omega^2 + \left(\frac{eB}{2mc}\right)^2}. \tag{2}$$

---

[1] The change in the electric field $\mathbf{E}$ connected with the change in the choice of the form of $\mathbf{A}$ would not occur if we had simultaneously changed the scalar potential by $\dot{B}xy/(2c)$ (gauge transformation).

The given vector potential defines a magnetic field which is symmetric with respect to the z-axis going through the centre of the oscillator. If we make a different choice for $\mathbf{A}$,

$$A_x = A_z = 0, \qquad A_y = xB(t), \tag{3}$$

we get, in fact, a different physical problem. The Lagrangian functions for these two problems differ by

$$\delta L = \frac{d}{dt}\left(\frac{e}{2c}Bxy\right) - \frac{e}{2c}\dot{B}xy; \tag{4}$$

that is, the difference is very small if we drop in (4) the inessential total derivative with respect to time.

In the preceding problem, where the motion was degenerate, just this extra term led to a complete change in the direction and drift velocity of the orbit. In the present case, however, the motion of the oscillator is not degenerate when $\mathbf{B} = \mathbf{const}$ and we can neglect the extra term δL (cf. problem 13.20). The relations (2) are thus valid also for a different choice of $\mathbf{A}$. When one passes through the degeneracy region ($\mathbf{B} = 0$), the equations (2) remains valid only when we choose the axially symmetric field given in the problem. For instance, the behaviour of the oscillator in the field (3) when $\mathbf{B}$ passes through zero requires additional study.

13.26. a) Using a canonical transformation, one can reduce the Hamiltonian function to a sum of the Hamiltonian functions for two independent oscillators (for X and Y; see problem 11.10). For each of the oscillators the ratio of the energy to the frequency is an adiabatic invariant. We remind ourselves that the oscillations of each of them correspond to motion along an ellipse (see problem 6.37). In terms of the amplitude a_k of the oscillations in the x-direction, for instance, the adiabatic invariants are equal to

$$I_k = \frac{ma_k^2}{2\Omega_k}\,\frac{\Omega_k^4 - \omega_1^2\omega_2^2}{\Omega_k^2 - \omega_2^2} \quad (k = 1, 2).$$

When the parameters of the system are changing, we must also add to the new Hamiltonian function the partial derivative with respect to the time of the generating function which is equal to $-\dot{\lambda}[m\omega_2 XY + P_X P_Y/(m\omega_2)]$ (see problem 11.18). This correction term is small ($\dot{\lambda} \ll \Omega_k$) and can be neglected provided the eigen-frequencies are not the same (cf. problem 13.20). One must consider separately the degenerate case when $\omega_1 = \omega_2$ and a magnetic field which can vanish.

When $\omega_1 = \omega_2$, the different choice of the adiabatic invariants is possible (see preceding problem).

b) To fix the ideas, let $\omega_1 > \omega_2$. The motion is along a circle of radius $a\sqrt{\omega_1/(\omega_B)}$ with a frequency ω_B, but the centre of the circle moves along an ellipse with semi-axes in the x- and y-direction and equal to

$$b\,\frac{\omega_2}{\sqrt{\omega_1\omega_B}}\quad\text{and}\quad b\sqrt{\frac{\omega_1}{\omega_B}}$$

with a frequency $\omega_1\omega_2/\omega_B$.

c) The oscillation will proceed almost in the y-direction; its amplitude is increased by a factor $\sqrt{\omega_1/\omega_2}$ (cf. problem 13.20).

13.28. a) The motion of the particle in the xy-plane takes place under the action of a magnetic field which is slowly varying as the particle moves along the z-axis. The adiabatic invariant

$$I_\perp = E_\perp\left[\frac{eB(z)}{mc}\right]^{-1}$$

is then conserved (see problem 13.24). From the energy conservation law, we have

$$\tfrac{1}{2}m\dot{z}^2 + I_\perp\frac{eB(z)}{mc} = E.$$

The particle moves in the z-direction as if it moved in a potential field

$$U(z) = I_\perp\frac{eB(z)}{mc}.$$

The period of the oscillations is (cf. problem 2b from [1], §11)

$$T = \frac{2\pi a}{v\sqrt{\lambda\sin^2\alpha - \cos^2\alpha}},$$

where α is the angle between the velocity $\mathbf{v}$ of a particle and the z-axis. Particles, for which $\tan^2\alpha < 1/\lambda$ are not contained in the trap. The condition for the applicability of the theory of adiabatic invariants consists in the requirement that the change in the magnetic field during one period of revolution of the particle be small. This gives $mc\lambda v_z \ll aeB_0$.

As an example of a magnetic trap, we can observe the radiation belts of the Earth.

b) $T = \dfrac{2\pi a}{v\sin\alpha}$.

13.29. a) $(\lambda E_\perp - E_z)a^2 = \text{const}$, $E_\perp/B_0 = \text{const}$, $E = E_\perp + E_z$;

b) $E_\perp/B_0 = \text{const}$, $aE_z\sqrt{B_0} = \text{const}$.

13.30. If we neglect in the Hamiltonian function

$$H = \frac{p_r^2}{2m} + \frac{p_\theta^2}{2mr^2} + \frac{p_\varphi^2}{2mr^2\sin^2\theta} - \frac{eBp_\varphi}{2cm} + \frac{e^2B^2r^2\sin^2\theta}{8mc^2} + U(r)$$

the term which is quadratic in B, we can separate the variables in the Hamilton–Jacobi equation.

The adiabatic invariants have the form

$$I_\varphi = p_\varphi, \quad I_\theta = \frac{1}{\pi} \int_{\theta_1}^{\theta_2} \sqrt{\beta - \frac{p_\varphi^2}{\sin^2\theta}}\, d\theta = I_\theta(p_\varphi, \beta),$$

$$I_r = \frac{1}{\pi} \int_{r_1}^{r_2} \sqrt{2m\left[E + \frac{eBp_\varphi}{2mc} - U(r)\right] - \frac{\beta}{r^2}}\, dr = I_r\left(E + \frac{eBp_\varphi}{2mc}, \beta\right).$$

When B is slowly changed, the quantities

$$p_\varphi, \quad \beta = p_\theta^2 + \frac{p_\varphi^2}{\sin^2\theta}, \quad \text{and } E + \frac{eBp_\varphi}{2mc}$$

thus remain constant.

13.31..

$$\frac{p_x^2}{2m} + \frac{1}{2} m\omega^2 x^2 = E_1, \quad \frac{p_y^2}{2m} + 2m\omega^2 y^2 = E_2,$$

$$(m^2\omega^2 x^2 - p_x^2)y + x p_x p_y = A,$$

$$(m^2\omega^2 x^2 - p_x^2)p_y - 4m\omega x p_x y = m\{E_2, A\}.$$

13.32. a) $w = \arctan\frac{m\omega x}{p}, \quad I = \frac{p^2}{2m\omega} + \frac{1}{2} m\omega x^2.$
These variables are convenient, for instance, to develop perturbation theory (see problem 13.10).

b) Let initially the particle moves to the right from the point $x = 0$, and we shall choose S such that $S = 0$ for $x = 0$. In that case,

$$S = \int_0^x |p|\, dx = \pi I - \pi a \left[\left(\frac{I}{a}\right)^{2/3} - Fx\right]^{3/2},$$

where

$$I = \frac{1}{\pi} \int_0^{x_m} |p|\, dx = aE^{3/2}, \quad a = \frac{2\sqrt{2m}}{3\pi F}, \quad x_m = \frac{E}{F}, \quad |p| = \sqrt{2m(E - xF)}.$$

If the motion is to the left,

$$S = \left(\int_0^{x_m} - \int_{x_m}^x\right)|p|\,dx = \pi I + \pi a\left[\left(\frac{I}{a}\right)^{2/3} - Fx\right]^{3/2}$$

and so on. For the nth oscillation,

$$S = (2n-1)\pi I \mp \pi a\left[\left(\frac{I}{a}\right)^{2/3} - Fx\right]^{3/2}$$

(the upper [lower]) sign corresponds to motion to the right [left]; Fig. 196).

One can use the $S(x, I)$ as a generating function to change to new canonical action and angle variables (see [1], § 49). The new variables are connected with the old ones in the following way:

$$x = \frac{1}{\pi^2 F}\left(\frac{I}{a}\right)^{2/3}\left\{\pi^2 - [(2n-1)\pi - w]^2\right\},$$

$$p = \frac{3}{2}aF\left(\frac{I}{a}\right)^{1/3}[(2n-1)\pi - w],$$

where x is a periodic function of w, while w in multivalued function of x (Fig. 197).

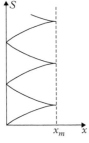

Figure 196

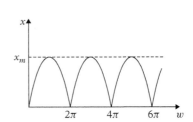

Figure 197

13.33. From the equation

$$P = \int_0^a \sqrt{2m(E - U)}\,dx$$

we find

$$E = \frac{P^2}{2ma^2} + \frac{1}{2}V + \frac{mV^2a^2}{8P^2}.$$

The abbreviated action is

$$S_0 = \int_0^x p \, dx = \begin{cases} \sqrt{2mE}(x - na) + nP, & \text{when } na < x < \left(n + \frac{1}{2}\right)a, \\ \sqrt{2m(E - V)}\left[x - (n + \frac{1}{2})a\right] + \\ \quad + \sqrt{2mE}\left(\frac{1}{2}a\right) + nP & \text{when } \left(n + \frac{1}{2}\right)a < x < (n + 1)a. \end{cases}$$

Eliminating E and introducing $A = \dfrac{ma^2 V}{2P^2}$, we get the generating function for the canonical transformation under consideration

$$S_0(x, P) = \frac{P}{a} \cdot \begin{cases} (1 + A)(x - na) + na, & \text{when } na < x < \left(n + \frac{1}{2}\right)a, \\ (1 - A)(x - na) + Aa + na, & \text{when } \left(n + \frac{1}{2}\right)a < x < (n + 1)a. \end{cases}$$

From the equations

$$p = \frac{\partial S_0}{\partial x}, \quad Q = \frac{\partial S_0}{\partial P},$$

we get

$$p(P, Q) = \frac{P}{a} \cdot \begin{cases} 1 + A, & \text{when } n < Q < n + Q_0, \\ 1 - A, & \text{when } n + Q_0 < Q < n + 1, \end{cases}$$

$$x(P, Q) = a \cdot \begin{cases} \dfrac{Q - nA}{1 - A}, & \text{when } n < Q < n + Q_0, \\ \dfrac{Q + (n + 1)A}{1 + A}, & \text{when } n + Q_0 < Q < n + 1, \end{cases}$$

where $Q_0 = \frac{1}{2}(1 - A)$. The variables P and Q are analogous to action and angle variables, and the quantity $a\dot{Q}$ is the average particle velocity.

Bibliography

[1] L. D. Landau, E. M. Lifshitz (1976) *Mechanics*, Pergamon Press, Oxford.

[2] L. D. Landau, E. M. Lifshitz (1980) *Classical Theory of Fields*, Butterworth-Heinenann, Oxford.

[3] L. D. Landau, E. M. Lifshitz (1984) *Electrodynamics of Continous Media*, Pergamon Press, Oxford.

[4] H. Goldstein (1960) *Classical mechanics*, Addison-Wesley, Reading, Mass.

[5] H. Goldstein, C. Poole, J. Safko (2000) *Classical Mechanics*, Addison-Wesley, Reading, Mass.

[6] D. ter Haar (1964) *Elements of Hamiltonian Mechanics*, North Holland, Amsterdam.

[7] G. L. Kotkin, V. G. Serbo, A. I. Chernykh (2010) *Lectures on Analytical Mechanics*, Regular and Chaotic Dynamics, Moscow-Izhevsk.

[8] V. I. Arnold (1989) *Mathematical Methods of Classical Mechanics*, Springer-Verlag, Berlin.

[9] L. I. Mandelstam (1972) *Lectures on Theory of Oscillations*, Nauka, Moscow.

[10] N. N. Bogolyubov, Yu. A. Metropolskii (1958) *Asymptotic Methods in the Non-linear Oscillations*, Noordhoff, Groningen.

[11] M. Abramowitz, I. A. Stegun (1965) *Handbook of Mathematical Functions*, Dover, New York.

[12] N. F. Mott, H. S. W. Massey (1985) *The Theory of Atomic Collisions*, Clarendon Press, Oxford.

[13] N. N. Bogolyubov, D. V. Shirkov (1980) *Introduction to the Theory of Quantized Fields*, Wiley, New York.

[14] P. Courant, D. Hilbert (1989) *Methods of Mathematical Physics*, Wiley, New York.

[15] J. V. Stratt (Lord Rayleigh) (1945) *Theory of Sound*, Vol. I., Chap. IV, Dover, New York.

[16] C. Kittel (1968) *Introduction to Solid State Physics*, Wiley, New York.

[17] J. P. Den Hartog (1985) *Mechanical Vibrations*, Dover, New York.

[18] E. Fermi (1971) *Scientific Papers*, Vol. I, p. 440, Nauka, Moscow.

[19] N. Bloembergen (1965) *Nonlinear Optics*, Appendix I, §§ 5,6, W. A. Benjamin, New York.

[20] L. D. Landau, E. M. Lifshitz (1980) *Statistical Physics*, Chap. XIV, Pergamon Press, Oxford.

[21] F. D. Stacey (1969) *Physics of the Earth*, Wiley, New York.

[22] A. I. Bazh, Ya. B. Zeldovich, A. M. Perelomov (1971) *Scattering, Reactions and Decays in Non-Relativistic Quantum Mechanics*, Nauka, Moscow.

[23] P. Calogero, O. Ragnisco (1975) *Lettere al Nuovo Cimento* **13**, 383.

[24] M. S. Ryvkin (1975) *Doklady Akademii Nauk, (USSR)* **221**, No 1.

[25] V. V. Sokolov *Nuclear Physics, (USSR)* **23**, 628 (1976); **26**, 427 (1977).

[26] F. Rafe (1974) *Scientific American*, **12**.

Index

The numbers refer to the problems; for instance, 4.12 is problem 12 of §4.